Recent Advances in Biomembrane Models for Studying Interactions with Bio-/Molecules

Recent Advances in Biomembrane Models for Studying Interactions with Bio-/Molecules

Guest Editors

Jadwiga Maniewska
Katarzyna Gębczak

MDPI

Basel • Beijing • Wuhan • Barcelona • Belgrade • Novi Sad • Cluj • Manchester

Guest Editors
Jadwiga Maniewska
Department of Medicinal Chemistry
Wroclaw Medical University
Wroclaw
Poland

Katarzyna Gębczak
Division of Basic Medical Sciences, Department of Basic Medical Sciences and Immunology
Wroclaw Medical University
Wroclaw
Poland

Editorial Office
MDPI AG
Grosspeteranlage 5
4052 Basel, Switzerland

This is a reprint of the Special Issue, published open access by the journal *Membranes* (ISSN 2077-0375), freely accessible at: https://www.mdpi.com/journal/membranes/special_issues/0X4XMUX5U7.

For citation purposes, cite each article independently as indicated on the article page online and as indicated below:

Lastname, A.A.; Lastname, B.B. Article Title. *Journal Name* **Year**, *Volume Number*, Page Range.

ISBN 978-3-7258-8285-4 (Hbk)
ISBN 978-3-7258-8286-1 (PDF)
https://doi.org/10.3390/books978-3-7258-8286-1

Contents

About the Editors vii

Preface ix

Jadwiga Maniewska and Katarzyna Gębczak
Recent Advances in Biomembrane Models for Studying Interactions with Bio-/Molecules
Reprinted from: *Membranes* **2026**, *16*, 174, https://doi.org/10.3390/membranes16050174 1

Jadwiga Maniewska, Katarzyna Gębczak, Łucja Cwynar-Zając, Żaneta Czyżnikowska and Berenika M. Szczęśniak-Sięga
New Oxicam Derivatives—Studies of Membrane Interactions, Cytotoxicity, Cyclooxygenase Inhibition and Molecular Docking
Reprinted from: *Membranes* **2026**, *16*, 166, https://doi.org/10.3390/membranes16050166 5

Ralitsa Veleva, Aneliya Kostadinova, Antoaneta Trendafilova, Viktoria Ivanova, Veselina Moskova-Doumanova, Kirilka Mladenova, et al.
Polyphenols from *Inula oculus-christi* L. Induced Cell-Specific Membrane and Cytoskeleton Reorganization
Reprinted from: *Membranes* **2025**, *15*, 357, https://doi.org/10.3390/membranes15120357 28

José Villalaín
Location and Dynamics of Nymphaeol A in a Complex Membrane
Reprinted from: *Membranes* **2025**, *15*, 163, https://doi.org/10.3390/membranes15060163 45

Berenika M. Szczęśniak-Sięga, Jadwiga Maniewska, Benita Wiatrak, Tomasz Janek, Paulina Nowotarska and Żaneta Czyżnikowska
Anti-Inflammatory Properties of Novel 1,2-Benzothiazine Derivatives and Their Interaction with Phospholipid Model Membranes
Reprinted from: *Membranes* **2024**, *14*, 274, https://doi.org/10.3390/membranes14120274 62

Florian Weber, Markus Axmann, Erdinc Sezgin, Mariana Amaro, Taras Sych, Armin Hochreiner, et al.
"Head-to-Toe" Lipid Properties Govern the Binding and Cargo Transfer of High-Density Lipoprotein
Reprinted from: *Membranes* **2024**, *14*, 261, https://doi.org/10.3390/membranes14120261 88

Raina Marie Seychell, Adam El Saghir and Neville Vassallo
Modulation of Biological Membranes Using Small-Molecule Compounds to Counter Toxicity Caused by Amyloidogenic Proteins
Reprinted from: *Membranes* **2024**, *14*, 231, https://doi.org/10.3390/membranes14110231 103

Kajetan Kiełbowski, Małgorzata Król, Estera Bakinowska and Andrzej Pawlik
The Role of ABCB1, ABCG2, and SLC Transporters in Pharmacokinetic Parameters of Selected Drugs and Their Involvement in Drug–Drug Interactions
Reprinted from: *Membranes* **2024**, *14*, 223, https://doi.org/10.3390/membranes14110223 123

Juan M. Giraldo-Lorza, Chad Leidy and Marcela Manrique-Moreno
The Influence of Cholesterol on Membrane Targeted Bioactive Peptides: Modulating Peptide Activity Through Changes in Bilayer Biophysical Properties
Reprinted from: *Membranes* **2024**, *14*, 220, https://doi.org/10.3390/membranes14100220 142

About the Editors

Jadwiga Maniewska

Jadwiga Maniewska is an Assistant Professor in the Department of Medicinal Chemistry, Faculty of Pharmacy, Wroclaw Medical University, Poland. Her research interests include the physicochemical characterization of bioactive compounds, particularly potential non-steroidal anti-inflammatory agents, and their interactions with model lipid membranes. Her work focuses on drug–membrane interactions and the role of membrane properties in drug activity, transport, and biological effects, combining approaches from medicinal chemistry and membrane biophysics.

Katarzyna Gębczak

Katarzyna Gębczak is an Assistant Professor in the Department of Basic Medical Sciences and Immunology, Faculty of Pharmacy, Wroclaw Medical University, Poland. Her research interests include natural and synthetic compounds with potential therapeutic applications, particularly their antioxidant, anti-inflammatory, and anticancer activities. She investigates the biological effects and mechanisms of action of bioactive molecules using cellular and in vitro models. Her work focuses on identifying compounds with potential biomedical relevance and evaluating their therapeutic properties.

Preface

Biological membranes are highly dynamic and structurally complex systems that regulate essential cellular processes and mediate interactions between cells and their environment. Understanding the mechanisms that shape membrane organization and membrane molecule interactions remains a major challenge in contemporary biophysics, pharmacology, and biomedical research. Increasing evidence demonstrates that membrane composition, lipid diversity, cholesterol content, and membrane heterogeneity strongly influence the activity of bioactive compounds, membrane proteins, and therapeutic agents.

This Reprint was prepared to provide a comprehensive overview of recent progress in biomembrane research, with particular emphasis on experimental and computational strategies used to investigate membrane-associated phenomena. The collected contributions present current approaches to studying interactions between model membranes and small molecules, peptides, lipoproteins, and natural compounds, while also addressing broader biological and pharmacological implications. Special attention is devoted to the relationship between membrane structure, physicochemical properties, and biological function.

The aim of this Reprint is to bring together interdisciplinary perspectives that connect membrane biophysics with medicinal chemistry, cell biology, computational studies, and drug development. The included studies highlight both the advantages and limitations of contemporary biomembrane models and demonstrate the importance of integrating multiple analytical methodologies to better reproduce the complexity of native biological membranes.

This Reprint is addressed to researchers, doctoral students, and professionals working in membrane science, biophysics, pharmaceutical sciences, molecular biology, computational studies, and related biomedical disciplines. We hope that the presented works will contribute to a deeper understanding of membrane-driven biological processes and inspire further research focused on the development of advanced membrane models and membrane-targeted therapeutic strategies.

Jadwiga Maniewska and Katarzyna Gębczak
Guest Editors

membranes

Editorial

Recent Advances in Biomembrane Models for Studying Interactions with Bio-/Molecules

Jadwiga Maniewska [1,*] and Katarzyna Gębczak [2,*]

1 Department of Medicinal Chemistry, Faculty of Pharmacy, Wroclaw Medical University, Borowska 211, 50-556 Wroclaw, Poland
2 Division of Basic Medical Sciences, Faculty of Pharmacy, Wroclaw Medical University, Borowska 211, 50-556 Wroclaw, Poland
* Correspondence: jadwiga.maniewska@umw.edu.pl (J.M.); katarzyna.gebczak@umw.edu.pl (K.G.)

1. Introduction

Biological membranes play a pivotal role in determining cellular organization and functionality, as they provide complex and dynamic environments for interactions with a diverse array of biomolecules and external compounds. The physicochemical properties of these membranes, shaped by factors such as lipid composition, membrane-associated proteins, and lateral heterogeneity, are critical for governing essential biological processes. Therefore, achieving a comprehensive understanding of these interactions is vital for advancing fields such as biophysics, pharmacology, and medicinal chemistry.

In recent years, significant progress has been made in developing biomembrane models and the methodologies employed to study them. Techniques such as calorimetry, fluorescence-based measurements, and high-resolution imaging, alongside molecular dynamics simulations, have opened avenues for increasingly precise characterization of membrane structures, dynamics, and mechanisms of interaction. At the same time, there has been a growing focus on how lipid diversity, domain formation, and cholesterol content influence membrane behavior and responsiveness to interacting molecules.

Nevertheless, several challenges persist. A key limitation is the difficulty in translating findings from simplified model systems to the inherently more complex environment of native biological membranes, which are characterized by compositional asymmetry, spatial heterogeneity, and continuous remodeling. Furthermore, the molecular mechanisms by which small molecules, peptides, and proteins disrupt membrane organization and initiate downstream biological effects remain inadequately understood, particularly regarding interconnected phenomena such as cytoskeletal rearrangement and lipid domain dynamics. Another unresolved issue is establishing predictive relationships between membrane-level interactions and pharmacological outcomes.

The Special Issue entitled "Recent Advances in Biomembrane Models for Studying Interactions with Bio-/Molecules," published in *Membranes* (MDPI), addresses these challenges by presenting eight contributions that integrate experimental and computational approaches to explore membrane–molecule interactions across various organizational levels. These studies enhance our understanding of how membrane composition, molecular structure, and biophysical properties influence biological activity.

2. Overview of Special Issue Contributions

Several papers in this collection focus on the interaction of small bioactive molecules with model lipid membranes, combining physicochemical characterization with biological evaluation. In Contribution 1, arylpiperazine oxicam derivatives were investigated using differential scanning calorimetry (DSC) in phospholipid bilayers composed of 1,2-dimyristoyl-sn-glycero-3-phosphocholine (DMPC). The results showed concentration-dependent decreases in the phase transition temperature, indicating incorporation of the compounds into the lipid bilayer and partial disruption of lipid organization. Biological evaluation using MTT assays in cancer (MCF-7, MCF-7/DX, LOVO, and LOVO/DX) and normal (V79) cell lines revealed moderate cytotoxic activity, particularly against colorectal cancer cells, whereas cyclooxygenase inhibition remained weak despite molecular docking, suggesting potential interactions with the COX-2 active site. These findings indicate that membrane interactions constitute a primary determinant of the observed biological activity.

In a related study (Contribution 4), another series of arylpiperazine oxicam derivatives was evaluated using a combined experimental and theoretical approach. Cyclooxygenase inhibition assays, cytokine mRNA expression analysis, and antioxidant activity tests based on DPPH and ABTS radical scavenging were complemented by density functional theory calculations to assess reactivity profiles. The results demonstrated moderate to high selectivity toward the COX-2 isoform, low cytotoxicity relative to reference compounds, and significant antioxidant capacity. The observed interactions with phospholipid membranes further emphasize the importance of membrane-related effects in determining pharmacological behavior.

Natural compounds and plant-derived molecules are addressed in detail in several contributions. In Contribution 2, polyphenolic fractions enriched in flavonoid glycosides and caffeoylquinic acids, derived from *Inula oculus-christi* L. (medicinal plant), were investigated using confocal microscopy. Membrane order was assessed with the fluorescent probe Di-4-ANEPPDHQ, while actin cytoskeleton organization was visualized using TRITC-phalloidin staining. The results revealed distinct cell-specific effects, with flavonoid glycosides inducing membrane ordering in cancerous A549 cells and membrane fluidization in non-malignant MDCK II cells, accompanied by pronounced reorganization of F-actin structures. These findings demonstrate a strong coupling between membrane biophysical properties and cytoskeletal dynamics, as well as the influence of molecular structure on membrane interactions.

Complementary insight at the molecular level is provided in Contribution 3, in which the author employed molecular dynamics simulations to investigate the behavior of nymphaeol A (isolated from propolis produced by honeybees) in complex lipid membranes. The simulations demonstrated spontaneous insertion of the molecule into the bilayer, preferential localization near the carbonyl region of phospholipids, and a tendency to adopt an extended conformation with an average tilt angle of approximately 35° relative to the membrane normal. The compound exhibited significant mobility along the membrane normal and showed preferential interactions with specific lipid species, including POPC and PSM, while slightly increasing membrane fluidity. These results provide detailed insight into the relationship between molecular structure, membrane localization, and functional activity.

The importance of lipid composition as a key determinant of membrane interactions is further emphasized in Contribution 5, in which the authors examined the interaction of high-density lipoprotein (HDL) particles with model membranes using fluorescence-based techniques. In the study, they analyzed lipid domain organization, membrane order, and both lateral and intramolecular mobility, with a particular focus on the glycerol region of lipids and the associated hydrogen bond network. The results demonstrated that both fluid-phase and certain gel-phase membranes interact with HDL particles, although with distinct functional outcomes: gel-phase membranes favor binding, whereas fluid-phase membranes facilitate cargo transfer. Increased glycerol region mobility and a strengthened hydrogen bond network were associated with enhanced anchoring interactions, while

https://doi.org/10.3390/membranes16050174

reduced water structuring promoted lipid transfer. The presence of cholesterol was shown to reduce HDL interaction by increasing membrane packing density and stiffness, thereby modulating membrane fluidity.

Two review articles provide broader perspectives on membrane-mediated biological processes and their pharmacological implications. Contribution 6 addressed the role of biological membranes in amyloidogenic protein toxicity, emphasizing how membrane physicochemical properties influence protein misfolding, aggregation, and cytotoxicity. In the review, the authors discuss small molecules capable of intercalating into lipid bilayers, including aminosterols, cholesterol, ubiquinone, and selected polyphenols, which can modulate membrane properties and reduce amyloid–membrane interactions. At the same time, inconsistencies in reported protective effects highlight the need for a deeper mechanistic understanding of these processes.

Contribution 7 focused on the role of membrane transporters, particularly ATP-binding cassette (ABC) and solute carrier (SLC) proteins, in drug pharmacokinetics. These transporters play a crucial role in drug absorption, distribution, and elimination. The complex interplay between transporter expression, physiological conditions, and co-administered compounds is identified as a key factor influencing drug efficacy and safety.

Finally, the authors of Contribution 8 examined the influence of cholesterol on membrane-targeted bioactive peptides. They highlight how cholesterol modulates lipid packing, membrane rigidity, and lateral mobility, thereby affecting peptide insertion, membrane disruption, and selectivity. The findings emphasize the importance of incorporating realistic membrane compositions, including cholesterol and sphingolipids, into experimental models in order to better replicate native biological systems.

3. Conclusions

The studies compiled in this Special Issue highlight the pivotal role of biomembrane models in elucidating the mechanisms governing membrane–molecule interactions. Taken together, the contributions demonstrate that membrane composition, structural organization, and dynamic properties critically influence molecular behavior and, consequently, biological function in systems ranging from small bioactive compounds to complex biological assemblies.

Importantly, the presented works contribute to narrowing several gaps that persist in the field. They provide mechanistic evidence that membrane interactions can act as primary drivers of biological activity, clarify the relationship between membrane organization and cellular processes, and emphasize the regulatory role of lipid composition, including cholesterol and specific lipid classes, in modulating these effects. Furthermore, the combined use of experimental techniques and computational modeling underscores the importance of multiscale approaches for achieving a more comprehensive description of membrane-associated phenomena.

However, several challenges remain. Future research should focus on developing membrane models that more accurately represent the structural and compositional complexities of native systems, including lipid asymmetry, heterogeneity, and curvature effects. Stronger integration of experimental findings with theoretical frameworks will also be essential for establishing predictive models that link molecular interactions to functional outcomes.

From a practical perspective, translating these fundamental insights into real-world applications is a key objective. Future studies should prioritize the rational design of membrane-active compounds, optimization of drug delivery strategies, and identification of therapeutic targets related to membrane interactions. Enhancing our understanding of membrane–molecule interactions in complex biological contexts will be vital for tackling issues such as multidrug resistance and targeted therapies.

https://doi.org/10.3390/membranes16050174

This Special Issue aims to serve as a valuable reference and inspire further research that connects model membrane systems with the intricacies of biological membranes, ultimately advancing the field of biomembrane science.

Author Contributions: J.M. and K.G. equally contributed to this Editorial. All authors have read and agreed to the published version of the manuscript.

Conflicts of Interest: The authors declare no conflict of interest.

List of Contributions:

1. Maniewska, J.; Gębczak, K.; Cwynar-Zając, Ł.; Czyżnikowska, Ż.; Szczęśniak-Sięga, B. New Oxicam Derivatives—Studies of Membrane Interactions, Cytotoxicity, Cyclooxygenase Inhibition and Molecular Docking. *Membranes* **2026**, *16*, 166. https://doi.org/10.3390/membranes16050166.
2. Veleva, R.; Kostadinova, A.; Trendafilova, A.; Ivanova, V.; Moskova-Doumanova, V.; Mladenova, K.; Doumanov, J.; Benkova, D.; Staneva, G.; Topouzova-Hristova, T. Polyphenols from *Inula oculus-christi* L. Induced Cell-Specific Membrane and Cytoskeleton Reorganization. *Membranes* **2025**, *15*, 357. https://doi.org/10.3390/membranes15120357.
3. Villalaín, J. Location and Dynamics of Nymphaeol A in a Complex Membrane. *Membranes* **2025**, *15*, 163. https://doi.org/10.3390/membranes15060163.
4. Szczęśniak-Sięga, B.; Maniewska, J.; Wiatrak, B.; Janek, T.; Nowotarska, P.; Czyżnikowska, Ż. Anti-Inflammatory Properties of Novel 1,2-Benzothiazine Derivatives and Their Interaction with Phospholipid Model Membranes. *Membranes* **2024**, *14*, 274. https://doi.org/10.3390/membranes14120274.
5. Weber, F.; Axmann, M.; Sezgin, E.; Amaro, M.; Sych, T.; Hochreiner, A.; Hof, M.; Schütz, G.; Stangl, H.; Plochberger, B. "Head-to-Toe" Lipid Properties Govern the Binding and Cargo Transfer of High-Density Lipoprotein. *Membranes* **2024**, *14*, 261. https://doi.org/10.3390/membranes14120261.
6. Seychell, R.; El Saghir, A.; Vassallo, N. Modulation of Biological Membranes Using Small-Molecule Compounds to Counter Toxicity Caused by Amyloidogenic Proteins. *Membranes* **2024**, *14*, 231. https://doi.org/10.3390/membranes14110231.
7. Kiełbowski, K.; Król, M.; Bakinowska, E.; Pawlik, A. The Role of ABCB1, ABCG2, and SLC Transporters in Pharmacokinetic Parameters of Selected Drugs and Their Involvement in Drug–Drug Interactions. *Membranes* **2024**, *14*, 223. https://doi.org/10.3390/membranes14110223.
8. Giraldo-Lorza, J.; Leidy, C.; Manrique-Moreno, M. The Influence of Cholesterol on Membrane Targeted Bioactive Peptides: Modulating Peptide Activity Through Changes in Bilayer Biophysical Properties. *Membranes* **2024**, *14*, 220. https://doi.org/10.3390/membranes14100220.

membranes

MDPI

Article

New Oxicam Derivatives—Studies of Membrane Interactions, Cytotoxicity, Cyclooxygenase Inhibition and Molecular Docking

Jadwiga Maniewska [1,*], Katarzyna Gębczak [2], Łucja Cwynar-Zając [2], Żaneta Czyżnikowska [3] and Berenika M. Szczęśniak-Sięga [1,*]

1 Department of Medicinal Chemistry, Faculty of Pharmacy, Wroclaw Medical University, Borowska 211, 50-556 Wroclaw, Poland
2 Division of Basic Medical Sciences, Department of Basic Medical Sciences and Immunology, Faculty of Pharmacy, Wroclaw Medical University, Borowska 211, 50-556 Wroclaw, Poland
3 Department of Basic Chemical Sciences, Faculty of Pharmacy, Wroclaw Medical University, Borowska 211a, 50-556 Wroclaw, Poland
* Correspondence: jadwiga.maniewska@umw.edu.pl (J.M.); berenika.szczesniak-siega@umw.edu.pl (B.M.S.-S.); Tel.: +48-71-7840397 (J.M. & B.M.S.-S.)

Abstract

Oxicam derivatives, a class of nonsteroidal anti-inflammatory drugs (NSAIDs), are important scaffolds for developing biologically active compounds. In this study, arylpiperazine oxicam derivatives (PR24–PR50) were examined for membrane interactions, cytotoxic activity, cyclooxygenase inhibition, and potential binding to COX-2 protein. Membrane interactions were examined using differential scanning calorimetry (DSC) in phospholipid bilayers formed from 1,2-dimyristoyl-*sn*-glycero-3-phosphocholine (DMPC). All compounds altered the thermotropic properties of the lipid bilayer, showing concentration-dependent decreases in phase transition temperature, indicating incorporation to bilayer and partial disruption of lipid organization. Cytotoxicity, assessed using the MTT assay in breast cancer (MCF-7, MCF-7/DX), colorectal cancer (LOVO, LOVO/DX), and normal V79 cell lines, showed moderate effects, particularly against colorectal cancer cells. Cyclooxygenase inhibition was rather weak, with IC_{50} values in the high micromolar range, indicating limited anti-inflammatory potential compared with reference COX inhibitors, although docking studies suggested possible interactions with the COX-2 active site. The obtained results indicate that the biological activity of the arylpiperazine oxicam derivatives is primarily associated with cytotoxicity and membrane effects rather than COX inhibition. These limitations should be considered in the design of future membrane-targeted bioactive compounds.

Keywords: model membranes; differential scanning calorimetry (DSC); oxicam derivatives; cyclooxygenase; ROS; molecular docking

1. Introduction

Oxicams are a subgroup of nonsteroidal anti-inflammatory drugs (NSAIDs) known for their ability to non-selectively inhibit cyclooxygenase 1 and 2 (COX-1 and COX-2) [1]. Since the discovery that inflammatory processes are involved in carcinogenesis, anti-inflammatory drugs have become a target of research to support many anticancer therapies [2]. Cancer is still a leading cause of death worldwide—ranking as the first or second cause of death before the age of 70 in 112 out of 183 countries [3]. The pathophysiology of carcinogenesis is very complex and still not fully understood. Cancer is a pathological

 https://doi.org/10.3390/membranes16050166

condition in which homeostasis in cell proliferation and differentiation is disturbed [4,5]. Less than 10% of cancers are caused by germline mutations, with most cases arising from somatic changes and environmental factors. Cancer is often associated with diet, obesity, inhaled pollutants, smoking, or an autoimmune reaction [6–8]. Moreover, chronic inflammation is recognized as a key pathological factor, representing a prolonged and dysregulated response to the loss of tissue homeostasis [9]. Cancer development may be viewed as a dysregulated repair response to tissue damage, where inflammatory and neoplastic processes co-exist, forming the concept of a "wound that does not heal" [10].

The link between inflammation and tumor development was first observed in 1863 by Rudolf Virchow, founder of modern pathology [11]. Virchow observed tumors in several patients where they had been struck or repeatedly irritated by shoes or work tools. Examining the tumor tissues under a microscope, he observed leukocyte infiltration. He hypothesized that cancer resulted from a failed attempt by the body's immune system to repair the wound. However, scientists did not take this concept seriously at the time. Only more than a century later, in 1986, Harold Dvorak, a pathologist at Harvard University, demonstrated a surprising similarity between the mechanisms naturally occurring in the inflammatory process and the development of tumor tissue [10]. Dvorak also noted that cancer is often directly related to chronic inflammation, a finding confirmed by the other subsequent research [12]. Inflammation arising from chronic infections is estimated to contribute to around 25% of the worldwide cancer burden [13]. Furthermore, chronic inflammatory processes triggered by chemical and physical agents, autoimmune factors, as well as reactions of unknown origin, are linked to an increased cancer risk [14].

Additionally, the use of NSAIDs for analgesic and anti-inflammatory purposes has revealed their chemopreventive effects in several cancers, including breast, lung, bladder, colon, and prostate cancers [15–31]. NSAIDs inhibit cyclooxygenase, an enzyme with at least two isoforms, COX-1 and COX-2. COX-1 is a constitutively expressed glycoprotein present in various tissues, including the stomach, intestines, ovaries, kidneys, and platelets, with important protective functions [32]. COX-2 is an inducible isoform upregulated by cytokines (IL-1β, IL-6, TNFα), mitogens, and growth factors during inflammation, pain, tissue damage, and carcinogenesis. Its overexpression, linked to increased prostaglandin E_2 (PGE_2) production, plays a key role in carcinogenesis and chronic inflammation [33]. COX-2 has been shown to participate in cancer development at multiple stages, from tumor initiation to metastasis [34]. The protumorigenic activity of COX-2 is thought to involve several mechanisms, including stimulation of mutant cell proliferation, inhibition of apoptosis, reduced epithelial cell adhesion, promotion of angiogenesis, increased invasiveness, modulation of immune surveillance, induction of DNA mutagenesis, and activation of carcinogenic xenobiotics through peroxidase activity [34].

Over the last three decades, numerous reports have emerged from experimental animal studies, epidemiological studies, and clinical trials showing that regular use of aspirin and other NSAIDs, including celecoxib, piroxicam, and sulindac, has been associated with a reduction in cancer risk of up to 50% [35]. Some anti-inflammatory drugs have even been approved for use in cancer treatment or as chemopreventive agents; for example, celecoxib is used to reduce the risk of colorectal adenomas in people with familial adenomatous polyposis [36]. Oxicams, such as piroxicam or meloxicam, are characterized by potent anti-inflammatory activity and a long duration of action, and were therefore also targeted for such studies. However, piroxicam was withdrawn from further studies due to its significant toxicity. To reduce its toxicity, new arylpiperazine oxicam derivatives were designed and synthesized. Some of them demonstrated anticancer or potential chemopreventive properties [37–41].

https://doi.org/10.3390/membranes16050166

A series of novel arylpiperazine oxicam derivatives (Table 1) was investigated for anti-inflammatory and cytotoxic properties. The anti-inflammatory potential was evaluated in vitro by measuring inhibition of COX-2, an enzyme involved in both inflammatory and pro-carcinogenic processes, whereas COX-1 inhibition was additionally assessed to define the COX-2/COX-1 selectivity profile. Due to the fact that cyclooxygenase is an enzyme anchored in the endoplasmic reticulum, drug permeation across biological membranes is necessary for interaction with the enzyme. For this purpose, studies of the interaction of oxicam derivatives with model biological membranes were also conducted. In addition, molecular docking studies focused on the COX-2 active site were conducted to investigate the binding mode of the compounds. The cytotoxic activity of the new oxicam derivatives was subsequently tested in two cancer cell lines, along with their doxorubicin-sensitive and doxorubicin-resistant sublines. What is more, to determine the compounds' safety, cytotoxicity was also tested on a normal cell line. The mechanism of cytotoxic action was determined by testing the compounds' activity on rhodamine-containing cells. In addition, the antioxidant activity of the new compounds was evaluated, as inflammation is associated with the generation of reactive oxygen and nitrogen species (ROS and RNS). These species may damage essential cellular components, including DNA, proteins, and lipids, thereby contributing—directly or indirectly—to neoplastic transformation.

Table 1. Chemical structures of studied oxicam derivatives. For synthesis route and experimental data please see [42,43].

Compound Abbreviation	Molecular Weight [g/mol]	Chemical Structure
PR24	535.63	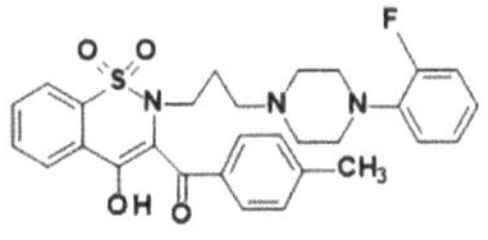
PR25	535.59	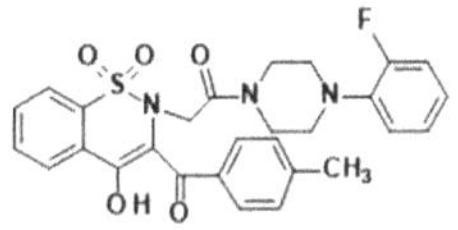
PR45	606.01	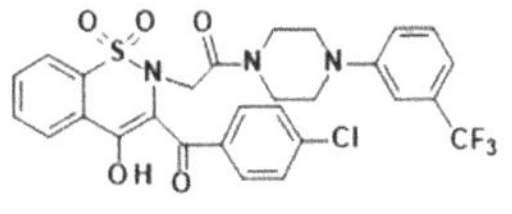
PR47	650.46	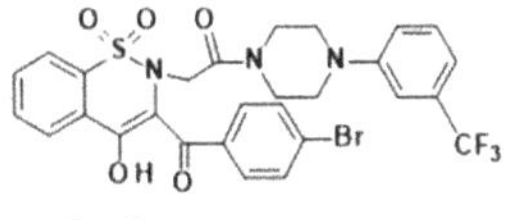
PR48	650.51	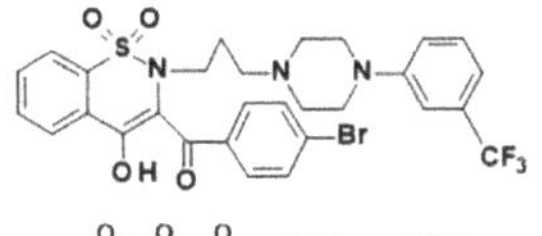
PR49	585.59	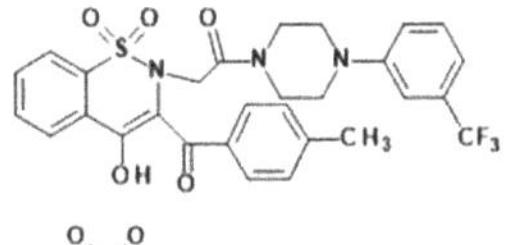
PR50	585.64	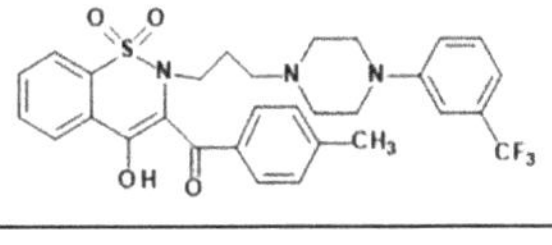

https://doi.org/10.3390/membranes16050166

2. Materials and Methods

2.1. Chemicals

Oxicam derivatives were obtained through synthesis performed at the Department of Medicinal Chemistry, Wroclaw Medical University, Poland. Chemical structures, molecular weights and abbreviations of studied oxicam derivatives are shown in Table 1. The synthesis route and spectral spectra confirming structures and purity of new compounds have been published previously [42,43]. Meloxicam and celecoxib (reference standards) were supplied by Thermo Fisher Scientific (Waltham, MA, USA). Given the lack of water solubility of the investigated compounds, they were dissolved in chloroform or dimethyl sulfoxide (DMSO) prior to use in the assays.

2.2. Differential Scanning Calorimetry (DSC)

The thermal response of phospholipid bilayers in the presence of oxicam derivatives was examined using 1,2-dimyristoyl-*sn*-glycero-3-phosphocholine (DMPC; Sigma-Aldrich, St. Louis, MO, USA) as a representative membrane lipid. The investigated compounds were initially dissolved in chloroform to obtain 5 mM stock solutions. Defined amounts of these solutions were subsequently combined with DMPC (2 mg per sample) to yield compound-to-lipid molar fractions ranging from 0.02 to 0.1. Following mixing, organic solvent was removed by evaporation under a stream of nitrogen, and samples were further subjected to vacuum drying for 2 h to eliminate residual traces. The resulting lipid films were rehydrated with 20 µL of Tris–EDTA buffer (20 mM, pH 7.4; Sigma-Aldrich), then heated above the main phase transition temperature of DMPC and vortexed to produce uniform dispersions. Prepared samples were hermetically sealed in aluminum *Concavus pans* (Netzsch GmbH & Co., Selb, Germany) and analyzed by differential scanning calorimetry using a DSC 214 *Polyma* instrument (Netzsch GmbH & Co., Selb, Germany) operating in heat-flow mode. Measurements were conducted over the temperature range of 10–35 °C at a constant heating rate of 1 °C/min under a nitrogen purge (25 mL/min). For each composition, two independently prepared samples were measured, each in duplicate. Transition enthalpy values (J/g) were calculated with respect to lipid mass using Netzsch Proteus software (version 7.1.0). The experimental protocol was based on our previously reported methodology [44].

2.3. Biological Assays

The biological activity of the investigated compounds was assessed using four complementary assays: MTT, cyclooxygenase (COX) inhibition, DCFH-DA, and Rhodamine 123. These assays were employed to evaluate cell viability, COX inhibitory activity, intracellular reactive oxygen species (ROS) generation, and mitochondrial membrane potential ($\Delta\Psi m$), respectively. Tests were performed for three cell lines: the normal V79 line, the breast cancer line MCF-7, and its doxorubicin-resistant variant MCF-7/DX, as well as the colorectal cancer line LOVO, and its resistant variant LOVO/DX. The V79 cell line is derived from the lung tissue of a young male Chinese hamster and displays fibroblast-like morphology [45]. The LOVO cell line originates from a metastatic lesion in the left supraclavicular region of a 56-year-old male patient with colorectal adenocarcinoma and is characterized by epithelial morphology [46]. The MCF-7 cell line was established in 1970 from pleural effusion obtained from a 69-year-old female patient with metastatic breast adenocarcinoma and is characterized by epithelial morphology [47]. The LOVO/DX and MCF-7/DX cell lines were created by cultivating parental counterparts in the presence of low concentrations of doxorubicin. Cells were seeded in 96-well plates at a density of 1–2 × 10^4 cells per well and cultured for 24 h under standard conditions (37 °C, 5% CO_2, humidified atmosphere) to allow cell adhesion. In all four assays, the cells were treated

https://doi.org/10.3390/membranes16050166

with the tested compounds at the same concentrations of 1, 10, and 50 μM. The starting concentration was chosen as 1 μM because doxorubicin, the reference drug, is often tested at this concentration, which facilitates comparison of the efficacy of the tested compounds. The subsequent concentrations—10 and 50 μM—were chosen after preliminary studies that demonstrated differences in the effects of the compounds between these concentrations. All measurements were performed using a Varioskan Microplate Reader (Thermo Fisher Scientific, Waltham, MA, USA).

2.3.1. Determination of Cell Proliferation (MTT Assay)

Cell viability was assessed using the MTT assay based on the reduction of tetrazolium salt to formazan by metabolically active cells, as previously described, with minor modifications [48]. Following 24 h exposure to the tested compounds, MTT solution (1 mg/mL) was added to each well, and the cells were incubated for 2 h at 37 °C. Subsequently, the medium was removed, and the formed formazan crystals were dissolved in dimethyl sulfoxide (DMSO). Absorbance was measured at 570 nm, and cell viability was calculated as a percentage relative to untreated control cells (100%) [49]. Non-viable cells do not reduce MTT and therefore do not generate a detectable signal.

2.3.2. Cyclooxygenase Inhibitory Activity

Cyclooxygenase (COX) inhibitory activity was determined using a colorimetric assay kit (Cayman Chemical Company, Ann Arbor, MI, USA) according to the manufacturer's instructions [50]. The assay is a high-throughput microplate-based method that measures COX peroxidase activity via colorimetric detection of oxidized *N,N,N′,N′*-tetramethyl-*p*-phenylenediamine (TMPD) at 590 nm, enabling assessment of compounds that inhibit COX activity in vitro. The assay was performed according to the manufacturer's protocol using 0.1 M Tris-HCl buffer (pH 8.0), heme solution in DMSO, COX-1 or COX-2 enzymes, arachidonic acid (100 μM), KOH (0.1 M), and TMPD solution. The reaction mixture consisted of 150 μL assay buffer, 10 μL heme, and 10 μL of COX-1 or COX-2 enzyme per well. Test samples were prepared by adding 10 μL of the investigated compounds at final concentrations of 1, 10, and 50 μM. For control wells, 10 μL of DMSO was used instead of the compound. The reaction was initiated by the addition of 20 μL TMPD, followed by the addition of arachidonic acid, and allowed to proceed for 2 min at 25 °C. The oxidation of TMPD was measured using a Varioskan Lux microplate reader (Thermo Scientific, USA) at 590 nm. Inhibitory activity was expressed as the concentration required to inhibit enzyme activity by 50% (IC_{50}). Selectivity was assessed using the selectivity index (SI), calculated as IC_{50} (COX-2)/IC_{50} (COX-1).

2.3.3. Reactive Oxygen Species Assay

Intracellular reactive oxygen species (ROS) levels were determined using the DCFH-DA assay (Sigma-Aldrich, St. Louis, MO, USA) as described elsewhere [48,51]. Following 24 h exposure to the tested compounds (1, 10, and 50 μM), the culture medium was removed and cells were washed with phosphate-buffered saline (PBS). Subsequently, DCFH-DA solution (20 μM in PBS) was added, and the cells were incubated for 30 min at 37 °C in the dark. After incubation, cells were washed with PBS, and fluorescence of the oxidized product (DCF) was measured using excitation and emission wavelengths of 485 nm and 530 nm, respectively. Results were expressed as a percentage of the untreated control (100%), with increased fluorescence corresponding to elevated ROS production.

2.3.4. Rhodamine 123 Assay

Mitochondrial membrane potential was evaluated using the Rhodamine 123 assay (Sigma-Aldrich, St. Louis, MO, USA) following a previously reported procedure [52].

https://doi.org/10.3390/membranes16050166

After 24 h exposure to the tested compounds (1, 10, and 50 μM), the culture medium was removed and cells were washed with phosphate-buffered saline (PBS). Cells were then incubated with Rhodamine 123 solution (5 μM in PBS) for 30 min at 37 °C in the dark. Following incubation, cells were washed with PBS, and fluorescence intensity was measured at excitation and emission wavelengths of 505 nm and 530 nm, respectively [53]. Results were expressed as a percentage of control fluorescence (100%), where a decrease in signal indicated loss of mitochondrial membrane potential.

2.3.5. Statistical Analysis

Statistical analysis was performed using Statistica 13.3 software (TIBCO Software Inc., Palo Alto, CA, USA). Data distribution normality was evaluated using the Shapiro–Wilk test. For normally distributed data, comparisons between independent groups were performed using Student's *t*-test. In cases where the assumption of normality was not satisfied, the nonparametric Mann–Whitney U test was applied [48,52]. Results are presented as mean ± standard deviation (SD). Statistical significance was defined as $p < 0.05$ (*), $p < 0.01$ (**), and $p < 0.001$ (***).

2.4. Computational Studies

The geometry of oxicam derivatives was optimized at the B3LYP/6-31++G** level of theory using the polarizable continuum model (PCM) to include the solvent effects. The calculations were performed by using the Gaussian 09 program [54–58]. The standard AutoDock 4.2 protocol was employed to obtain the binding modes of the compounds within the COX-2 binding cavity [59]. The crystal structure of cyclooxygenase-2 (COX-2; PDB ID: 4M11), co-crystallized with meloxicam, was retrieved from the Protein Data Bank [60]. The structure of the molecular target (chain A) was prepared by adding polar hydrogen atoms and assigning appropriate solvation parameters. Gasteiger partial charges were assigned to all atoms. The binding site was defined using a grid box of 60 × 60 × 60 points centered on the active site based on the position of the co-crystallized ligand, with a grid spacing of 0.375 Å. Visualization was performed using Discovery Studio Visualizer 2021 (Dassault Systèmes, Waltham, MA, USA; https://discover.3ds.com/, accessed on 1 October 2025).

3. Results

3.1. Differential Scanning Calorimetry

Differential Scanning Calorimetry (DSC) is one of the techniques used in studying drug–membrane interactions by measurements of the thermotropic phase behavior of phospholipids building the membrane [61]. In our research, this tool was used to characterize the influence of oxicam derivatives on model phospholipid membranes formed from DMPC, because this technique makes it possible to follow the effects of any additives to model membrane on cooperativity and enthalpy of the lipid phase transition. The dependencies of the main transition temperature, half-height width and enthalpy change on the oxicam derivative concentration are shown in Figure 1.

The investigated compounds had previously been studied on model biological membranes obtained from dipalmitoylphosphatidylcholine (DPPC) [42,43]. Herein they were examined using DMPC, a phospholipid characterized by shorter hydrocarbon chains than DPPC. Uniform molar ratios were applied to allow direct comparison of the obtained results. All tested compounds induced a concentration-dependent decrease in the main phase transition temperature of DMPC, indicating a destabilization of the ordered gel phase. In parallel, incorporation of the compounds into the phospholipid bilayer resulted in a broadening of the phase transition peaks, suggesting increased heterogeneity of the lipid environment and reduced cooperativity of the phase transition. Furthermore, a decrease in the enthalpy of the main DMPC phase transition was observed for all compounds, reflecting weakened lipid–lipid interactions. These findings demonstrate that the examined compounds interact with the phosphatidylcholine model membrane and significantly modify its thermotropic properties. The most pronounced effects on the transition temperature were observed for compounds PR25, PR49, and PR50. The nature of the observed changes suggests that oxicam derivatives affect both the polar headgroup region and the hydrocarbon chains of DMPC, leading to a reduction in the stability of the gel phase and weakened intermolecular interactions within the lipid bilayer [62].

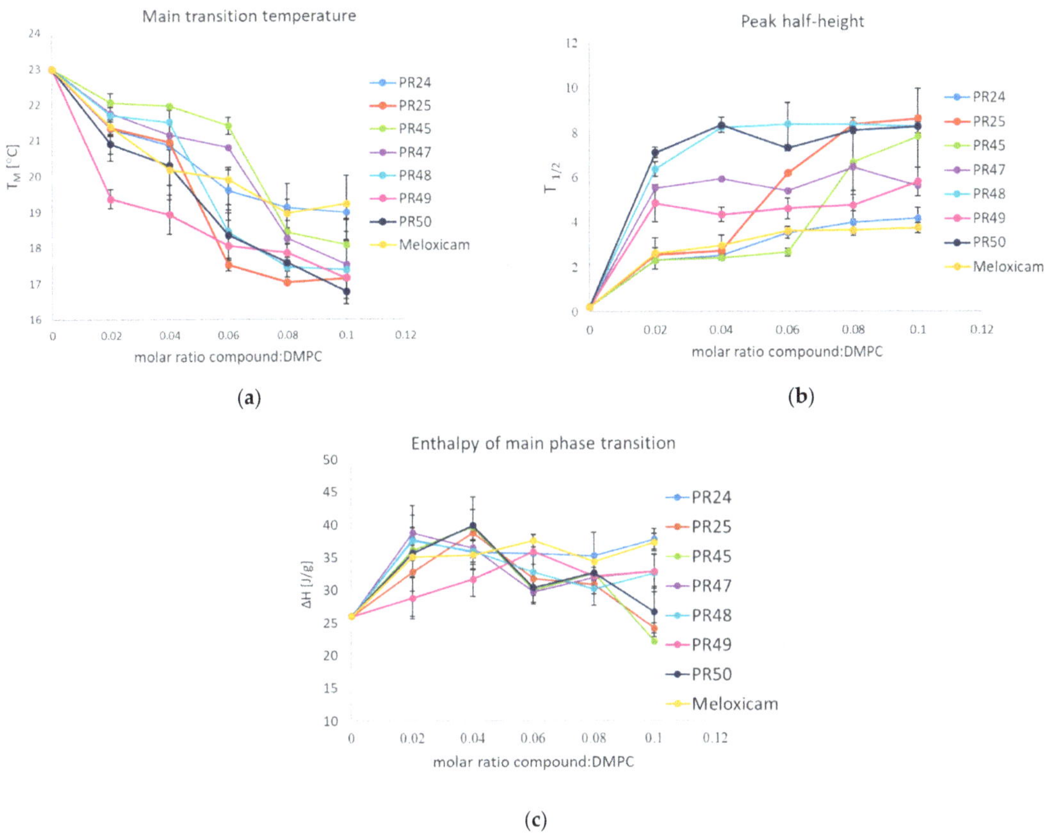

Figure 1. Influence of studied compounds on the DMPC: (**a**) main phase transition temperature; (**b**) main phase transition peak half-height width; (**c**) main phase transition enthalpy; mean ± SD, $n = 4$.

3.2. Biological Assays

3.2.1. Determination of Cell Proliferation

In Figure 2, the results of the MTT assay are presented, determining the effect of the tested compounds PR24–PR50 on the viability of cancer cells (MCF-7, MCF-7/DX, LOVO, LOVO/DX) and normal cells (V79) after 24 h of incubation at three concentrations: 1, 10, and 50 μM. Normal cell line V79 is a Chinese hamster lung fibroblast cell commonly used in toxicology studies. Cancer cell line MCF-7 is a breast cancer cell, and LOVO is a colorectal cancer cell. Cancer cell lines MCF-7/DX and LOVO/DX are their doxorubicin-resistant variants.

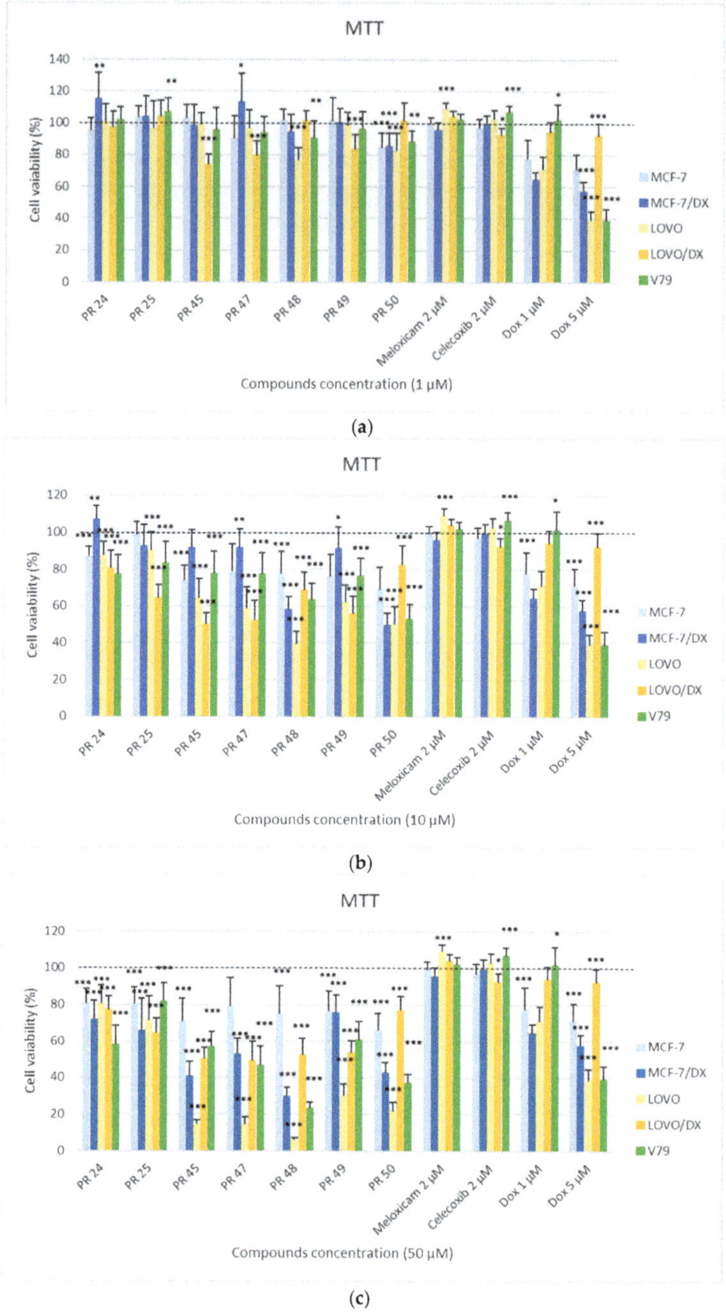

Figure 2. Effect of the tested compounds PR24–PR50 on the viability of cancer (MCF-7, MCF-7/DX, LOVO, LOVO/DX) and normal (V79) cells at a concentration of 1 µM (**a**), 10 µM (**b**), 50 µM (**c**), evaluated by the MTT assay after 24 h of incubation. Results are presented as the percentage of cell viability relative to the untreated control (100%). The presence of statistical significance markers (* $p \leq 0.05$; ** $p < 0.01$; *** $p < 0.001$) indicates significant differences compared to the control group.

https://doi.org/10.3390/membranes16050166

The MTT assay showed differentiated cytotoxicity of the tested derivatives depending on the cell type, the applied concentration, as well as the structure of tested compound. With increasing concentration, a clear decrease in cell viability was observed, confirming the dose-dependent nature of cytotoxicity. The strongest cytotoxic effect occurred at the concentration of 50 μM (Figure 2c), especially against the cancer cells LOVO and LOVO/DX (colon cancer cell lines). For most compounds (PR45, PR47, PR48, PR49, PR50), the viability of LOVO cells decreased below 40–60% compared to the control (100%). At the same time, the resistant LOVO/DX cells were characterized by higher viability, indicating persistent resistance to the tested derivatives.

In the breast cancer cell lines MCF-7 and MCF-7/DX, moderate cytotoxicity of most derivatives was observed. The most active compounds were again PR48 and PR50, causing a reduction in the viability of MCF-7/DX cells to about 60–70% at 50 μM. The MCF-7 (non-resistant) cells generally showed lower sensitivity to the tested compounds than LOVO cells, which may result from metabolic differences between cancer types.

In the case of normal V79 cells, the effect of the tested derivatives was varied and depended on the type of compound. Some derivatives, such as PR47 and PR50, showed moderate cytotoxicity. The lowest toxicity toward normal cells was exhibited by compound PR25, while PR48 showed the strongest effect in this line.

At lower concentrations of 1 and 10 μM (Figure 2a,b), most compounds showed a smaller effect on cell viability, although statistically significant results were obtained in most cases. It is worth emphasizing that compounds PR48 and PR50, which were most cytotoxic to cancer cells, were equally toxic to healthy cells. Compounds PR25, PR45, PR47, and PR49, on the other hand, were more cytotoxic to resistant colon cancer cells (LOVO/DX) than to healthy cells (V79).

The IC_{50} values (μM) for the tested derivatives PR24–PR50 were determined based on the results of the MTT assay using the linear regression method (y = ax + b). Calculations were performed for five cell lines: the normal cell line (V79), and two cancer lines (MCF-7, MCF-7/DX, LOVO, LOVO/DX). The results are presented in Table 2. The symbol nd (not determined) indicates that the IC_{50} value could not be determined within the tested concentration range.

Table 2. Estimated IC_{50} values determined by the MTT assay.

Compound	MCF-7	MCF-7/DX	LOVO	LOVO/DX	V79
			[μM]		
PR24	nd	nd	nd	nd	108.0
PR25	nd	nd	nd	110.0	nd
PR45	nd	54.4	12.3	28.7	123.0
PR47	nd	87.4	11.0	28.5	55.6
PR48	nd	15.3	4.7	54.2	13.7
PR49	nd	nd	17.5	47.3	178.0
PR50	nd	18.3	8.8	nd	16.2

nd (not determined) indicates that IC_{50} value could not be determined within the tested concentration range.

The analysis of the obtained IC_{50} values confirmed the variable sensitivity of the tested cell lines to the studied compounds. The greatest cytotoxic activity against cancer cells was demonstrated by derivatives PR48 and PR50, which showed the lowest IC_{50} values—4.7 μM and 8.8 μM, respectively, for LOVO cells. High activity of these compounds was also confirmed in the resistant lines LOVO/DX (PR48—54.2 μM) and MCF-7/DX

(PR48—15.3 μM and PR50—18.3 μM). Unfortunately, these compounds were also the most toxic to healthy cells V79 (PR48—13.7 μM and PR50—16.2 μM).

Compounds PR45, PR47 and PR49 exhibited moderate cytotoxicity toward cancer cell lines, with particularly high activity against LOVO cells (IC_{50} values of 12.3, 11.0, and 17.5 μM, respectively) and their resistant variant LOVO/DX (IC_{50} values of 28.7, 28.5, and 47.3 μM, respectively). However, against healthy cells V79 they showed very low toxicity (IC_{50} values of 123.0, 55.6, and 178.0 μM, respectively) which indicates their greater selectivity towards cancer cells.

In the case of the MCF-7 cell line, it was not possible to determine IC_{50} values for any of the tested compounds. This was due to the fact that cell viability across the entire tested concentration range (1–50 μM) remained at approximately 80–90%, which prevented the establishment of a dose–response relationship required to calculate IC_{50} values using the linear regression method. The obtained results indicate that MCF-7 cells exhibited low sensitivity to the tested derivatives.

A comparison of IC_{50} values between sensitive and resistant colon cancer cell lines showed that the resistant cells (LOVO/DX) generally exhibited higher IC_{50} values, suggesting reduced sensitivity to the tested compounds.

3.2.2. Structure–Activity Relationship (SAR) Analysis

The two most cytotoxic compounds, both against cancer and healthy cell lines, PR48 and PR50, differ from the moderately active compounds PR25, PR45, PR47, and PR49 in the type of linker between the thiazine and piperazine nitrogen atoms. The former possesses a propylene linker, while the latter possesses an acetyl linker, indicating the importance of this linker not only for the compound's overall toxicity but also for its activity profile. Compounds PR48 and PR50 also possess an *m*-trifluoromethylphenyl substituent on the piperazine moiety. The most cytotoxic compound, PR48, also has a bromine atom in its structure, while the less cytotoxic PR50 has a methyl group instead of bromine. This indicates that both the bromine atom and the *m*-trifluoromethyl substituent significantly enhance the compound's cytotoxicity.

On the other hand, compounds with the lowest cytotoxicity are PR24 and PR25. Structurally, they are distinguished by the presence of *o*-fluorophenyl substituent on the piperazine. The remaining compounds tested possess *m*-trifluoromethyl substituent, demonstrating that this substituent significantly enhances their cytotoxicity. The three compounds that demonstrated the highest cytotoxicity against LOVO cell line were PR45, PR47, and PR48. In addition to *m*-trifluoromethyl substituent mentioned above, these compounds also contain chlorine atom (PR45) or bromine atom (PR47 and PR48), suggesting that these substituents affect cytotoxicity more selectively. Due to the limited number of compounds, it is not possible to draw more detailed conclusions on the structure–activity relationship.

3.2.3. Cyclooxygenase Inhibitory Activity

The inhibitor concentration causing 50% inhibition of COX-1 and COX-2 activity (IC_{50}) was calculated based on the experimental data for the tested compounds at three concentrations: 1 μM, 10 μM, 50 μM and is presented in Table 3. The selectivity index (SI) was calculated as the quotient of IC_{50} COX-2/IC_{50} COX-1.

All tested compounds showed much lower effect on COX-1 and COX-2 activity compared to the reference compound, celecoxib. Compounds PR24, PR45, PR47, and PR48 exhibited a detectable reduction in COX-1 and COX-2 activity only at higher concentrations than celecoxib, with a somewhat stronger reduction in COX-1 activity than COX-2. Compound PR49 affected only COX-2 activity. Compounds PR25 and PR50 showed minimal or no detectable effect on either COX-1 or COX-2 within the tested concentration range.

https://doi.org/10.3390/membranes16050166

Table 3. IC_{50} values estimated for COX-1 and COX-2 activities at 2 min of incubation with the tested compounds [mean; SD].

Compound	Cyclooxygenase Inhibition Assay IC_{50} [µM]		SI IC_{50}(COX-2)/ IC_{50}(COX-1)
	COX-1	COX-2	
PR24	115.50 (9.6)	359.80 (58.2)	3.10
PR25	nd	nd	
PR45	241.05 (81.6)	507.80 (72.1)	2.10
PR47	41.55 (9.05)	206.20 (31.6)	4.96
PR48	103.20 (10)	377.35 (22.1)	3.66
PR49	nd	87.50 (0.49)	
PR50	nd	nd	
celecoxib	6.53 (0.01)	0.72 (0.01)	0.11

nd (not determined) indicates that IC_{50} value could not be determined within the tested concentration range.

3.2.4. Reactive Oxygen Species Assay

In Figure 3, the results of the DCFH-DA assay are presented, determining the effect of the tested compounds PR24–PR50 on the level of reactive oxygen species (ROS) in cancer cells (MCF-7, MCF-7/DX, LOVO, LOVO/DX) and normal cells (V79) after 24 h of incubation at three concentrations: 1, 10, and 50 µM.

The analysis of the obtained results showed that the effect of the tested compounds on ROS levels depended on both the cell type and the applied concentration. In most cases, a decrease in fluorescence intensity was observed with increasing compound concentration, indicating a reduced level of reactive oxygen species and, consequently, the potential antioxidant activity of the tested compounds.

The strongest antioxidant effect was observed at a concentration of 50 µM (Figure 3c), particularly in the LOVO and LOVO/DX cell lines, where for compounds PR47 and PR48 a clear decrease in fluorescence intensity was noted compared to the control. In the LOVO/DX line, this effect was slightly weaker than in LOVO cells, which may confirm the lower sensitivity of the drug-resistant colorectal cancer line.

In breast cancer cell lines MCF-7 and MCF-7/DX, a moderate but statistically significant decrease in ROS levels was observed in response to most of the tested derivatives. The most pronounced antioxidant effect was observed for compounds at the highest concentration (50 µM), particularly for compound PR45. In the MCF-7/DX line (doxorubicin-resistant), for most of the tested compounds, a tendency toward a smaller decrease in fluorescence signal was observed compared to the sensitive line, which may confirm the lower susceptibility of the resistant line, similarly to the LOVO/DX line.

In the case of normal V79 cells, the response to the tested derivatives was varied. At the highest concentration (50 µM), most of the tested compounds caused a decrease in ROS levels compared to the control. The most pronounced decrease in fluorescence intensity was observed for compound PR47. This effect may indicate the antioxidant activity of all the tested compounds also in normal cells.

At lower concentrations 1 and 10 µM (Figure 3a,b), smaller but statistically significant changes in ROS levels were observed compared to the control. In some cases (particularly for PR48 and PR49 in MCF-7/DX and PR49 in LOVO/DX), a statistically significant increase in fluorescence was observed, indicating that these compounds induce oxidative stress at lower concentrations.

https://doi.org/10.3390/membranes16050166

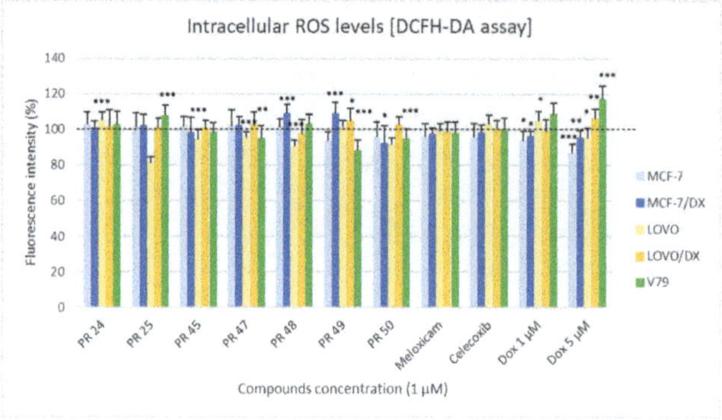

(**a**)

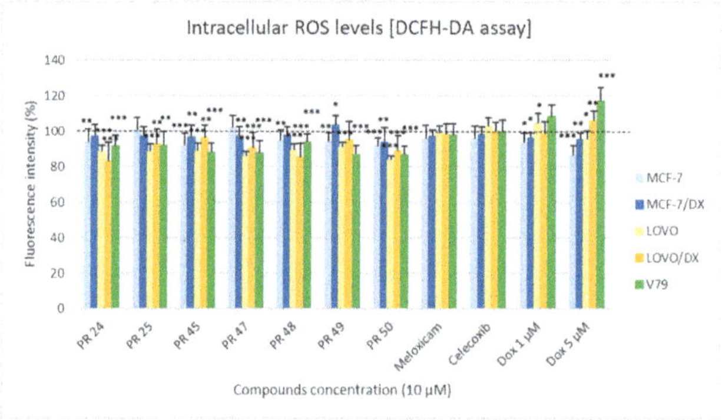

(**b**)

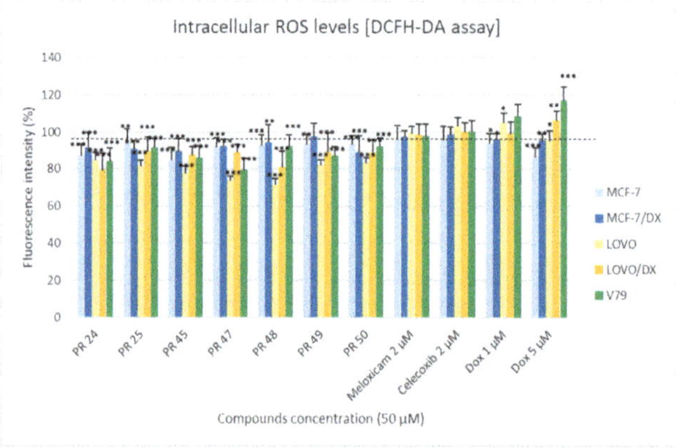

(**c**)

Figure 3. Effect of the tested compounds PR24–PR50 on the viability of cancer (MCF-7, MCF-7/DX, LOVO, LOVO/DX) and normal (V79) cells at a concentration of 1 μM (**a**), 10 μM (**b**) and 50 μM (**c**) evaluated by the DCFH-DA assay after 24 h of incubation. Results are presented as the percentage of cell viability relative to the untreated control (100%). The presence of statistical significance markers (* $p \leq 0.05$; ** $p < 0.01$; *** $p < 0.001$) indicates significant differences compared to the control group.

3.2.5. Rhodamine 123 Assay

In Figure 4, the results of the Rhodamine 123 (Rh123) assay are presented, determining the effect of the tested compounds PR24–PR50 on the mitochondrial membrane potential

https://doi.org/10.3390/membranes16050166

in cancer cells (MCF-7, MCF-7/DX, LOVO, LOVO/DX) and normal cells (V79) after 24 h of incubation at three concentrations: 1, 10, and 50 μM.

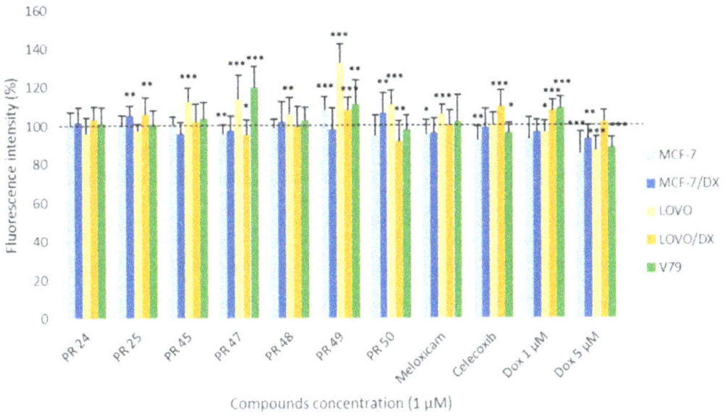

(**a**)

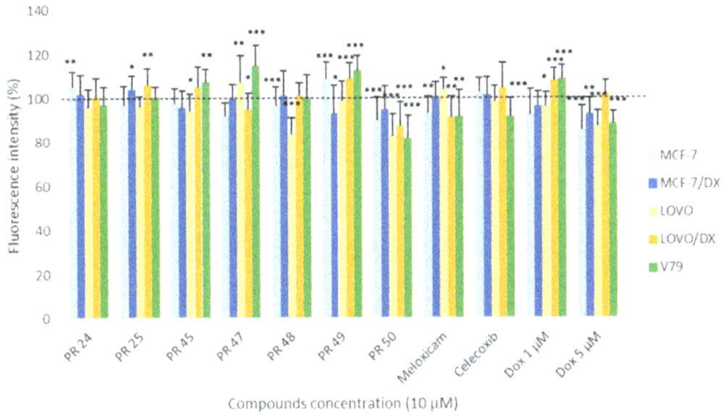

(**b**)

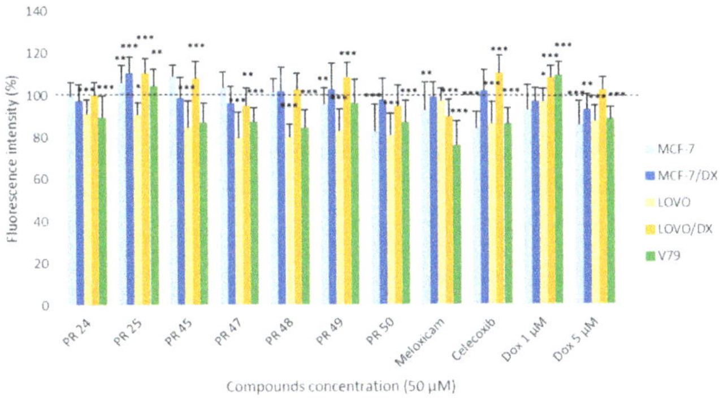

(**c**)

Figure 4. Effect of the tested compounds PR24–PR50 on the viability of cancer (MCF-7, MCF-7/DX, LOVO, LOVO/DX) and normal (V79) cells at a concentration of 1 μM (**a**), 10 μM (**b**) and 50 μM (**c**) evaluated by the Rhodamine 123 assay after 24 h of incubation. Results are presented as the percentage of cell viability relative to the untreated control (100%). The presence of statistical significance markers (* $p \leq 0.05$; ** $p < 0.01$; *** $p < 0.001$) indicates significant differences compared to the control group.

https://doi.org/10.3390/membranes16050166

The obtained results indicate that the effect of the tested derivatives on the mitochondrial membrane potential depended on both the type of cells and the concentration of the compound. In most cases, a decrease in Rhodamine 123 fluorescence intensity was observed with increasing concentration, which indicates a reduction in mitochondrial membrane potential and disruption of mitochondrial function, potentially activating processes leading to apoptosis.

The strongest decrease in fluorescence signal was observed at the concentration of 50 μM (Figure 4c), particularly in the LOVO sublines, where all compounds caused a clearly significant reduction in fluorescence intensity compared to the control. In drug-resistant colorectal cancer cells (LOVO/DX), this effect was slightly weaker for compound PR47 and PR50 while for the other tested compounds the fluorescence intensity increased, which may result from greater stability of mitochondrial membranes in doxorubicin-resistant cells.

In breast cancer cell lines MCF-7 and MCF-7/DX, no reduction in mitochondrial potential was observed after incubation with most of the tested derivatives, while a moderate decrease in fluorescence intensity was noted for PR50 in sensitive MCF-7 subline at various concentrations. The smaller decrease in mitochondrial potential in resistant cells compared to the sensitive line may indicate the activation of compensatory mechanisms associated with increased resistance to stress factors.

In normal V79 cells, a decrease in fluorescence intensity was observed at the highest concentration (50 μM) for compounds PR45, PR47, and PR48, as well as at 10 μM for PR50. This suggests that most of the analyzed compounds within the tested concentration range did not cause significant disturbances in the mitochondrial membrane potential of normal cells.

At lower concentrations of 1 and 10 μM (Figure 4a,b), changes in mitochondrial membrane potential were less pronounced, although in some cases (PR50 at 10 μM) a statistically significant decrease in fluorescence intensity was still observed.

Considering the results of the Rhodamine 123 assay, it can be concluded that the tested compounds act most effectively on sensitive colorectal cancer cells subline (LOVO), reducing their mitochondrial membrane potential. This may result from mitochondrial damage and activation of apoptosis-related processes, as confirmed by the results of the cell viability assay (MTT). It also indicates the presence of cellular stress.

3.2.6. Reference Compounds

In the study, reference compounds—meloxicam, celecoxib, and doxorubicin—were also used to compare their effects with the tested derivatives PR24–PR50. Both meloxicam and celecoxib did not cause significant changes in cell viability in most of the tested cell lines in the MTT assay, nor in fluorescence levels in the DCFH-DA assay, indicating the absence of a distinct cytotoxic effect at the applied concentrations (2 μM).

The IC_{50} values for meloxicam reported in the literature indicate its low cytotoxic activity. In studies conducted on MCF-7 cells, IC_{50} values exceeded 1.4 mM, confirming the very weak anticancer effect of this drug [63]. Similar results were obtained for C32 melanoma cells, where IC_{50} values were above 100 μM [64].

For celecoxib, the IC_{50} values reported in the literature were markedly lower but still indicative of moderate cytotoxicity. In colorectal cancer HT-29 cells, the IC_{50} value was approximately 30.4 μM [65]. These data confirm that celecoxib shows cytotoxic activity comparable to that of most of the tested derivatives PR24–PR50.

In contrast, doxorubicin exhibited high cytotoxic activity, as confirmed by both literature data and experimental results. In the present study, doxorubicin caused a significant decrease in cell viability in most of the examined cell lines in the MTT assay, both in concentrations of 1 and 5 μM.

https://doi.org/10.3390/membranes16050166

The obtained results confirm that meloxicam and celecoxib are characterized by low cytotoxic potential, whereas doxorubicin, as expected, shows strong cytotoxic activity, confirming the accuracy of the applied methodology and the reliability of the results obtained for the tested derivatives PR24–PR50.

3.3. Molecular Docking Studies

The binding modes were determined by using molecular docking. The obtained data are presented in Figures 5 and 6. Compound PR45, which exhibited best selectivity index COX-2/COX-1, and compound PR49, which was active exclusively towards COX-2, were selected for detailed analysis of their binding mode within the COX-2 binding site. Docking protocol validation was performed by re-docking the co-crystalized ligand (meloxicam) into the crystal structure of COX-2 binding site using the same docking parameters as those applied for the designed compounds. The accuracy of the docking protocol was evaluated by calculating the root mean square deviation (RMSD) between the docked pose and the crystallographic conformation. The obtained RMSD value below 1 Å, calculated using the LigRMSD web server, confirmed the reliability of the docking procedure [66]. Moreover, the docking protocol employed in this study has been previously validated and successfully applied in earlier work [67]. The binding score obtained for PR45 and PR49 were −10.83 and −10.00 kcal/mol, respectively. For comparison, the reference inhibitor meloxicam exhibited a docking score of −9.90 kcal/mol. These results indicate comparable predicted affinities towards the COX-2 active site, with PR45 showing negligibly stronger binding than both PR49 and meloxicam. It should be noted, however, that docking scores obtained using empirical scoring functions provide only semi-quantitative estimates of binding affinity and should be interpreted primarily in a comparative rather than an absolute manner.

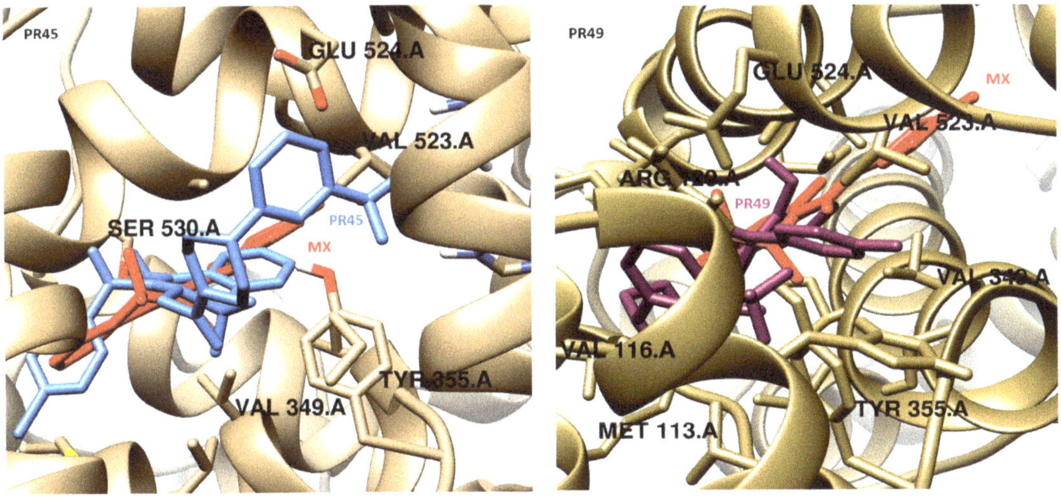

Figure 5. Comparison of the docking poses of PR45 (**left**, light blue) and PR49 (**right**, purple) with the reference inhibitor meloxicam (MX, red) within the COX-2 active site.

As illustrated in Figures 5 and 6, PR45 binds within the COX-2 active site like meloxicam, occupying the main catalytic channel and extending into the selectivity side pocket [60]. While meloxicam predominantly occupies the main catalytic channel. The observed selectivity of COX-2 inhibitors originates from subtle structural modifications within the enzyme binding site. Of particular importance is an additional selectivity pocket that contains Leu352, Ser353, Tyr355, Phe518, and Val523. This structural feature, resulting from the substitution of Ile523 with Val523 and the reorientation of the side chain of Tyr355, is consistently employed in the development of selective COX-2 inhibitors. In contrast, PR49 adopts a distinct binding orientation, which differs from that of both PR45 and meloxicam. Instead of penetrating deeply into the COX-2 selectivity pocket, PR49 is oriented towards an alternative region and extends in the opposite region to meloxicam. This altered binding mode may be related to structural differences between the two compounds, as PR49 has a methyl substituent instead of the chlorine atom present in PR45. The presence of the bulkier and more flexible CH_3 group may impose steric constraints that favor an alternative accommodation within the binding cavity, thereby influencing the overall orientation of PR49. Designed compounds form key hydrogen bonds with Ser530 (PR45) and Arg120 (PR49). While interactions with Arg120 contribute to the general anchoring of the ligands within the catalytic domain and are also observed for meloxicam, they do not determine the selectivity of the compounds. In contrast to meloxicam the elongated aromatic substituents of PR45 penetrate deeper into the COX-2 selectivity side pocket, resulting in additional stabilizing interactions. This region is absent in COX-1 due to the presence of the previously mentioned bulkier residuum of Ile523. These fragments establish additional π-alkyl/alkyl interactions, van der Waals contacts and hydrogen bonds with His90, Leu93, Arg,513, Val523, and Met535, which are absent or less pronounced in the meloxicam-COX-2 complex.

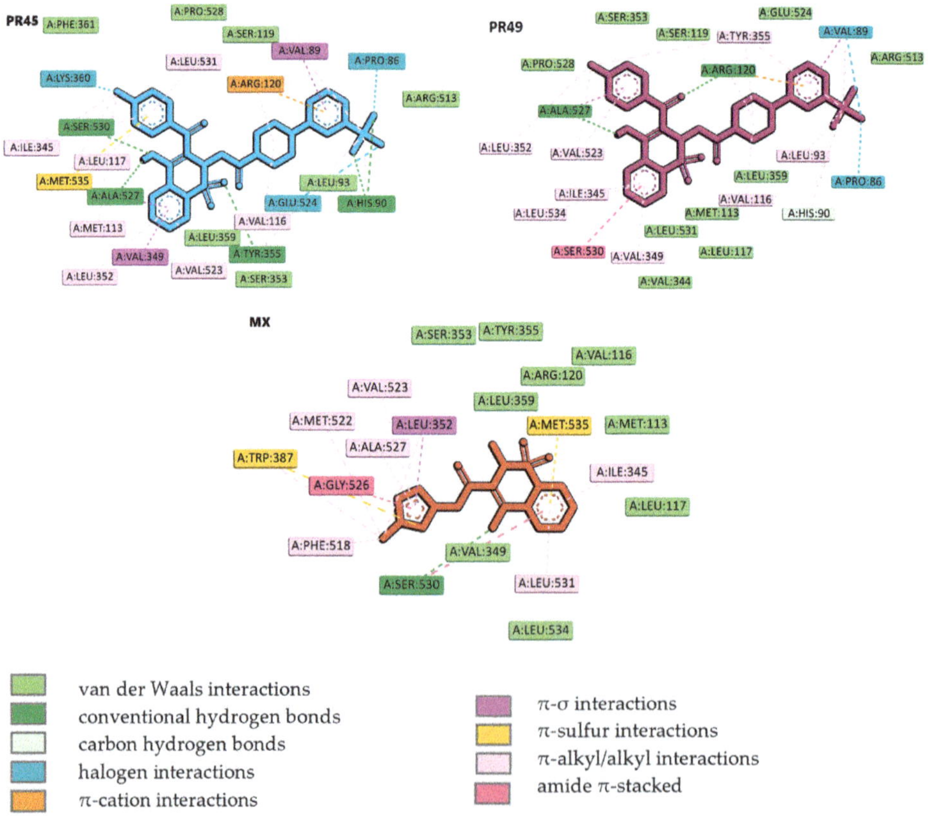

Figure 6. Intermolecular interactions of compounds PR45, PR49, and meloxicam (MX) in the binding cavity of COX-2.

https://doi.org/10.3390/membranes16050166

Compound PR45 forms numerous nonpolar interactions that contribute to its high binding affinity. The trifluoromethylphenyl moiety is stabilized by π-cation, π-σ and halogen interactions with Arg120, Val89 and Pro86 amino acid residues, respectively. The 1,2-benzothiazine scaffold forms two hydrogen bonds with Ala527 and Ser530. Meanwhile, the chlorobenzoyl substituent is stabilized by numerous interactions with Met535 (π-sulfur), Lys350 (halogen), Ile345, Leu117, Leu531 (alkyl/π-alkyl) amino acid residues. In comparison, meloxicam forms fewer stabilizing contacts within the COX-2 selectivity pocket and does not engage the same extent of π-type and halogen interactions observed for PR45. Compound PR49 is anchored by hydrogen bonding interactions with Arg120, and Ala527. Further stabilization is achieved through halogen interactions with Pro86 and Val89. The occupancy of the pocket is increased by the van der Waals interactions of the 1,2-benzothiazine fragment with Met113, Leu117, Val344, Ser353, Leu359, and Leu531. The alternative orientation leads to less pronounced occupation of the canonical selectivity pocket, but enables PR49 to establish a complementary interaction network along the periphery of the COX-2 binding site. Molecular docking results indicate that both compounds are potent COX-2 inhibitors.

4. Discussion

Cytotoxicity studies demonstrated that all tested compounds reduced cancer cell viability in a cell line-dependent manner, with the most pronounced effects observed in the LOVO and LOVO/DX colon cancer cell lines. This observation is consistent with previous reports indicating that nonsteroidal anti-inflammatory drugs (NSAIDs) exhibit particularly strong anticancer activity in colorectal cancer models [68–80]. For instance, celecoxib, a selective COX-2 inhibitor, was shown to significantly reduce the number of duodenal polyps in patients with familial adenomatous polyposis (FAP) when administered at high doses (400 mg twice daily) [81]. The observed cytotoxic effects are further supported by the results of the Rhodamine 123 assay, which revealed a decrease in mitochondrial membrane potential in the same cell lines, indicating mitochondrial damage and activation of apoptosis-related mechanisms. Similar mitochondrial dysfunction and selective cytotoxicity toward colorectal cancer cells have been reported for other NSAIDs, suggesting that apoptosis-related pathways and mitochondrial impairment may contribute to their anticancer activity independently of cyclooxygenase inhibition [82].

Compounds PR48 and PR50 exhibited the highest cytotoxicity toward cancer cells; however, these compounds also showed substantial toxicity toward healthy cells, which limits their therapeutic potential. In contrast, PR24 and PR25 displayed low toxicity toward healthy cells but were also the least effective against cancer cells, resulting in no clear therapeutic advantage. Among the remaining compounds, PR45, PR47, and PR49 demonstrated a more favorable balance between anticancer activity and toxicity toward normal cells. Notably, PR45 and PR49 were the least toxic to healthy V79 cells, with estimated IC_{50} values of 123.0 µM and 178.0 µM, respectively, making them the most promising candidates within the tested series. Both compounds exhibited moderate cytotoxicity against cancer cell lines, with particularly strong activity against LOVO cells (IC_{50} = 12.3 and 17.5 µM, respectively) and their doxorubicin-resistant variant LOVO/DX (IC_{50} = 28.7 and 47.3 µM, respectively). Moreover, PR45 at a concentration of 1 µM inhibited LOVO/DX cell proliferation more effectively than the reference drug doxorubicin at the same concentration. Given the chemoresistant nature of the LOVO/DX cell line, this finding highlights the potential relevance of PR45 for further investigation.

Since overexpression of inducible cyclooxygenase-2 (COX-2), accompanied by increased production of prostaglandin E_2 (PGE_2), plays a key role in carcinogenesis, the ability of the oxicams derivatives to inhibit COX-2 was evaluated. The COX inhibition

https://doi.org/10.3390/membranes16050166

assay revealed that five out of seven tested compounds inhibited COX-2 activity, although their inhibitory potency was markedly lower than that of the reference compound celecoxib, indicating that COX-2 inhibition by the tested compounds is very weak. As expected for oxicam-derived structures, all tested compounds inhibited COX-1 to a greater extent than COX-2, reflecting a non-selective cyclooxygenase inhibition profile. Although the studied compounds retain the characteristic 1,2-benzothiazine core of oxicams, substantial structural modifications were introduced at the 2- and 3-positions of the thiazine ring. Specifically, the small methyl substituent at the 2-position was replaced with a bulky arylpiperazine moiety, while the arylcarboxamide group at the 3-position was substituted with a benzoyl group. These modifications likely altered the molecular geometry and steric properties of the compounds, limiting their optimal accommodation within the narrow COX active site and thereby reducing COX-2 inhibitory activity compared with classical oxicams.

Among the tested compounds, PR49 exhibited the strongest COX-2 inhibition, whereas PR45 showed the highest COX-2/COX-1 selectivity, justifying their selection for molecular docking studies. Docking analysis indicated that both compounds are capable of binding to the COX-2 active site with predicted affinities comparable to that of the reference drug meloxicam. It should be emphasized, however, that docking scores provide only semi-quantitative estimates of binding affinity and cannot be directly translated into biological potency. Accordingly, the docking results were used to support structure–activity trends rather than to predict absolute inhibitory activity.

The COX inhibition data are further supported by membrane interaction studies, which demonstrated that all tested compounds were able to penetrate DMPC model membranes. This finding suggests that the compounds can access membrane-associated cyclooxygenase enzymes localized in the endoplasmic reticulum. Among the tested derivatives, PR49 exerted one of the strongest effects on the thermotropic properties of the phospholipid bilayer. This observation is consistent with previous studies indicating that the presence of an acetyl linker facilitates deeper penetration into phospholipid membranes and enhances their impact on membrane organization [42]. Moreover, earlier work demonstrated that oxicam derivatives containing an acetyl linker were effective modulators of multidrug resistance in LOVO/DX colon cancer cells [38].

From a structural perspective, PR45 and PR49 are closely related compounds. Both contain an acetyl linker connecting the thiazine and piperazine nitrogen atoms and an *m*-trifluoromethyl substituent on the phenylpiperazine moiety at the 2-position of the thiazine ring. The only structural difference between these compounds lies in the substituent at the 3-position of the thiazine ring: PR45 contains a *p*-chlorobenzoyl group, whereas PR49 bears a *p*-methylbenzoyl substituent. Structure–activity relationship (SAR) analysis indicated that the presence of an acetyl linker reduced the overall toxicity of the compounds, while the most toxic derivatives, PR48 and PR50, contained a propylene linker. Furthermore, the *p*-methylbenzoyl substituent in PR49 appeared to enhance COX-2 selectivity, whereas the *p*-chlorobenzoyl group in PR45 increased cytotoxicity, particularly against the LOVO/DX cell line. In summary, the structural modifications introduced into the oxicams scaffold resulted in reduced COX-2 inhibitory activity but enhanced cytotoxicity toward both cancerous and healthy cells. Importantly, certain substituents were found to reduce toxicity toward healthy cells (e.g., the acetyl linker between the thiazine and piperazine nitrogen atoms) while increasing cytotoxicity toward cancer cells (e.g., chlorine or bromine substituents and the *m*-trifluoromethyl group).

https://doi.org/10.3390/membranes16050166

5. Conclusions

Although the compounds evaluated in this study exhibit lower COX inhibitory activity than clinically available agents and most do not demonstrate superior cytotoxic potency compared to established drugs, the results provide a rational basis for further structural optimization. Taking into account all studies conducted, compounds PR45 and PR49 appear to be the most promising candidates for further investigation. Both compounds exhibited significant cytotoxic activity against the LOVO and LOVO/DX cell lines while showing considerably lower toxicity toward healthy cells. Notably, PR45 at a concentration of 1 μM inhibited LOVO/DX cell proliferation more effectively than the reference compound doxorubicin at the same concentration. PR45 demonstrated the highest COX-2/COX-1 selectivity, whereas PR49 showed the strongest COX-2 inhibitory activity. The dual activity of these compounds, combining COX inhibition and cytotoxic effects, may be particularly relevant for cancer chemoprevention, especially in inflammation-associated malignancies such as colorectal cancer, where COX-2 is a key therapeutic target and may contribute to disease prevention and progression control. However, COX-2 inhibition is weak and unlikely to be the primary mechanism underlying the antitumor effects of these compounds. These limitations should be explicitly taken into account when interpreting their potential applications. Overall, both PR45 and PR49 exhibit promising pharmacological potential; nevertheless, further mechanistic studies are required to clarify their precise mode of action and evaluate their suitability for future therapeutic development.

Author Contributions: Conceptualization, J.M.; methodology, J.M.; validation, J.M., B.M.S.-S., K.G., Ł.C.-Z., Ż.C.; formal analysis, J.M., B.M.S.-S., K.G., Ł.C.-Z., Ż.C.; investigation, J.M., B.M.S.-S., K.G., Ł.C.-Z., Ż.C.; resources, J.M., B.M.S.-S., K.G., Ł.C.-Z., Ż.C.; data curation, J.M., B.M.S.-S., K.G., Ł.C.-Z., Ż.C.; writing—original draft preparation, J.M., B.M.S.-S., K.G., Ł.C.-Z., Ż.C.; writing—review and editing, J.M. and B.M.S.-S.; visualization, J.M., B.M.S.-S., K.G., Ł.C.-Z., Ż.C.; supervision, J.M.; project administration, J.M.; funding acquisition, B.M.S.-S. All authors have read and agreed to the published version of the manuscript.

Funding: This research was funded by the Ministry of Health subvention according to the number SUBZ.D070.26.006 from the IT Simple System of Wroclaw Medical University, Poland.

Institutional Review Board Statement: Not applicable.

Informed Consent Statement: Not applicable.

Data Availability Statement: The original contributions presented in this study are included in the article. Further inquiries can be directed to the corresponding authors.

Acknowledgments: DSC measurements were carried out in the Laboratory of Elemental Analysis and Structure Research, Faculty of Pharmacy, Wroclaw Medical University. Calculations for molecular modeling were carried out in Wroclaw Center of Networking and Supercomputing (WCSS).

Conflicts of Interest: The authors declare no conflicts of interest.

Abbreviations

The following abbreviations are used in this manuscript:

COX	cyclooxygenase
COX-1	cyclooxygenase 1
COX-2	cyclooxygenase 2
DCFH-DA	2′,7′-dichlorodihydrofluorescein diacetate
DMPC	1,2-dimyristoyl-*sn*-glycero-3-phosphocholine
DMSO	dimethyl sulfoxide
DPPC	dipalmitoylphosphatidylcholine
DSC	differential scanning calorimetry

https://doi.org/10.3390/membranes16050166

IC_{50}	half-maximal inhibitory concentration
LOVO	colorectal cancer cell line
LOVO/DX	doxorubicin-resistant colorectal cancer cell subline
MCF-7	breast cancer cell line
MCF-7/DX	doxorubicin-resistant breast cancer cell subline
MTT	3-(4,5-dimethylthiazol-2-yl)-2,5-diphenyltetrazolium bromide
MX	meloxicam
NSAIDs	nonsteroidal anti-inflammatory drugs
Rh123	Rhodamine 123
RNS	reactive nitrogen species
ROS	reactive oxygen species
SI	selectivity index
TMPD	*N,N,N′,N′*-tetramethyl-*p*-phenylenediamine
V79	normal cell line (fibroblast)

References

1. Xu, S.; Rouzer, C.A.; Marnett, L.J. Oxicams, a Class of Nonsteroidal Anti-Inflammatory Drugs and Beyond. *IUBMB Life* **2014**, *66*, 803–811. [CrossRef]
2. Antonucci, L.; Karin, M. The Past and Future of Inflammation as a Target to Cancer Prevention. *Cancer Prev. Res.* **2024**, *17*, 141–155. [CrossRef]
3. Sung, H.; Ferlay, J.; Siegel, R.L.; Laversanne, M.; Soerjomataram, I.; Jemal, A.; Bray, F. Global Cancer Statistics 2020: GLOBOCAN Estimates of Incidence and Mortality Worldwide for 36 Cancers in 185 Countries. *CA Cancer J. Clin.* **2021**, *71*, 209–249. [CrossRef] [PubMed]
4. Larrea Murillo, L.; Green, M.; Mahon, N.; Saiani, A.; Tsigkou, O. Modelling Cancer Pathophysiology: Mechanisms and Changes in the Extracellular Matrix During Cancer Initiation and Early Tumor Growth. *Cancers* **2025**, *17*, 1675. [CrossRef]
5. Zhang, S.; Xiao, X.; Yi, Y.; Wang, X.; Zhu, L.; Shen, Y.; Lin, D.; Wu, C. Tumor Initiation and Early Tumorigenesis: Molecular Mechanisms and Interventional Targets. *Signal Transduct. Target. Ther.* **2024**, *9*, 149. [CrossRef] [PubMed]
6. Sun, M.; da Silva, M.; Bjørge, T.; Fritz, J.; Mboya, I.B.; Jerkeman, M.; Stattin, P.; Wahlström, J.; Michaëlsson, K.; van Guelpen, B.; et al. Body Mass Index and Risk of over 100 Cancer Forms and Subtypes in 4.1 Million Individuals in Sweden: The Obesity and Disease Development Sweden (ODDS) Pooled Cohort Study. *Lancet Reg. Health Eur.* **2024**, *45*, 101034. [CrossRef] [PubMed]
7. Marino, P.; Mininni, M.; Deiana, G.; Marino, G.; Divella, R.; Bochicchio, I.; Giuliano, A.; Lapadula, S.; Lettini, A.R.; Sanseverino, F. Healthy Lifestyle and Cancer Risk: Modifiable Risk Factors to Prevent Cancer. *Nutrients* **2024**, *16*, 800. [CrossRef]
8. De Guzman, R.; Schiller, J. Air Pollution and Its Impact on Cancer Incidence, Cancer Care and Cancer Outcomes. *BMJ Oncol.* **2025**, *4*, e000535. [CrossRef]
9. Medzhitov, R. Origin and Physiological Roles of Inflammation. *Nature* **2008**, *454*, 428–435. [CrossRef]
10. Dvorak, H.F. Tumors: Wounds That Do Not Heal-Redux. *Cancer Immunol. Res.* **2015**, *3*, 1–11. [CrossRef]
11. Virchow, R. An Address on the Value of Pathological Experiments. *Br. Med. J.* **1881**, *2*, 198–203. [CrossRef] [PubMed]
12. Balkwill, F.; Mantovani, A. Inflammation and Cancer: Back to Virchow? *Lancet* **2001**, *357*, 539–545. [CrossRef]
13. Wu, Y.; Antony, S.; Meitzler, J.L.; Doroshow, J.H. Molecular Mechanisms Underlying Chronic Inflammation-Associated Cancers. *Cancer Lett.* **2014**, *345*, 164–173. [CrossRef] [PubMed]
14. Furman, D.; Campisi, J.; Verdin, E.; Carrera-Bastos, P.; Targ, S.; Franceschi, C.; Ferrucci, L.; Gilroy, D.W.; Fasano, A.; Miller, G.W.; et al. Chronic Inflammation in the Etiology of Disease across the Life Span. *Nat. Med.* **2019**, *25*, 1822–1832. [CrossRef]
15. Mansouri, D.; McMillan, D.C.; Roxburgh, C.S.D.; Crighton, E.M.; Horgan, P.G. The Impact of Aspirin, Statins and ACE-Inhibitors on the Presentation of Colorectal Neoplasia in a Colorectal Cancer Screening Programme. *Br. J. Cancer* **2013**, *109*, 249–256. [CrossRef]
16. Cook, N.R.; Lee, I.M.; Zhang, S.M.; Moorthy, M.V.; Buring, J.E. Alternate-Day, Low-Dose Aspirin and Cancer Risk: Long-Term Observational Follow-up of a Randomized Trial. *Ann. Intern. Med.* **2013**, *159*, 77–85. [CrossRef]
17. Rothwell, P.M.; Fowkes, F.G.R.; Belch, J.F.; Ogawa, H.; Warlow, C.P.; Meade, T.W. Effect of Daily Aspirin on Long-Term Risk of Death Due to Cancer: Analysis of Individual Patient Data from Randomised Trials. *Lancet* **2011**, *377*, 31–41. [CrossRef] [PubMed]
18. Rothwell, P.M.; Wilson, M.; Elwin, C.E.; Norrving, B.; Algra, A.; Warlow, C.P.; Meade, T.W. Long-Term Effect of Aspirin on Colorectal Cancer Incidence and Mortality: 20-Year Follow-up of Five Randomised Trials. *Lancet* **2010**, *376*, 1741–1750. [CrossRef]
19. Sahin, I.H.; Hassan, M.M.; Garrett, C.R. Impact of Non-Steroidal Anti-Inflammatory Drugs on Gastrointestinal Cancers: Current State-of-the Science. *Cancer Lett.* **2014**, *345*, 249–257. [CrossRef]

https://doi.org/10.3390/membranes16050166

20. Arantes-Rodrigues, R.; Pinto-Leite, R.; Ferreira, R.; Neuparth, M.J.; Pires, M.J.; Gaivão, I.; Palmeira, C.; Santos, L.; Colaço, A.; Oliveira Collins, P. Meloxicam in the Treatment of in Vitro and in Vivo Models of Urinary Bladder Cancer. *Biomed. Pharmacother.* **2013**, *67*, 277–284. [CrossRef]
21. Serrano, D.; Lazzeroni, M.; Bonanni, B. Cancer Chemoprevention: Much Has Been Done, but There Is Still Much to Do. State of the Art and Possible New Approaches. *Mol. Oncol.* **2015**, *9*, 1008–1017. [CrossRef] [PubMed]
22. Steward, W.P.; Brown, K. Cancer Chemoprevention: A Rapidly Evolving Field. *Br. J. Cancer* **2013**, *109*, 1–7. [CrossRef] [PubMed]
23. Gravitz, L. Chemoprevention: First Line of Defence. *Nature* **2011**, *471*, 5–7. [CrossRef]
24. Ghosh, N.; Chaki, R.; Mandal, V.; Mandal, S.C. COX-2 as a Target for Cancer Chemotherapy. *Pharmacol. Rep.* **2010**, *62*, 233–244. [CrossRef] [PubMed]
25. Boghossian, S.; Hawash, A. Chemoprevention in Colorectal Cancer—Where We Stand and What We Have Learned from Twenty Year's Experience. *Surgeon* **2012**, *10*, 43–52. [CrossRef]
26. Elwood, P.C.; Gallagher, A.M.; Duthie, G.G.; Mur, L.A.; Morgan, G. Aspirin, Salicylates, and Cancer. *Lancet* **2009**, *373*, 1301–1309. [CrossRef]
27. Abdalla, L.F.; Chaudhry Ehsanullah, R.; Karim, F.; Oyewande, A.; Khan, S. Role of Using Nonsteroidal Anti-Inflammatory Drugs in Chemoprevention of Colon Cancer in Patients with Inflammatory Bowel Disease. *Cureus* **2020**, *12*, e8240. [CrossRef] [PubMed]
28. Ulrich, C.M.; Bigler, J.; Potter, J.D. Non-Steroidal Anti-Inflammatory Drugs for Cancer Prevention: Promise, Perils and Pharmacogenetics. *Nat. Rev. Cancer* **2006**, *6*, 130–140. [CrossRef]
29. Rostom, A.; Dubé, C.; Lewin, G.; Tsertsvadze, A.; Barrowman, N.; Code, C.; Sampson, M.; Moher, D. Non-steroidal Anti-Inflammatory Drugs and Cyclooxygenase-2 Inhibitors for Primary Prevention of Colorectal Cancer: A Systematic Review Prepared for the U.S. Preventive Services Task Force. *Ann. Intern. Med.* **2007**, *146*, 376–389. [CrossRef]
30. Vinogradova, Y.; Hippisley-Cox, J.; Coupland, C.; Logan, R.F. Risk of Colorectal Cancer in Patients Prescribed Statins, Nonsteroidal Anti-Inflammatory Drugs, and Cyclooxygenase-2 Inhibitors: Nested Case-Control Study. *Gastroenterology* **2007**, *133*, 393–402. [CrossRef]
31. Harris, R.E.; Beebe-Donk, J.; Alshafie, G.A. Similar Reductions in the Risk of Human Colon Cancer by Selective and Nonselective Cyclooxygenase-2 (COX-2) Inhibitors. *BMC Cancer* **2008**, *8*, 237. [CrossRef]
32. Dhingra, A.; Chopra, B.; Dass, R.; Mittal, S. An Update on Anti-Inflammatory Compounds: A Review. *Anti-Inflamm. Anti-Allergy Agents Med. Chem.* **2015**, *14*, 81–97. [CrossRef] [PubMed]
33. Liu, B.; Qu, L.; Tao, H. Cyclo-Oxygenase 2 up-Regulates the Effect of Multidrug Resistance. *Cell Biol. Int.* **2009**, *34*, 21–25. [CrossRef]
34. Hull, M.A. Cyclooxygenase-2: How Good Is It as a Target for Cancer Chemoprevention? *Eur. J. Cancer* **2005**, *41*, 1854–1863. [CrossRef]
35. Maniewska, J.; Jeżewska, D. Non-Steroidal Anti-Inflammatory Drugs in Colorectal Cancer Chemoprevention. *Cancers* **2021**, *13*, 594. [CrossRef] [PubMed]
36. Arber, N.; Eagle, C.J.; Spicak, J.; Rácz, I.; Dite, P.; Hajer, J.; Zavoral, M.; Lechuga, M.J.; Gerletti, P.; Tang, J.; et al. Celecoxib for the Prevention of Colorectal Adenomatous Polyps. *N. Engl. J. Med.* **2006**, *355*, 885–895. [CrossRef] [PubMed]
37. Środa-Pomianek, K.; Wesołowska, O.; Szczęśniak-Sięga, B.; Puła, B.; Dzięgiel, P.; Maniewska, J.; Malinka, W.; Palko-Łabuz, A.; Michalak, K. Effect of New Oxicam Derivatives on Efflux Pumps Overexpressed in Resistant a Human Colorectal Adenocarcinoma Cell Line. *Anticancer Res.* **2015**, *35*, 2835–2840.
38. Środa-Pomianek, K.; Michalak, K.; Palko-Łabuz, A.; Uryga, A.; Szczęśniak-Sięga, B.; Wesołowska, O. Simvastatin Strongly Augments Proapoptotic, Anti-Inflammatory and Cytotoxic Activity of Oxicam Derivatives in Doxorubicin-Resistant Colon Cancer Cells. *Anticancer Res.* **2019**, *39*, 727–734. [CrossRef]
39. Szczęśniak-Sięga, B.M.; Zaręba, N.; Czyżnikowska, Ż.; Janek, T.; Kepinska, M. Rational Design, Synthesis, Molecular Docking, and Biological Evaluations of New Phenylpiperazine Derivatives of 1,2-Benzothiazine as Potential Anticancer Agents. *Molecules* **2024**, *29*, 4282. [CrossRef]
40. Lewandowska, P.; Szczuka, I.; Bednarz-Misa, I.; Szczęśniak-Sięga, B.M.; Neubauer, K.; Mierzchała-Pasierb, M.; Zawadzki, M.; Witkiewicz, W.; Krzystek-Korpacka, M. Modulating Properties of Piroxicam, Meloxicam and Oxicam Analogues against Macrophage-Associated Chemokines in Colorectal Cancer. *Molecules* **2021**, *26*, 7375. [CrossRef]
41. Krzystek-Korpacka, M.; Szczęśniak-Sięga, B.; Szczuka, I.; Fortuna, P.; Zawadzki, M.; Kubiak, A.; Mierzcha-ła-Pasierb, M.; Fleszar, M.G.; Lewandowski, Ł.; Serek, P.; et al. L-Arginine/Nitric Oxide Pathway Is Altered in Colorectal Cancer and Can Be Modulated by Novel Derivatives from Oxicam Class of Non-Steroidal Anti-Inflammatory Drugs. *Cancers* **2020**, *12*, 2594. [CrossRef]
42. Maniewska, J.; Gąsiorowska, J.; Szczęśniak-Sięga, B.; Michalak, K. The Interaction of New Oxicam Derivatives with Lipid Bilayers as Measured by Calorimetry and Fluorescence Spectroscopy. *Acta Biochim. Pol.* **2018**, *65*, 185–191. [CrossRef] [PubMed]
43. Maniewska, J.; Gąsiorowska, J.; Czyżnikowska, Ż.; Michalak, K.; Szczęśniak-Sięga, B.M. New Meloxicam Derivatives—Synthesis and Interaction with Phospholipid Bilayers Measured by Differential Scanning Calorimetry and Fluorescence Spectroscopy. *Membranes* **2023**, *13*, 416. [CrossRef]

44. Szczęśniak-Sięga, B.M.; Maniewska, J.; Wiatrak, B.; Janek, T.; Nowotarska, P.; Czyżnikowska, Ż. Anti-Inflammatory Properties of Novel 1,2-Benzothiazine Derivatives and Their Interaction with Phospholipid Model Membranes. *Membranes* **2024**, *14*, 274. [CrossRef] [PubMed]
45. ATCC. Animal Cells V79. Available online: https://www.atcc.org/products/ccl-93 (accessed on 4 May 2024).
46. ATCC. Human Cells LoVo. Available online: https://www.atcc.org/products/ccl-229 (accessed on 29 July 2025).
47. ATCC. Human Cells MCF-7. Available online: https://www.atcc.org/products/htb-22 (accessed on 29 July 2025).
48. Gębczak, K.; Cwynar-Zając, Ł.; Sapeta-Nowińska, M.; Barg, E. Additive Anticancer and Antioxidant Effects of Metformin and Luteolin in Breast and Colorectal Cancer Cell Lines. *Pharmaceuticals* **2025**, *18*, 1660. [CrossRef] [PubMed]
49. Mosmann, T. Rapid colorimetric assay for cellular growth and survival: Application to proliferation and cytotoxicity assays. *J. Immunol. Methods* **1983**, *65*, 55–63. [CrossRef]
50. *COX Colorimetric Inhibitor Screening Assay Kit—Instruction Manual*; Cayman Chemical Company: Ann Arbor, MI, USA, 2025.
51. Wang, H.; Joseph, J.A. Quantifying cellular oxidative stress by dichlorofluorescein assay using microplate reader. *Free Radic. Biol. Med.* **1999**, *27*, 612–616. [CrossRef]
52. Gębarowski, T.; Wiatrak, B.; Gębczak, K.; Tylińska, B.; Gąsiorowski, K. Effect of new olivacine derivatives on p53 protein level. *Pharmacol. Rep.* **2020**, *72*, 214–224. [CrossRef]
53. Johnson, L.V.; Walsh, M.L.; Chen, L.B. Localization of mitochondria in living cells with rhodamine 123. *Proc. Natl. Acad. Sci. USA* **1980**, *77*, 990–994. [CrossRef]
54. Becke, A.D. Density-Functional Exchange-Energy Approximation with Correct Asymptotic Behavior. *Phys. Rev. A* **1988**, *38*, 3098–3100. [CrossRef]
55. Tomasi, J.; Mennucci, B.; Cammi, R. Quantum Mechanical Continuum Solvation Models. *Chem. Rev.* **2005**, *105*, 2999–3093. [CrossRef]
56. Cancès, E.; Mennucci, B.; Tomasi, J. A New Integral Equation Formalism for the Polarizable Continuum Model: Theoretical Background and Applications to Isotropic and Anisotropic Dielectrics. *J. Chem. Phys.* **1997**, *107*, 3032–3041. [CrossRef]
57. Frisch, M.J.; Trucks, G.W.; Schlegel, H.B.; Scuseria, G.E.; Robb, M.A.; Cheeseman, J.R.; Scalmani, G.; Barone, V.; Mennucci, B.; Petersson, G.A. *Gaussian 09*; revision A.02; Gaussian, Inc.: Wallingford, CT, USA, 2016.
58. Potemkin, A.V.; Grishina, M.A.; Potemkin, V.A. Grid-Based Continual Analysis of Molecular Interior for Drug Discovery, QSAR and QSPR. *Curr. Drug Discov. Technol.* **2017**, *14*, 181–205. [CrossRef]
59. Morris, G.M.; Ruth, H.; Lindstrom, W.; Sanner, M.F.; Belew, R.K.; Goodsell, D.S.; Olson, A.J. Software News and Updates AutoDock4 and AutoDockTools4: Automated Docking with Selective Receptor Flexibility. *J. Comput. Chem.* **2009**, *30*, 2785–2791. [CrossRef]
60. Xu, S.; Hermanson, D.J.; Banerjee, S.; Ghebreselasie, K.; Clayton, G.M.; Garavito, R.M.; Marnett, L.J. Oxicams Bind in a Novel Mode to the Cyclooxygenase Active Site via a Two-Water-Mediated h-Bonding Network. *J. Biol. Chem.* **2014**, *289*, 6799–6808. [CrossRef]
61. Zambrano, P.; Manrique-Moreno, M.; Petit, K.; Colina, J.R.; Jemiola-Rzeminska, M.; Suwalsky, M.; Strzalka, K. Differential Scanning Calorimetry in Drug-Membrane Interactions. *Biochem. Biophys. Res. Commun.* **2024**, *709*, 149806. [CrossRef]
62. Jain, M.K.; Wu, N.M. Effect of Small Molecules on the Dipalmitoyl Lecithin Liposomal Bilayer: III. Phase Transition in Lipid Bilayer. *J. Membr. Biol.* **1977**, *34*, 157–201. [CrossRef]
63. Marinov, L.; Georgieva, A.; Toshkova, R.; Kostadinova, I.; Mangarov, I.; Toshkova-Yotova, T.; Nikolova, I. The Effects of Meloxicam, Lornoxicam, Ketoprofen, and Dexketoprofen on Human Cervical, Colorectal, and Mammary Carcinoma Cell Lines. *Pharmacia* **2024**, *71*, 1–12. [CrossRef]
64. Karkoszka-Stanowska, M.; Rzepka, Z.; Wrześniok, D. The Evaluation of Potential Anticancer Activity of Meloxicam—In Vitro Study on Amelanotic and Melanotic Melanoma. *Int. J. Mol. Sci.* **2025**, *26*, 5985. [CrossRef]
65. Azandaryani, A.H.; Barati, A.; Shahlaei, M.; Shamsipur, M.; Moradi, S.; Arkan, E. Epirubicin-Calf Thymus DNA Interaction: A Comprehensive Investigation Using Molecular Docking, Spectroscopy and Fluorescent Quantum Dots. *Cell. Mol. Biol.* **2018**, *64*, 7. [CrossRef]
66. Luis Velázquez-Libera, J.; Durán-Verdugo, F.; Valdés-Jiménez, A.; Núñez-Vivanco, G.; Caballero, J. LigRMSD: A Web Server for Automatic Structure Matching and RMSD Calculations among Identical and Similar Compounds in Protein-Ligand Docking. *Bioinformatics* **2020**, *36*, 2912–2914. [CrossRef]
67. Glomb, T.; Wiatrak, B.; Gębczak, K.; Gębarowski, T.; Bodetko, D.; Czyżnikowska, Ż.; Świątek, P. New 1,3,4-Oxadiazole Derivatives of Pyridothiazine-1,1-Dioxide with Anti-Inflammatory Activity. *Int. J. Mol. Sci.* **2020**, *21*, 9122. [CrossRef]
68. Lichtenberger, L.M. Using Aspirin to Prevent and Treat Cancer. *Inflammopharmacology* **2024**, *32*, 903–908. [CrossRef]
69. Kawai, N.; Tsujii, M.; Tsuji, S. Cyclooxygenases and Colon Cancer. *Prostaglandins Other Lipid Mediat.* **2002**, *68–69*, 187–196. [CrossRef] [PubMed]
70. Pathi, S.; Li, X.; Safe, S. Tolfenamic Acid Inhibits Colon Cancer Cell and Tumor Growth and Induces Degradation of Specificity Protein (Sp) Transcription Factors. *Mol. Carcinog.* **2014**, *53*, E53–E61. [CrossRef]

https://doi.org/10.3390/membranes16050166

71. Tinsley, H.N.; Grizzle, W.E.; Abadi, A.; Keeton, A.; Zhu, B.; Xi, Y.; Piazza, G.A. New NSAID Targets and Derivatives for Colorectal Cancer Chemoprevention. *Recent. Results Cancer Res.* **2013**, *191*, 105. [CrossRef]
72. Ouyang, N.; Ji, P.; Williams, J.L. A Novel NSAID Derivative, Phospho-Ibuprofen, Prevents AOM-Induced Colon Cancer in Rats. *Int. J. Oncol.* **2012**, *42*, 643–650. [CrossRef] [PubMed]
73. Ghanghas, P.; Jain, S.; Rana, C.; Sanyal, S.N. Chemoprevention of Colon Cancer through Inhibition of Angiogenesis and Induction of Apoptosis by Nonsteroidal Anti-Inflammatory Drugs. *J. Environ. Pathol. Toxicol. Oncol.* **2016**, *35*, 273–289. [CrossRef] [PubMed]
74. Tsioulias, G.J.; Go, M.F.; Rigas, B. NSAIDs and Colorectal Cancer Control: Promise and Challenges. *Curr. Pharmacol. Rep.* **2015**, *1*, 295–301. [CrossRef]
75. Mishra, T.; Goswami, A.; Bajpai, A.; Basu, S. Detouring NSAID into Mitochondria to Induce Apoptosis in Cancer Cells. *Chem. Asian J.* **2024**, *19*, e202400732. [CrossRef]
76. Turchanowa, L.; Dauletbaev, N.; Milovic, V.; Stein, J. Nonsteroidal Anti-Inflammatory Drugs Stimulate Spermidine/Spermine Acetyltransferase and Deplete Polyamine Content in Colon Cancer Cells. *Eur. J. Clin. Investig.* **2001**, *31*, 887–893. [CrossRef]
77. Chen, J.; Stark, L.A. Aspirin Prevention of Colorectal Cancer: Focus on NF-KB Signalling and the Nucleolus. *Biomedicines* **2017**, *5*, 43. [CrossRef]
78. Gu, M.; Nishihara, R.; Chen, Y.; Li, W.; Shi, Y.; Masugi, Y.; Hamada, T.; Kosumi, K.; Liu, L.; da Silva, A.; et al. Aspirin Exerts High Anti-Cancer Activity in PIK3CA-Mutant Colon Cancer Cells. *Oncotarget* **2017**, *8*, 87379. [CrossRef] [PubMed]
79. Srivastava, S.; Dewangan, J.; Mishra, S.; Divakar, A.; Chaturvedi, S.; Wahajuddin, M.; Kumar, S.; Rath, S.K. Piperine and Celecoxib Synergistically Inhibit Colon Cancer Cell Proliferation via Modulating Wnt/β-Catenin Signaling Pathway. *Phytomedicine* **2021**, *84*, 153484. [CrossRef]
80. Setia, S.; Nehru, B.; Sanyal, S.N. The PI3K/Akt Pathway in Colitis Associated Colon Cancer and Its Chemoprevention with Celecoxib, a Cox-2 Selective Inhibitor. *Biomed. Pharmacother.* **2014**, *68*, 721–727. [CrossRef] [PubMed]
81. Yang, P.; Zuo, X.; Advani, S.; Wei, B.; Malek, J.; Day, R.S.; Shureiqi, I. Celecoxib Colorectal Bioavailability and Chemopreventive Response in Patients with Familial Adenomatous Polyposis. *Cancer Prev. Res.* **2022**, *15*, 217–223. [CrossRef]
82. Tegeder, I.; Pfeilschifter, J.; Geisslinger, G. Cyclooxygenase-independent actions of cyclooxygenase inhibitors. *FASEB J.* **2001**, *15*, 2057–2072. [CrossRef] [PubMed]

https://doi.org/10.3390/membranes16050166

membranes

MDPI

Article

Polyphenols from *Inula oculus-christi* L. Induced Cell-Specific Membrane and Cytoskeleton Reorganization

Ralitsa Veleva [1,2], Aneliya Kostadinova [3], Antoaneta Trendafilova [2,4], Viktoria Ivanova [4], Veselina Moskova-Doumanova [1], Kirilka Mladenova [1], Jordan Doumanov [1], Dayana Benkova [3], Galya Staneva [3] and Tanya Topouzova-Hristova [1,2,*]

1 Faculty of Biology, Sofia University 'St. Kliment Ohridski', 8 Dragan Tzankov Blvd., 1164 Sofia, Bulgaria; ralitsa_veleva@biofac.uni-sofia.bg (R.V.); moskova@biofac.uni-sofia.bg (V.M.-D.); k_mladenova@biofac.uni-sofia.bg (K.M.); doumanov@biofac.uni-sofia.bg (J.D.)

2 Centre of Competence "Sustainable Utilization of Bio-resources and Waste of Medicinal and Aromatic Plants for Innovative Bioactive Products" (BIORESOURCES BG), 1000 Sofia, Bulgaria; antoaneta.trendafilova@orgchm.bas.bg

3 Institute of Biophysics and Biomedical Engineering, Bulgarian Academy of Sciences, Acad. G. Bonchev Str., Bl. 21, 1113 Sofia, Bulgaria; aneliakk@yahoo.com (A.K.); dayanabenkova369@abv.bg (D.B.); gstaneva@bio21.bas.bg (G.S.)

4 Institute of Organic Chemistry with Centre of Phytochemistry, Bulgarian Academy of Sciences, 1113 Sofia, Bulgaria; viktoria.genova@orgchm.bas.bg

* Correspondence: topouzova@biofac.uni-sofia.bg

Abstract

Interrelations between the plasma membrane and cytoskeleton are of crucial importance for essential cellular processes such as endocytosis, formation of intercellular junctions, cell morphology, etc. Many studies validate the beneficial effects of polyphenols as antioxidant and protective agents, but a molecular mechanism of their interaction and transition through the plasma membranes of different cell lines is still missing. In this study, we examined the affinity of fractions enriched in flavonoid glycosides (FGs) and caffeoylquinic acids (CQAs), obtained from the methanol extract of the medicinal plant *Inula oculus-christi* L., to reorganize the plasma membrane structure and actin cytoskeleton by using confocal microscopy. Assessment of the degree of membrane ordering aiming to distinguish the ordered from disordered regions of the cellular membranes was performed using the fluorescent dye Di-4-ANEPPDHQ, and visualization of F-actin was by TRITC-phalloidin. Two epithelial cell lines with clear differences in their origin and plasma membrane organization were chosen: the non-malignant MDCK II and the cancerous A549. Our results showed that flavonoid glycosides exhibited an ordering effect on plasma membranes of cancerous cells and fluidized one on non-malignant cells. Different patterns of actin reorganization were observed for both cell lines after treatment. Our results indicate the potential of plant-derived polyphenols as modulators of the membrane's structural organization, offering valuable insights for the development of membrane-targeted therapeutic strategies.

Keywords: *Inula oculus-christi* L.; polyphenols; caffeoylquinic acids; flavonoid glycosydes plasmalaemma

1. Introduction

Plants of the genus *Inula* have long been utilized in ethnomedicine across Asia, Europe, and North America due to their diverse therapeutic benefits [1]. Notably, *Inula* sp. plants possess pronounced anti-inflammatory activity which has led to their widespread empirical application as expectorants and antitussives in the treatment of respiratory ailments, including lung inflammation, asthma, cough, sputum production, and wheezing [2].

The health-promoting properties of *Inula* plants are largely attributed to their high phenolic content. Of particular phytopharmacological interest are phenolic acids (PAs) and flavonoid glycosides (FGs). Both compound classes exhibit well-documented biological activities, including antioxidant [3–8], anti-inflammatory, and anticancer effects [9–13]. The cell-protective effects of these polyphenols have been partially attributed to their interactions with cellular compartments, thereby modulating key cellular processes. In particular, the cell membrane represents a primary target for the molecular therapeutic actions of various plant-derived polyphenols.

The eukaryotic cellular membrane is composed of a phospholipid bilayer with embedded proteins, creating a heterogeneous landscape with important structural and functional roles [14]. Specific molecular and cooperative lipid–lipid and lipid–protein interactions give rise to lateral segregation and formation of specialized and highly ordered membrane domain "rafts" in a Lo-like (liquid-ordered) phase state enriched in cholesterol, sphingomyelin, and raft-associated proteins. These domains co-exist with a milieu of fluid and disordered Ld-like (liquid-disordered) phases composed predominately of saturated phospholipids [15–18]. Structurally, membrane rafts range from nanoscale domains that can cluster upon extracellular stimulation to larger assemblies spanning hundreds of nanometers to several microns in size [17,19]. Functional raft macrodomains play crucial roles in diverse cellular processes, including signal transduction, protein sorting, membrane trafficking, and cell adhesion [20–23].

The implementation of these vital cellular processes requires synergetic interaction between membrane rafts and the cell cytoskeleton. Cortical actin filaments criss-cross with the cytoplasmic membrane surface, forming a dynamic mesh-like structure. These interactions drive the compartmentalization of the entire plasma membrane ($\sim$100 nm and in the range of 30–230 nm depending on the cell type), within which nanoscale lipid rafts ($\sim$2–200 nm) facilitate the organization and temporal regulation of membrane proteins [17]. Membrane rafts are particularly enriched with cytoskeletal proteins, modulators of the cortical actin cytoskeleton, and with factors that anchor actin filaments to the plasma membrane [24]. Activation signals induce enrichment of actin filaments in the cortical cytoskeleton, which anchor to membrane rafts to modulate the diffusion of adjacent proteins and lipids, increasing molecular order and driving the clustering of initially dispersed nanoclusters into functional platforms [17,24–26]. Thus, the interplay between membrane rafts and the cytoskeleton governs both raft assembly and cytoskeletal dynamics, with actin-driven raft clusters exerting multifaceted effects on cellular responses. These include modulation of raft-associated signal transduction pathways, facilitation of extracellular-to-intracellular communication via cytoskeletal components, establishment of connections with the extracellular matrix (ECM), generation of cell membrane polarity, and formation of cell junctions [26–28]. Disruptions in this synergism—or alterations induced by bioactive compounds—can trigger remodeling of the cytoskeletal architecture, thereby influencing cell migration, morphology, mechanotransduction, and cell proliferation [24,26].

Given the remarkable therapeutic potential of both phenolic acids and flavonoid glycosides, continued investigation of their effects on membrane dynamics and lipid rafts is essential. Previous studies have examined their interactions with the membrane bilayer, including biophysical and biochemical properties and their modulatory effects on raft formation [29–34]. However, the literature remains limited regarding their influence on raft–actin cytoskeleton synergism, despite its critical role in maintaining cell viability and function.

In our previous study, we observed selective cytotoxicity of extracts from the Bulgarian plant *Inula oculus-christi* and examined two fractions enriched in flavonoid glycosides (FGs) and phenolic acids (PAs) derived from the methanol extract on membrane properties using model membranes [35]. We observed specific molecular effects, including alterations in lipid order, fluidity, polarity, and raft domain dynamics. In the present study, we aim to extend our investigation to their effects on raft–actin cytoskeleton interactions in cells. We focus on two epithelial cell lines (one cancerous and one non-cancerous) with distinctly different membrane polarity, raft and actin cytoskeleton organization, as well as the capability to form tight junctions in a monolayer. These findings may provide deeper insights into the mechanisms underlying their biological activities and potential anticancer therapeutic applications.

2. Materials and Methods

2.1. Plant Material and Plant Extracts

Wild-growing *I. oculus-christi* L. plants were collected in full flowering stage in Bulgaria. Dr. Ina Aneva (Institute of Biodiversity and Ecosystem Research, BAS, Sofia, Bulgaria) identified the plant, and a voucher specimen (SOM 1360) has been deposited in the Herbarium of the Institute of Biodiversity and Ecosystem Research, BAS, Sofia, Bulgaria. Air-dried leaves of *I. oculus-christi* L. were successively extracted with chloroform and methanol at room temperature. Crude extracts were obtained after filtration and evaporation of the solvent. Methanol extract was further fractioned by column chromatography on Sephadex LH-20 using methanol as an eluent as described in [36,37]. Fractions enriched in flavonoid glycosides (FFGs) and in caffeoylquinic acids (FCQAs) were used in this study. The presence of chlorogenic and caffeoylquinic acids, or flavonoid glucosides, in the fractions was confirmed by Thin-Layer Chromatography (TLC) comparison with the isolated compounds (proven by NMR) and authentic standards, as previously reported [36].

2.2. Cell Cultures

Two mammalian epithelial cell lines with distinct membrane lipid order were used in this study. MDCK II cells, derived from immortalized normal renal epithelium, possess well-defined ordered membrane regions (lipid rafts), are engaged in tight junctions, and a large number of intercellular contacts, making them a good model for studying membrane polarization. In contrast, A549 cells, derived from non-small-cell lung carcinoma, exhibit a more fluid membrane, do not form tight junctions with each other, and do not polarize in a monolayer.

Cells were maintained under standard conditions at 37 °C in a humidified atmosphere containing 5% CO_2 in Dulbecco's Modified Eagle Medium (DMEM), supplemented with 10% fetal bovine serum (FBS) and 1% (v/v) antibiotic-antimycotic solution (penicillin 100 U.mL^{-1}, streptomycin 100 µg.mL^{-1} and amphotericin B 0.25 µg·mL^{-1}). Cells were cultured in 25 cm^2 flasks before experiments. For fluorescence imaging, cells were seeded on sterile coverslips in a 24-well plate at an initial density of 2×10^5 cells·mL^{-1}(or 88,106 cells per cm^2). Cells were treated with 100 or 200 µg·mL^{-1} for 24 or 48 h and then proceeded onto fluorescent analyses as described below.

2.3. Evaluation of the Cytotoxicity of FFGs and FCQAs

To assess the cytotoxicity of the extracts used, we applied crystal violet staining and spectrophotometric measurement as previously described in [38]. This approach circumvents potential complications associated with oxidative enzyme-based assays, which may be affected by plant secondary metabolites. The intensity of crystal violet staining depends on the amount of DNA and proteins in the cells, which is proportional to the total number of cells in each sample. Both FFG and FCQA fractions were tested in increasing concentrations—50, 100, 250, and 300 $\mu g \cdot mL^{-1}$. Cytotoxicity was evaluated at 24 and 48 h post-treatment.

2.4. Di-4-ANEPPDHQ Microscopy Measurements

The degree of membrane order was assessed using an optimized protocol with the fluorescent probe Di-4-ANEPPDHQ [39]. This styryl dye, originally developed as a voltage-sensitive probe to detect electrical activity in cells and tissues, also serves as an environmentally sensitive indicator capable of discriminating between lipid phases in both model membranes and intact cells. The probe localizes deep into the membrane and is cholesterol-dependent. Di-4-ANEPPDHQ was purchased from Thermo Fisher Scientific (Sofia, Bulgaria).

Fluorophore was solubilized in dimethyl sulfoxide (DMSO) to a stock concentration of 5 mM and stored at room temperature in the dark. Stock solutions were used for no longer than one month before being replaced with fresh solution. The optimized protocol described in [38] was followed to obtain micrographs of both cell lines. Prior to microscopy experiments, cells were incubated in serum-free medium for 30 min during which fluorophore was added to achieve 5 μM final concentration. Di-4-ANEPPDHQ staining was visualized using a widefield fluorescent Leica TCS SPE microscope (Leica microsystems, Wetzlar, Germany) equipped with a 63× immersion objective and Leica microsystems LAS AF software version 2.2.1. Generalized Polarization (GP) calculations were based on the equation $GP = (I_{500-580} - I_{620-750})/(I_{500-580} + I_{620-750})$, as previously described [39]. Pseudo-colored GP images were merged with mean fluorescence intensity images to preserve structural information. We used the plugin published in Owen et al., 2011 [39], adapted for the CellTool 3.0 software package by Georgi Danovski (http://dnarepair.bas.bg/software/CellTool/ (accessed on 23 June 2023)). The images were thresholders before GP calculation to optimize the signal-to-noise ratio. Photomultiplier tube (PMT) gains were adjusted to avoid saturation in either channel and to ensure comparable intensity across channels. Because membrane packing assessment is a ratiometric measurement between the two channels, PMT gains were kept constant between treated and untreated polyphenol-exposed cells. Under these conditions, the G factor for sensitivity between channels was close to unity. GP values were calculated from raw fluorescence images and averaged from a minimum of 30 cells per experiment. Each experiment was performed at least three times. Values are expressed as mean ± SEM; the p value was computed using Student's t-test. Data with $p < 0.05$ are considered significant [38].

2.5. Fluorescent Staining of F-Actin

The cells were treated as described above. Cell monolayers were washed with 1× PBS containing 100 μM $CaCl_2$ and 1 mM $MgCl_2$, followed by fixation with 4% formaldehyde in PBS for 20 min. Cells were then permeabilized with 0.5% Triton X-100 in PBS for 5 min and blocked with 1.5% bovine serum albumin (BSA) in PBS with 0.1% Tritin X-100 for 30 min. After washing with PBS, cells were stained with TRITC-phalloidin in PBS for 1 h to visualize the actin cytoskeleton. The cell nuclei were stained with 5 $\mu g \cdot mL^{-1}$ DAPI (4′,6-diamidino-2-phenylindole), (excitation: 358 nm; emission: 461 nm) in Fluoromount.

2.6. Statistical Analysis

The statistical analysis was performed using OriginPro 9.0. The mean values of each measured parameter at 37 °C were compared using Student's t-test. Statistically significant differences based on the Tukey test linked to the control group and between the two investigated PPs (FGs and CQAs) were determined at $p < 0.05$ (*), and no significant differences were marked as ns.

3. Results

3.1. Dominant Compounds in the Fractions Enriched in Phenolic Acids (FCQAs) and Flavonoid Glycosydes (FFGs) from Inula oculus-christi L.

The phytochemical study of the methanol extract of *Inula oculus-christi L.* has already been reported [36,37]. Two fractions were used in this study, namely fractions enriched in phenolic acides (FCQAs) and flavonoid glycosides (FFGs). TLC comparison with previously isolated compounds as standards revealed the presence of chlorogenic (5-caffeoylquinic acid, 5-CQA), 1,5-, 3,5-, 4,5-, and 3,4-dicaffeoylquinic (DCQA) acids in FCQAs and nepitrin (nepetin-7-O-glucoside) and hispidulin-7-O-glucoside (see Figure 1).

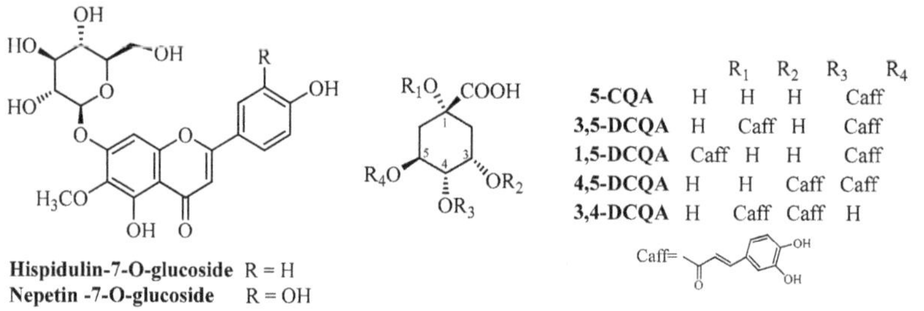

Figure 1. Structures of the dominant compounds in FFGs and FCQAs.

3.2. Cytotoxicity of the Fractions Enriched in Caffeoylquinic Acids (FCQAs) and Flavonoid Glycosydes (FFGs) from Inula oculus-christi L.

Both tested fractions, FFGs and FCQAs, exhibited dose- and time-dependent effects on the cell viability of both cell lines. In the non-cancerous MDCKII cell line, FFG treatment resulted in consistent cell responses at both early (24 h) and late (48 h) time points. At lower concentrations (50–100 $\mu g \cdot mL^{-1}$), cell viability remained high (~100%), while higher concentrations (250–300 $\mu g \cdot mL^{-1}$) induced a modest reduction in viability (~75%). No further cytotoxic effects were observed within the tested range, indicating good biocompatibility of FFGs with MDCKII cells (Figure 2). In contrast, FFGs exerted a pronounced dose- and time-dependent cytotoxic effect on the cancerous A549 cell line. At concentrations of 250–300 $\mu g \cdot mL^{-1}$, a significant decrease in cell viability was detected, with an IC_{50} value of 295 $\mu g \cdot mL^{-1}$ at 24 h post-treatment. This effect became more substantial over time, with viability dropping to ~50% at 250 $\mu g \cdot mL^{-1}$ and below 30% at 300 $\mu g \cdot mL^{-1}$ after 48 h, resulting in a lower IC_{50} value of 256 $\mu g \cdot mL^{-1}$.

The FCQA fraction demonstrated comparable effects in both MDCKII and A549 cells during early exposure. At lower concentrations (50–100 $\mu g \cdot mL^{-1}$), minimal to no cytotoxicity was observed. However, at higher concentrations (250–300 $\mu g \cdot mL^{-1}$), a moderate reduction in viability (65–70%) was noted in both cell lines, indicating moderate cytotoxicity (Figure 2b,d). These results suggest that FCQAs are generally well-tolerated across both cell types within the tested concentration range. At 48 h post-treatment, a clear dose- and time-dependent effect was evident for FCQAs in both cell lines. While MDCKII cells exhibited a slight reduction in viability (maintaining ~60%), a more pronounced cytotoxic response was observed in A549 cells, with an estimated IC_{50} value of 280 $\mu g \cdot mL^{-1}$ (Figure 2d). Overall, these findings demonstrate that both FFG and FCQA fractions exhibit selective dose- and time-dependent cytotoxicity toward the cancerous A549 cell line, while maintaining favorable compatibility with non-cancerous MDCKII cells.

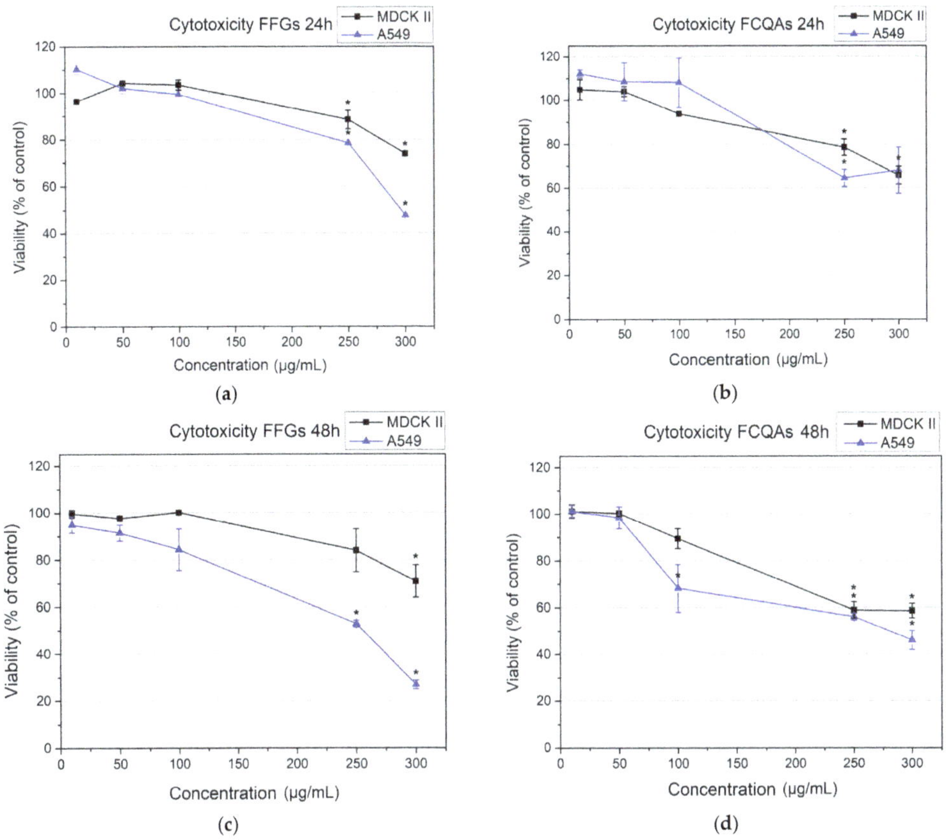

Figure 2. Cytotoxicity of FFGs (**a**,**c**) and FCQAs (**b**,**d**) on MDCK II and A549 cell lines after 24 h (**a**,**b**) and 48 h (**c**,**d**). The data were analyzed with OriginPro 9.0 and presented as a mean value ± SE. Statistical significance (*) is according to one-way ANOVA at the 0.05 level.

3.3. Membrane Order Imaging of Cells with the Fluorescent Probe Di-4-ANEPPDHQ Through Confocal Microscopy

The membrane lipid order of both MDCKII and A549 cells was visualized using the environment-sensitive fluorescent probe Di-4-ANEPPDHQ. This probe exhibits a fluorescence emission spectrum that is sensitive to the lipid packing and polarity of the membrane environment, properties that are indirectly influenced by the presence of water molecules in more disordered membrane regions. Di-4-ANEPPDHQ partitions evenly between liquid-ordered (L_o) and liquid-disordered (L_d) membrane phases but emits from two distinct excited states: a relaxed state associated with loosely packed (disordered) lipids, and a non-relaxed state corresponding to tightly packed (ordered) bilayers. Upon excitation at 488 nm, the probe emits with a peak around 560 nm when located in ordered (L_o) domains (visualized in green in Figures 3 and 4) and around 620 nm when localized in disordered (L_d) regions (visualized in red in Figures 3 and 4). This dual-emission behavior enables direct visualization and assessment of membrane organization and lipid order in mammalian cells, including human-derived cell lines [38,39].

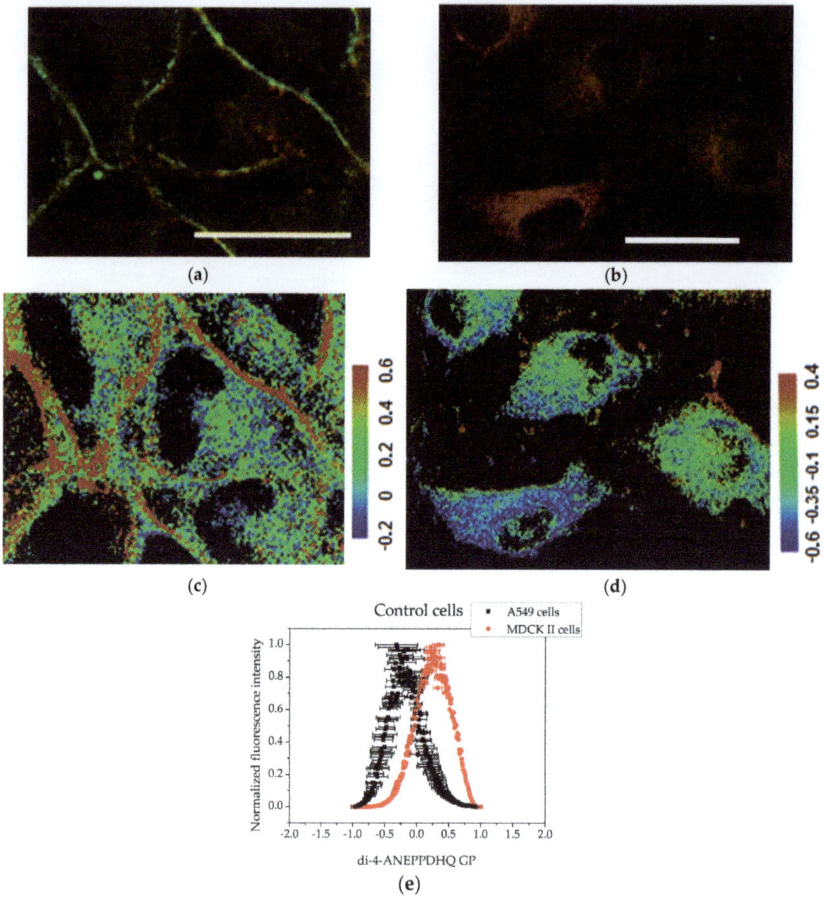

Figure 3. Assessment of the degree of cell membrane order using the fluorescent probe Di-4-ANEPPDHQ excited at 488 nm: (**a**) MDCK II cells merged image of emission at 560 nm (green) and 620 nm (red); (**b**) A549 cells merged image of emission at 560 nm (green) and 620 nm (red). Scale bar—25 μm. GP images of control cells: (**c**) MDCK II cells GP image; (**d**) A549 cells GP image. Scale bar—blue range corresponds to a more fluid lipid state and the red range to the ordered phase, respectively. Summarized data for the generalized polarization of membranes. (**e**) Histograms obtained with the CellTool and Origin programs from GP images of control cells. Mean values of all measurements and standard errors (bars).

In the micrographs of control (untreated) MDCKII cells, a strong green fluorescence signal—corresponding to the liquid-ordered (L_o) phase—is observed outlining the plasma membrane. As MDCKII cells are polarized epithelial cells, they form well-defined tight junctions and possess raft-enriched plasma membranes. The predominance of the L_o phase can be attributed to the high degree of intercellular contacts, which serve as sites for cell–cell adhesion and for anchoring the actin cytoskeleton to the plasma membrane (Figure 3a). The cancerous cell line, A549 (Figure 3b), exhibits a cytoplasm rich in organelles and lamellar bodies, reflecting high metabolic activity and active vesicular trafficking. In contrast to MDCKII cells, the Di-4-ANEPPDHQ fluorescence in A549 cells is predominantly distributed throughout the cytoplasmic membranes, suggesting a lower degree of membrane lipid order at the plasma membrane and possibly higher endocytic or vesicular membrane activity.

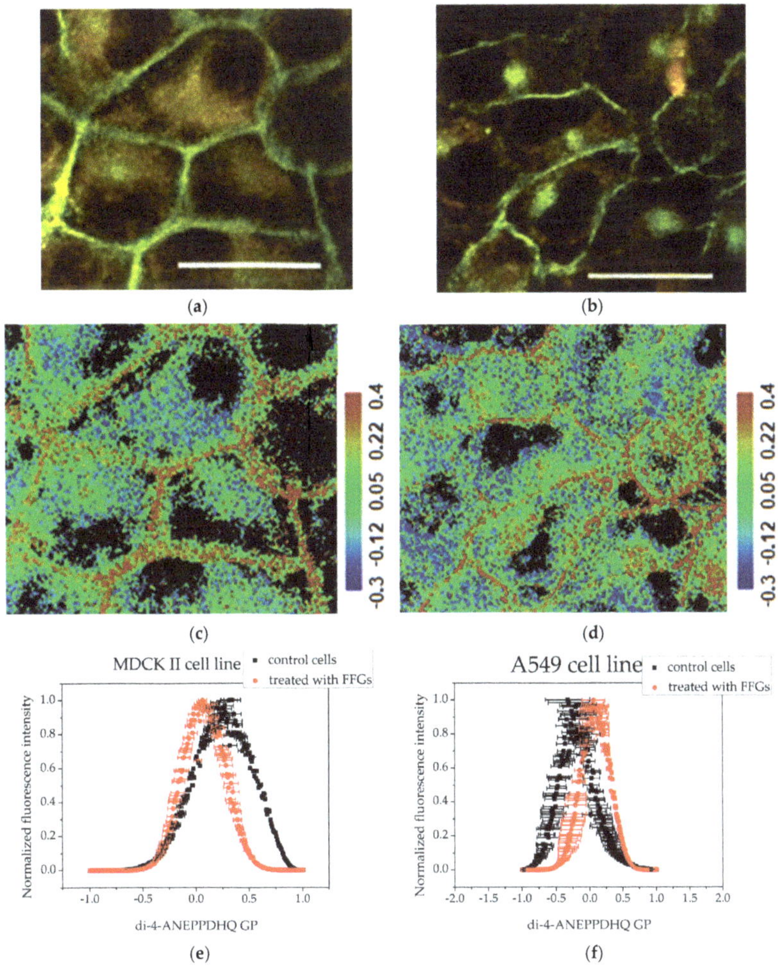

Figure 4. Assessment of the degree of membrane ordering after treatment with 200 μg·mL^{-1} fraction enriched in FFGs using the fluorescent probe Di-4-ANEPPDHQ excited at 488 nm: (**a**) MDCK II cells merged image of emission at 560 nm (green) and 620 nm (red); (**b**) A549 cells—merged image of emission at 560 nm (green) and 620 nm (red). Scale bar—25 μm. (**c**) MDCK II GP image; (**d**) A549 cells GP image. Scale bar—blue range corresponds to more fluid lipid state and the red range to the ordered phase, respectively. Summarized data for the generalized polarization of membranes—(**e**,**f**): histograms obtained with the CellTool and Origin programs from GP images of treated cells. Mean values of all measurements and standard errors (bars).

Treatment with fractions enriched in FFGs at a concentration of 200 μg·mL^{-1} led to changes in membrane organization in both cell lines. The membranes of MDCK II cells (Figure 4a) retained their distinct line, but noticeable changes in cell shape were observed. The cells appeared more roundish compared to the controls, which is likely a consequence of the loss of some intercellular contacts. Additionally, a more intense fluorescence in the cytoplasm was noticeable. After treatment of the A549 cell line with FFGs (Figure 4b), cells displayed a pronounced fluorescence in the plasma membranes in contrast to the control. The cytoplasm signal from Di-4-ANEPPDHQ was densely concentrated in a specific manner in an area located next to the nucleus.

Microscopic analysis of MDCKII cells treated with FCQAs revealed notable morphological changes, including a more elongated shape compared to the control group. The cell membranes appeared thinner. In A549 cancer cells, FCQAs caused significant damage. The treated cells exhibited highly altered morphology and extensive vesicle formation, suggesting membrane disintegration (Figure 5).

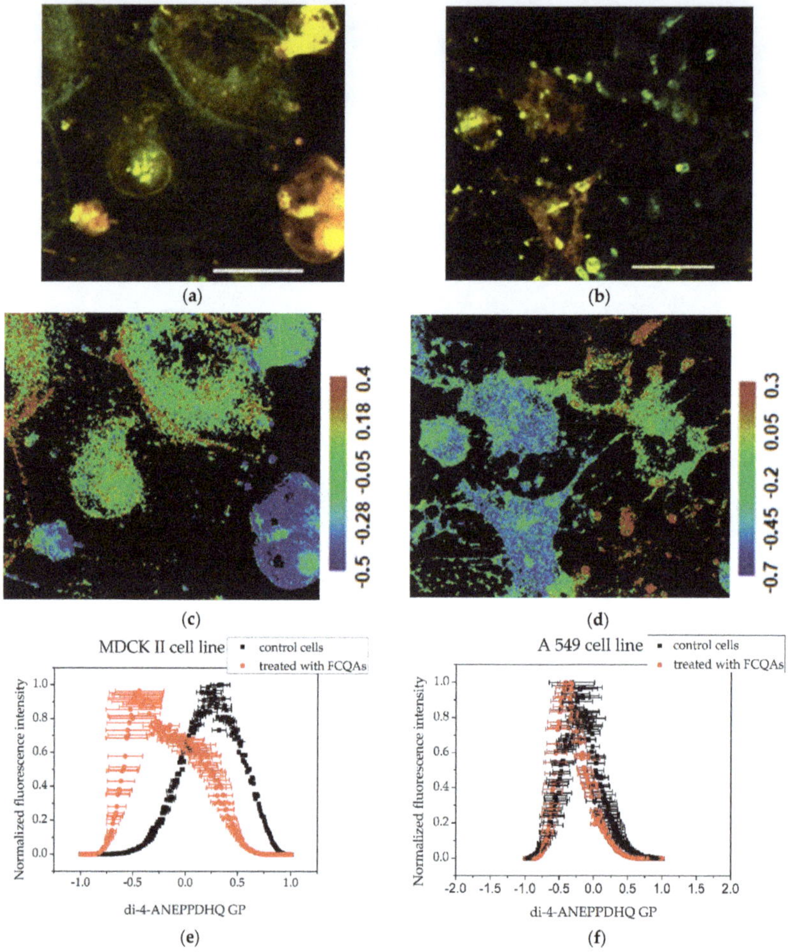

Figure 5. Assessment of the degree of membrane ordering after treatment with 200 μg·mL^{-1} fraction enriched in FCQAs using the fluorescent probe Di-4-ANEPPDHQ excited at 488 nm: (**a**) MDCK II cells merged image of emission at 560 nm (green) and 620 nm (red); (**b**) A549 cells—merged images of emission at 560 nm (green) and 620 nm (red). Scale bar—25 μm. (**c**) MDCK II GP images; (**d**) A549 cells GP images. Scale bar—blue range corresponds to more fluid lipid state and the red range to the ordered phase, respectively. Summarized data for the generalized polarization of membranes—(**e**,**f**): histograms obtained with the CellTool and Origin programs from GP images of treated cells. Mean values of all measurements and standard errors (bars).

3.4. GP Image Acquisition

The shift in emission spectra of Di-4-ANEPPDHQ can be quantified by Generalized Polarization (GP) analysis. This function (GP) represents normalized fluorescence intensity ratio. The spectral channels corresponding to L_o and L_d phases are used for the GP equation. Di-4-ANEPPDHQ microscopy measurements and GP calculations are described in Materials and Methods. Quantitative representation of the processed data from the fluorescence micrographs is shown in Figures 3–5 and Table 1.

Table 1. Peak of histograms obtained from GP images of control and treated cells from MDCK II and A549 cell lines.

Cells	MDCK II—GP Peak at	A549—GP Peak at
Control	0.331 (SE 0.09026)	−0.323 (SE 0.33225)
Treated with FFGs	0.063 (SE 0.04053)	0.047 (SE 0.16069)
Treated with FCQAs	−0.441 (SE 0.23289)	−0.338 (SE 0.07277)

GP images of controls and treated cells are presented in Figures 3–5. The blue color range corresponds to a more fluid lipid state and the red color range to a more ordered phase. The generated histograms from the GP images are summarized in Figures 3–5. The control MDCK II (Figure 3c,e) cells had higher GP values compared to A549 controls (Figure 3d,e), indicating a more ordered membrane of the MDCK II cells.

Treatment of MDCK II cells with FFG (Figure 4) and FCQA fractions (Figure 5) resulted in increased membrane fluidity compared to control cells (Figure 3), as evidenced by lower GP values. Among the two, FCQAs induced a more pronounced fluidizing effect, as shown in Figure 5 and quantified in Table 1. In contrast, A549 cells exhibited a different response. The FFG had a membrane-ordering effect in these cancer cells (Figure 3), opposite to its effect in non-cancerous MDCK II cells. On the other hand, FCQA treatment led to increased membrane fluidity in A549 cells (Figure 5).

3.5. Actin Cytoskeleton Reorganization

The cell lines used for this study are epithelial. Fluorescent images of control cells from both cell lines (Figure 6a,b) displayed normal morphology, characteristic features of the respective cell type with confluent monolayer and active cell division. In MDCKII cells (Figure 6a), the fluorescence signal was most intense at the cell periphery and along the plasma membrane. They form many adherent junctions and tight junctions, which ensure the separation of the membrane surface into basolateral and apical. In A549 cells (Figure 6b), a well-structured actin cytoskeleton was observed with pronounced stress fibrils.

Following treatment with FFGs, MDCK cells (Figure 6c,d) continued to exhibit a confluent monolayer with active cell division. Similarly to the control, the fluorescence signal remained most intense in the cell membrane area. In contrast, FFGs reduced the monolayer confluency of the A549 cell line (Figure 6d). After 24 h post-treatment with 100 $\mu g \cdot mL^{-1}$ FFGs, a redistribution of actin was observed in the cortex of lung carcinoma cells even in areas lacking intercellular contacts. This redistribution persisted after 48 h of treatment (Figure 6e,f). At higher concentrations, the actin rearrangement progressed into distinct cytoplasmic clumping near the nucleus. Comparable clumps were observed in the same cells stained with Di-4-ANEPPDHQ (Figure 4), and GP analysis confirmed that these regions exhibit increased membrane order.

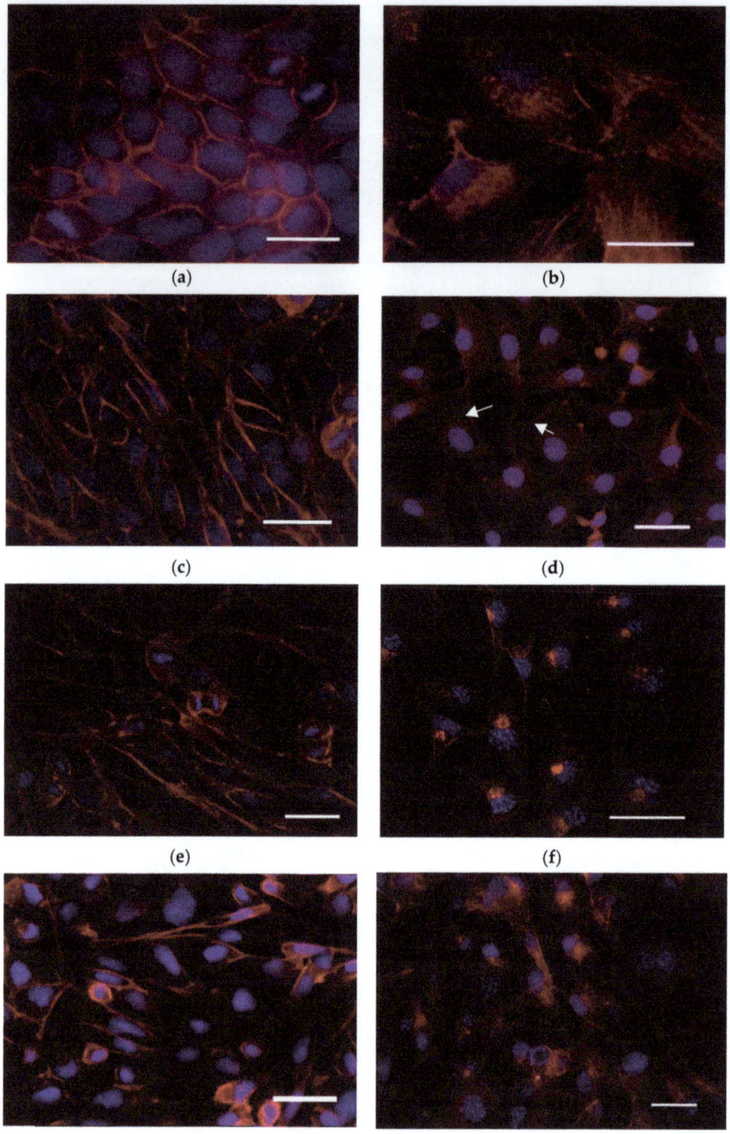

(**a**) (**b**) (**c**) (**d**) (**e**) (**f**) (**g**) (**h**)

Figure 6. Assessment of the actin cytoskeleton using the TRITC-phalloidine (red) and DAPI (blue): control cells; (**a**) MDCK II cells; (**b**) A549 cells. Cells treated with 100 μg·mL^{-1} FFGs: MDCK II cells at 24 (**c**) and 48 h of treatment (**e**); A549 cells at 24 (**d**) and 48 h (**f**) of treatment. Cells treated with 100 μg·mL^{-1} FCQAs: (**g**) MDCK II cells; (**h**) A549 cells. Staining: TRITC-phalloidine (red) and DAPI (blue). The preparations were observed and imaged using a Nikon TiU confocal laser scanning microscope (Nikon, Tokyo, Japan) and EZC1 software v.3.90. Magnification 400×. Scale bar—25 μm, arrows—actin aggregates.

Treatment with fraction enriched in FCQAs resulted in reduced intensity of the signal of F-actin in MDCKII cells (Figure 6g). Actin filaments were generally preserved in the membrane area, where actin aggregates could be found. A549 cells also exhibited decreased fluorescence intensity following FCQA treatment (Figure 6h). These cells were characterized by a branched actin meshwork and the presence of prominent stress fibers.

In general, the tested fractions showed a weaker effect on the actin cytoskeleton of cells from the MDCK II cell line. The A549 line showed great sensitivity to the fractions, as we observed different patterns of disorganization of the actin cytoskeleton upon treatment with the extracts.

4. Discussion

The cytotoxicity and biocompatibility of plant-derived polyphenols are largely influenced by their chemical structures and lipophilicity. An increase in the lipophilicity of chlorogenic acids (CGAs) is positively correlated with enhanced cytotoxic effects due to improved membrane permeability and intracellular accumulation [40]. Conversely, naturally occurring CGAs such as 5-caffeoylquinic acid are inherently hydrophilic because of multiple hydroxyl (-OH) and carboxyl (-COOH) groups, which enhance water solubility and contribute to their favorable biocompatibility and selective bioavailability [41]. The presence of a glycosidic moiety at the 7-*O* position imparts amphiphilic properties to the examined flavonoid glycosides [42], which has been shown to correlate exponentially with decreased cytotoxicity, primarily due to limited passive membrane diffusion [43]. In this study, two plant-derived fractions of methanol extracts from Bulgarian *Inula oculus-christi* were used: one enriched in flavonoid glycosides (FFGs) and the other predominantly containing caffeoylquinic acids (FCQAs). The FFG fraction was characterized by flavonoid glycosides such as hispidulin-7-O-glucoside and nepetin-7-O-glucoside, while the FCQA fraction comprised primarily chlorogenic acids (CGAs), including 5-CQA and dicaffeoylquinic acid (diCQA) isomers (1,5-; 3,5-; 4,5-; and 3,4-). According to our results from the cytotoxicity assay, both FFGs and FCQAs fractions demonstrated selective dose- and time-dependent cytotoxicity toward the cancerous A549 cell line, while exhibiting favorable biocompatibility with the non-cancerous MDCK II cells. These findings suggest a degree of cell-type selectivity, which is consistent with previous studies indicating that both chlorogenic acids (CGAs) and flavonoid glycosides (FGs) can exert cell-specific cytotoxic effects towards cancer cells [44–46].

At sub-toxic concentrations, both polyphenolic fractions induced alterations in membrane organization and notable rearrangements of the actin cytoskeleton in the examined cell lines. MDCKII membranes exhibited high GP values (+0.331), reflecting enrichment in cholesterol- and sphingomyelin-rich L_o domains. This high lipid order aligns with their origin from epithelial cells with many intercellular junctions, including tight junctions [47]. Cortical actin formed a dense mesh beneath the plasma membrane, reinforced by adherent and tight junctions that maintain epithelial integrity. In contrast, A549 cells, derived from lung carcinoma, displayed less ordered, more fluid membranes (GP = −0.323). Elevated membrane fluidity is characteristic of cancer cells and contributes to enhanced motility, invasiveness, and metastatic potential [48]. In these cells, the actin cytoskeleton was organized into prominent stress fibers spanning the cytoplasm, consistent with a migratory phenotype and weaker cell–cell adhesion [49,50]. These clearly distinct differences make both cell lines good models for tracking changes in membrane organization and the actin cytoskeleton in living cells.

The flavonoid glycoside-enriched fraction demonstrated a cell-specific effect on plasmalaema. FFGs exhibited a slightly fluidizing effect on non-cancerous cells (GP = 0.063), but the opposite effect on cancerous cells. However, the membranes of MDCK II cells retained their relatively ordered state without significant alterations in the actin cytoskeleton pattern, and the cellular monolayer was confluent. In contrast, treatment of A549 cells with FFGs caused pronounced actin redistribution in regions lacking intercellular contacts, progressing into highly ordered cytoplasmic clumps near the nucleus. At the same time, the FCQAs induced persistent membrane fluidization in both cell lines: for MDCK II, GP values were −0.441, and in A549 cells we detected minor fluidization (GP = −0.338). The cortical actin filaments in MDCK II cells were mostly preserved, with occasional small actin aggregates near the membrane, while changes in actin organization in A549 lung cells were insignificant.

Our previous findings showed that the same FFGs and FCQAs modulate the phase heterogeneity of model membranes through differential penetration into the lipid bilayer, governed by their chemical structures and degree of lipophilicity [35]. CGAs largely remain at the membrane surface, whereas flavonoid glycosides integrate into the bilayer. These data support the findings of other authors of significant proportions of membrane-incorporated flavonoids compared to those represented in the cytosol when treating colon carcinoma cell line HT-29 with tea flavonoids [51]. The amphipathic FGs embed their hydrophobic aglycone into the membrane core while positioning the polar glycoside near the glycerol backbone [35,52,53], enhancing fluidity in ordered L_o domains and increasing order in fluid L_d regions [35]. These findings are consistent with our observations on cell membranes. In ordered membranes of non-cancerous MDCK II cells, FFGs could remain near the headgroup region, where they increase local polarity and disrupt lipid packing at this membrane level. In contrast, in more fluid membranes of the cancerous A549 cells, they could penetrate deeper into the hydrophobic core, inducing tighter lipid packing and a subsequent increase in membrane order. On the other hand, hydrophilic CGAs remain primarily in the outer leaflet of the membrane, interacting with lipid headgroups. This increases local polarity at the membrane–water interface and weakens lipid–cholesterol interactions [54,55], as reflected by the elevated GP values observed in both cell lines.

The organization of the actin cytoskeleton, particularly the peripheral cortical actin, is closely linked to the domain organization of the plasma membrane [24–26]. Therefore, alterations in membrane order can potentially affect the anchoring and spatial arrangement of actin filaments at the membrane interface. To date, there is limited data on the effects of polyphenols on both membrane order and the actin cytoskeleton simultaneously, with most studies focusing on either membrane properties or actin organization independently [39,56,57].

Lipid raft L_o domains are enriched in proteins that tether directly or indirectly to actin filaments, including caveolins, flotillins, integrins, ERM proteins, and GPI-anchored proteins [24,26,27,58]. The tightly packed, ordered lipid environment of these domains enhances the affinity of raft-associated proteins for actin, thereby stabilizing cortical cytoskeleton organization. Disruption or fluidization of the plasma membrane can induce lateral dispersion of lipid raft-associated proteins, weakening their connections to the actin cytoskeleton and leading to actin redistribution or aggregation [24]. A similar phenomenon was observed in MDCK II cells treated with FCQAs. This treatment resulted in a decrease in membrane order, accompanied by partial actin aggregation near the membrane, likely due to disrupted affinity between raft-associated proteins and the actin cytoskeleton. The incorporation of amphiphilic FFGs into the lipid bilayer stiffens the membrane and generates mechanical load, which can be sensed by mechanosensitive complexes, such as integrin-based focal adhesions and cytoskeletal linkers (talin, vinculin, α-actinin) [27,59]. These complexes transmit mechanical cues to the actin cytoskeleton, activating Rho GTPases (RhoA, Rac1, Cdc42), which regulate actin polymerization [60]. The clustering of actin filaments near the nucleus likely represents a mechanosensitive compensatory response to maintain intracellular tension under increased membrane order.

These findings highlight how the initial membrane order, raft composition, and compound chemistry collectively shape lipid raft dynamics and actin organization. They underscore the unique ability of polyphenols to modulate membrane properties, revealing a direct link between membrane order and cytoskeletal architecture that governs cellular responses.

5. Conclusions

This study demonstrates distinct effects on membrane organization and cytoskeletal architecture of two polyphenol fractions of methanolic extracts derived from Bulgarian *Inula oculus-christi*—enriched in flavonoid glycosides (FFGs) and caffeoylquinic acids (FCQAs) in both cancerous (A549) and non-cancerous (MDCK II) cell lines. FFGs exhibited selective cytotoxicity toward cancer cells while maintaining biocompatibility with healthy epithelial cells, suggesting potential for therapeutic application.

At a sub-toxic concentration (200 $\mu g \cdot mL^{-1}$), both polyphenolic fractions induced notable alterations in cell membrane order and actin cytoskeleton dynamics. Prominent effects were observed in the cancerous A549 cell line, where FFGs caused membrane ordering, which was associated with the formation of perinuclear actin aggregates.

The distinct effects of both fractions on membrane–cytoskeleton dynamics across different cell types are likely driven by their specific chemical structures, which determine their mode of interaction with the lipid bilayer. The more hydrophilic nature of FCQAs favors their interaction with the membrane surface of MDCK II cells, where they likely disrupt lipid–lipid and lipid–protein interactions. This leads to increased membrane polarity and a reduction in membrane order. In contrast, the amphipathic properties of FFGs enable deeper insertion into the more disordered plasma membranes of A549 cancer cells, potentially imposing mechanical stress on the cell membrane.

The findings in this study highlight the intricate interplay between membrane biophysical properties, lipid raft–actin cytoskeleton dynamics, and the chemical nature of exogenous compounds. They underscore the potential of plant-derived polyphenols as modulators of membrane-associated signaling and structural organization, offering valuable insights for the development of membrane-targeted therapeutic strategies.

Author Contributions: Conceptualization, T.T.-H., A.K. and G.S.; methodology, T.T.-H. and G.S.; software, G.S.; validation, V.M.-D., J.D. and R.V.; formal analysis, R.V., D.B., T.T.-H. and A.T.; investigation, R.V., K.M. and A.K.; resources, A.T. and V.I.; writing—original draft preparation, R.V., T.T.-H., A.T., G.S. and D.B.; writing—review and editing, T.T.-H., A.T., G.S., V.M.-D., J.D., K.M. and A.K.; visualization, R.V. and K.M.; supervision, T.T.-H., A.T. and G.S.; project administration, T.T.-H., A.T. and G.S.; funding acquisition, T.T.-H., A.T. and G.S. All authors have read and agreed to the published version of the manuscript.

Funding: This research received no external funding.

Institutional Review Board Statement: Not applicable.

Informed Consent Statement: Not applicable.

Data Availability Statement: Data are contained within the article.

Acknowledgments: The support of the Centre of Competence "Sustainable Utilization of Bio-resources and Waste of Medicinal and Aromatic Plants for Innovative Bioactive Products" (BIORESOURCES BG), project BG16RFPR002-1.014-0001, funded by the Program "Research, Innovation and Digitization for Smart Transformation" 2021–2027, co-funded by the EU, is greatly acknowledged. Ralitsa Veleva was supported by the Bulgarian Ministry of Education and Science under the National Program "Young Scientists and Postdoctoral Students—2".

Conflicts of Interest: The authors declare no conflicts of interest. The funders had no role in the design of the study; in the collection, analyses, or interpretation of data; in the writing of the manuscript; or in the decision to publish the results.

Abbreviations

The following abbreviations are used in this manuscript:

diCQA	Dicaffeoylquinic acid isomers
PA	Phenolic acids
FCQAs	Fraction of caffeoylquinic acids
CGAs	Chlorogenic acids
FFGs	Fraction enriched in flavonoid glycosides
GP	Generalized polarization

References

1. Seca, A.M.; Grigore, A.; Pinto, D.C.; Silva, A.M. The genus *Inula* and their metabolites: From ethnopharmacological to medicinal uses. *J. Ethnopharmacol.* **2014**, *154*, 286–310. [CrossRef]
2. Yang, L.; Wang, X.; Hou, A.; Zhang, J.; Wang, S.; Man, W.; Yu, H.; Zheng, S.; Wang, Q.; Jiang, H.; et al. A review of the botany, traditional uses, phytochemistry, and pharmacology of the Flos Inulae. *J. Ethnopharmacol.* **2021**, *276*, 114125. [CrossRef] [PubMed]
3. Spiegel, M.; Kapusta, K.; Kołodziejczyk, W.; Saloni, J.; Żbikowska, B.; Hill, G.A.; Sroka, Z. Antioxidant activity of selected phenolic acids—Ferric reducing antioxidant power assay and QSAR analysis of the structural features. *Molecules* **2020**, *25*, 3088. [CrossRef] [PubMed]
4. Rogozinska, M.; Lisiecki, K.; Czarnocki, Z.; Biesaga, M. Antioxidant activity of sulfate metabolites of chlorogenic acid. *Appl. Sci.* **2023**, *13*, 2192. [CrossRef]
5. Belew, A.A.; Hanan, G.G.M.W.; Meshesha, D.S.; Akele, M.L. Evaluation of total phenolic, flavonoid contents, antioxidant and antibacterial activity of leaf extracts from *Rhus vulgaris*. *Discov. Plants* **2025**, *2*, 141. [CrossRef]
6. Xie, L.; Deng, Z.; Zhang, J.; Dong, H.; Wang, W.; Xing, B.; Liu, X. Comparison of flavonoid O-glycoside, C-glycoside and their aglycones on antioxidant capacity and metabolism during in vitro digestion and in vivo. *Foods* **2022**, *11*, 882. [CrossRef]
7. Gumisiriza, H.; Olet, E.A.; Mwikali, L.; Akatuhebwa, R.; Omara, T.; Lejju, J.B.; Sesaazi, D.C. Antibacterial and antioxidant activities of flavonoids, phenolic and flavonoid glycosides from *Gouania longispicata* leaves. *Microbiol. Res.* **2024**, *15*, 2085–2101. [CrossRef]
8. Ghozzi, I.; Fontaine, J.-X.; Molinié, R.; Elboutachfaiti, R.; Akkouche, L.; Sebei, K.; Mathiron, D.; Hano, C.; Garros, L.; Choque, E.; et al. Relationship between the structure of the flavone C-glycosides of linseed (*Linum usitatissimum* L.) and their antioxidant activity. *Molecules* **2024**, *29*, 5829. [CrossRef]
9. Xie, J.; Xiong, S.; Li, Y.; Xia, B.; Li, M.; Zhang, Z.; Shi, Z.; Peng, Q.; Li, C.; Lin, L.; et al. Phenolic acids from medicinal and edible homologous plants: A potential anti-inflammatory agent for inflammatory diseases. *Front. Immunol.* **2024**, *15*, 1345002. [CrossRef]
10. Ekowati, J.; Tejo, B.A.; Maulana, S.; Kusuma, W.A.; Fatriani, R.; Ramadhanti, N.S.; Norhayati, N.; Nofianti, K.A.; Sulistyowaty, M.I.; Zubair, M.S.; et al. Potential utilization of phenolic acid compounds as anti-inflammatory agents through TNF-α convertase inhibition mechanisms: A network pharmacology, docking, and molecular dynamics approach. *ACS Omega* **2023**, *8*, 43243–43254. [CrossRef]
11. Liu, W.; Cui, X.; Zhong, Y.; Ma, R.; Liu, B.; Xia, Y. Phenolic metabolites as therapeutic in inflammation and neoplasms: Molecular pathways explaining their efficacy. *Pharmacol. Res.* **2023**, *193*, 106812. [CrossRef]
12. Wu, J.; Huang, H.; Gong, L.; Tian, X.; Peng, Z.; Zhu, Y.; Wang, W. A flavonoid glycoside compound from *Siraitia grosvenorii* with anti-inflammatory and hepatoprotective effects in vitro. *Biomolecules* **2024**, *14*, 450. [CrossRef]
13. Maleki, S.J.; Crespo, J.F.; Cabanillas, B. Anti-inflammatory effects of flavonoids. *Food Chem.* **2019**, *299*, 125124. [CrossRef]
14. Zhukov, A.; Vereshchagin, M. Polar glycerolipids and membrane lipid rafts. *Int. J. Mol. Sci.* **2024**, *25*, 8325. [CrossRef] [PubMed]
15. Yang, J.; Jin, J.; Li, S. Role of polyunsaturated phospholipids in liquid-ordered and liquid-disordered phases. *RSC Adv.* **2021**, *11*, 27115–27120. [CrossRef] [PubMed]
16. Pike, L.J. Rafts defined: A report on the Keystone Symposium on lipid rafts and cell function. *J. Lipid Res.* **2006**, *47*, 1597–1598. [CrossRef] [PubMed]
17. Kusumi, A.; Tsunoyama, T.A.; Tang, B.; Hirosawa, K.M.; Morone, N.; Fujiwara, T.K.; Suzuki, K.G.N. Cholesterol- and actin-centered view of the plasma membrane: Updating the Singer–Nicolson fluid mosaic model to commemorate its 50th anniversary. *Mol. Biol. Cell* **2023**, *34*, pl1. [CrossRef]
18. Brown, D.A.; London, E. Structure and function of sphingolipid- and cholesterol-rich membrane rafts. *J. Biol. Chem.* **2000**, *275*, 17221–17224. [CrossRef]
19. Sharma, P.; Varma, R.; Sarasij, R.C.; Ira; Gousset, K.; Krishnamoorthy, G.; Rao, M.; Mayor, S. Nanoscale organization of multiple GPI-anchored proteins in living cell membranes. *Cell* **2004**, *116*, 577–589. [CrossRef]

20. Trybus, M.; Hryniewicz-Jankowska, A.; Wójtowicz, K.; Trombik, T.; Czogalla, A.; Sikorski, A.F. EFR3A: A new raft domain organizing protein? *Cell. Mol. Biol. Lett.* **2023**, *28*, 86. [CrossRef]
21. Castello-Serrano, I.; Heberle, F.A.; Diaz-Rohrer, B.; Ippolito, R.; Shurer, C.R.; Lujan, P.; Campelo, F.; Levental, K.R.; Levental, I. Partitioning to ordered membrane domains regulates the kinetics of secretory traffic. *eLife* **2024**, *12*, RP89306. [CrossRef] [PubMed]
22. Lietha, D.; Izard, T. Roles of membrane domains in integrin-mediated cell adhesion. *Int. J. Mol. Sci.* **2020**, *21*, 5531. [CrossRef] [PubMed]
23. Daumann, I.M.; Hiesinger, P.R. Lipid rafts, Rab GTPases, and a late endosomal checkpoint for plasma membrane recycling. *Proc. Natl. Acad. Sci. USA* **2023**, *120*, e2302320120. [CrossRef]
24. Chichili, G.R.; Rodgers, W. Cytoskeleton-membrane interactions in membrane raft structure. *Cell. Mol. Life Sci.* **2009**, *66*, 2319–2328. [CrossRef]
25. Liu, A.P.; Fletcher, D.A. Actin polymerization serves as a membrane domain switch in model lipid bilayers. *Biophys. J.* **2006**, *91*, 4064–4070. [CrossRef]
26. Head, B.P.; Patel, H.H.; Insel, P.A. Interaction of membrane/lipid rafts with the cytoskeleton: Impact on signaling and function: Membrane/lipid rafts, mediators of cytoskeletal arrangement and cell signaling. *Biochim. Biophys. Acta* **2014**, *1838*, 532–545. [CrossRef]
27. Warda, M.; Tekin, S.; Gamal, M.; Khafaga, N.; Çelebi, F.; Tarantino, G. Lipid rafts: Novel therapeutic targets for metabolic, neurodegenerative, oncological, and cardiovascular diseases. *Lipids Health Dis.* **2025**, *24*, 147. [CrossRef]
28. Kalappurakkal, J.M.; Anilkumar, A.A.; Patra, C.; van Zanten, T.S.; Sheetz, M.P.; Mayor, S. Integrin mechano-chemical signaling generates plasma membrane nanodomains that promote cell spreading. *Cell* **2019**, *177*, 1738–1756.e23. [CrossRef]
29. Veiko, A.G.; Sekowski, S.; Lapshina, E.A.; Wilczewska, A.Z.; Markiewicz, K.H.; Zamaraeva, M.; Zhao, H.; Zavodnik, I.B. Flavonoids modulate liposomal membrane structure, regulate mitochondrial membrane permeability and prevent erythrocyte oxidative damage. *Biochim. Biophys. Acta Biomembr.* **2020**, *1862*, 183442. [CrossRef]
30. Tarahovsky, Y.S.; Kim, Y.A.; Yagolnik, E.A.; Muzafarov, E.N. Flavonoid–Membrane Interactions: Involvement of Flavonoid–Metal Complexes in Raft Signaling. *Biochim. Biophys. Acta Biomembr.* **2014**, *1838*, 1235–1246. [CrossRef]
31. Wang, R.; Zhu, W.; Dang, M.; Li, X.; Zhang, Y.; Chen, J. Targeting lipid rafts as a rapid screening strategy for potential antiadipogenic polyphenols along with the structure-activity relationship and mechanism elucidation. *J. Agric. Food Chem.* **2022**, *70*, 3872–3885. [CrossRef] [PubMed]
32. Kanga, Y.; Kimb, B.; Kima, S.; Leec, Y.; Yoona, Y. Inhibitory potential of flavonoids on PtdIns(3,4,5)P3 binding with the phosphoinositide-dependent kinase 1 pleckstrin homology domain. *Bioorganic Med. Chem. Lett.* **2017**, *27*, 420–426. [CrossRef] [PubMed]
33. Wang, R.; Shi, X.; Li, C. Insights into the surface binding and structural interference of polyphenols with the membrane raft domains in relation to their distinctive ability to inhibit preadipocyte differentiation in 3T3-L1 cells. *J. Agric. Food Chem.* **2023**, *71*, 19845–19855. [CrossRef] [PubMed]
34. Weng, Z.; Zeng, F.; Wang, M.; Guo, S.; Tang, Z.; Itagaki, K.; Lin, Y.; Shen, X.; Cao, Y.; Duan, J.; et al. Antimicrobial activities of lavandulylated flavonoids in *Sophora flavences* against methicillin-resistant *Staphylococcus aureus* via membrane disruption. *J. Adv. Res.* **2024**, *57*, 197–212. [CrossRef]
35. Veleva, R.; Topouzova-Hristova, T.; Kostadinova, A.; Benkova, D.; Trendafilova, A.; Ivanova, V.; Moskova-Doumanova, V.; Mladenova, K.; Doumanov, J.; Yordanova, V.; et al. Flavonoid Glycosides and Phenolic Acids from *Inula oculus-christi* Modulate Membrane Organization and Provide Antioxidant Protection. *Molecules* **2025**, *30*, 2740. [CrossRef]
36. Trendafilova, A.; Ivanova, V.; Rangelov, M.; Todorova, M.; Ozek, G.; Yur, S.; Ozek, T.; Aneva, I.; Veleva, R.; Moskova-Doumanova, V.; et al. Caffeoylquinic acids, cytotoxic, antioxidant, acetylcholinesterase and tyrosinase enzyme inhibitory activities of six *Inula* species from Bulgaria. *Chem Biodivers.* **2020**, *17*, e2000051. [CrossRef]
37. Trendafilova, A.; Todorova, M.; Ivanova, V.; Aneva, I. Phenolic constituents and antioxidant capacity of *Inula oculus-christi* from Bulgaria. *Bulg. Chem. Commun.* **2017**, *49*, 176–180. Available online: http://www.bcc.bas.bg/BCC_Volumes/Volume_49_Special_D_2017/BCC-49-SE-D-176-180.pdf (accessed on 23 November 2025).
38. Kostadinova, A.; Staneva, G.; Topouzova-Hristova, T.; Moyankova, D.; Yordanova, V.; Veleva, R.; Nikolova, B.; Momchilova, A.; Djilianov, D.; Hazarosova, R. Myconoside Affects the Viability of Polarized Epithelial MDCKII Cell Line by Interacting with the Plasma Membrane and the Apical Junctional Complexes. *Separations* **2022**, *9*, 239. [CrossRef]
39. Owen, D.M.; Rentero, C.; Magenau, A.; Abu-Siniyeh, A.; Gaus, K. Quantitative imaging of membrane lipid order in cells and organisms. *Nat Protoc.* **2011**, *7*, 24–35. [CrossRef]
40. Garrido, J.; Gaspar, A.; Garrido, E.M.; Miri, R.; Tavakkoli, M.; Pourali, S.; Saso, L.; Borges, F.; Firuzi, O. Alkyl esters of hydroxycinnamic acids with improved antioxidant activity and lipophilicity protect PC12 cells against oxidative stress. *Biochimie* **2012**, *94*, 961–967. [CrossRef]
41. Delmondes, P.; Stefani, R. Computational study of natural phenolic acid solubility and their interactions with chitosan. *MOL2NET* **2017**, *2*, 1–5. [CrossRef]

42. Yang, B.; Liu, H.; Jiali, Y.; Gupta, V.; Jiang, Y. New insights on bioactivities and biosynthesis of flavonoid glycosides. *Trends Food Sci. Technol.* **2018**, *79*, 116–124. [CrossRef]
43. Plochmann, K.; Korte, G.; Koutsilieri, E.; Richling, E.; Riederer, P.; Rethwilm, A.; Schreier, P.; Scheller, C. Structure–activity relationships of flavonoid-induced cytotoxicity on human leukemia cells. *Arch. Biochem. Biophys.* **2007**, *460*, 1–9. [CrossRef] [PubMed]
44. Kalinowska, M.; Świsłocka, R.; Wołejko, E.; Jabłońska-Trypuć, A.; Wydro, U.; Kozłowski, M.; Koronkiewicz, K.; Piekut, J.; Lewandowski, W. Structural characterization and evaluation of antimicrobial and cytotoxic activity of six plant phenolic acids. *PLoS ONE* **2024**, *19*, e0299372. [CrossRef] [PubMed]
45. Chang, C.-J.; Hung, Y.-L.; Chen, T.-C.; Li, H.-J.; Lo, Y.-H.; Wu, N.-L.; Chang, D.-C.; Hung, C.-F. Anti-proliferative and anti-migratory activities of hispidulin on human melanoma A2058 cells. *Biomolecules* **2021**, *11*, 1039. [CrossRef]
46. Zeng, A.; Liang, X.; Zhu, S.; Liu, C.; Wang, S.; Zhang, Q.; Zhao, J.; Song, L. Chlorogenic acid induces apoptosis, inhibits metastasis and improves antitumor immunity in breast cancer via the NF-κB signaling pathway. *Oncol. Rep.* **2021**, *45*, 717–727. [CrossRef]
47. Yeager, A.N.; Weber, P.K.; Kraft, M.L. Three-dimensional imaging of cholesterol and sphingolipids within a Madin–Darby canine kidney cell. *Biointerphases* **2016**, *11*, 02A309. [CrossRef]
48. Nicolson, G.L. Cell membrane fluid–mosaic structure and cancer metastasis. *Cancer Res.* **2015**, *75*, 1169–1176. [CrossRef]
49. Xu, X.; Xu, S.; Wan, J.; Wang, D.; Pang, X.; Gao, Y.; Ni, N.; Chen, D.; Sun, X. Disturbing cytoskeleton by engineered nanomaterials for enhanced cancer therapeutics. *Bioact. Mater.* **2023**, *29*, 50–71. [CrossRef]
50. Nurmagambetova, A.; Mustyatsa, V.; Saidova, A.; Vorobjev, I. Morphological and cytoskeleton changes in cells after EMT. *Sci. Rep.* **2023**, *13*, 22164. [CrossRef]
51. Deusser, H.; Groh, I.; Bakuradze, T.; Simson, N.; Kaiser, E.; Barth, H.; Richling, E. Are compounds membrane-associated or present in the cytosol? A study using polyphenols in a colon carcinoma cell line model. *Curr. Pharmacol. Rep.* **2020**, *6*, 213–224. [CrossRef]
52. Selvaraj, S.; Krishnaswamy, S.; Devashya, V.; Sethuraman, S.; Krishnan, U.M. Influence of membrane lipid composition on flavonoid–membrane interactions: Implications on their biological activity. *Prog. Lipid Res.* **2015**, *58*, 1–13. [CrossRef]
53. Oteiza, P.I.; Erlejman, A.G.; Verstraeten, S.V.; Keen, C.L.; Fraga, C.G. Flavonoid–membrane interactions: A protective role of flavonoids at the membrane surface? *Clin. Dev. Immunol.* **2005**, *12*, 19–25. [CrossRef] [PubMed]
54. Bonarska-Kujawa, D.; Cyboran-Mikołajczyk, S.; Kleszczyńska, H. Molecular mechanism of action of chlorogenic acid on erythrocyte and lipid membranes. *Mol. Membr. Biol.* **2015**, *32*, 46–54. [CrossRef] [PubMed]
55. Filipe, H.A.L.; Sousa, C.; Marquês, J.T.; Vila-Viçosa, D.; de Granada-Flor, A.; Viana, A.S.; Santos, M.S.C.S.; Machuqueiro, M.; de Almeida, R.F.M. Differential targeting of membrane lipid domains by caffeic acid and its ester derivatives. *Free Radic. Biol. Med.* **2018**, *115*, 232–245. [CrossRef] [PubMed]
56. Carrano, R.; Grande, M.; Leti Maggio, E.; Zucca, C.; Bei, R.; Palumbo, C.; Focaccetti, C.; Nardozi, D.; Lucarini, V.; Angiolini, V.; et al. Dietary polyphenols effects on focal adhesion plaques and metalloproteinases in cancer invasiveness. *Biomedicines* **2024**, *12*, 482. [CrossRef]
57. Wang, R.; Peng, J.; Shi, X.; Cao, S.; Xu, Y.; Xiao, G.; Li, C. Change in membrane fluidity induced by polyphenols is highly dependent on the position and number of galloyl groups. *Biochim. Biophys. Acta Biomembr.* **2022**, *1864*, 184015. [CrossRef]
58. Zhou, A.; Xue, B.; Zhong, J.; Liu, J.; Peng, R.; Wang, F.; Zhou, Y.; Tang, J.; Yang, Q.; Chen, X. Cholesterol-rich lipid rafts mediate endocytosis as a common pathway for respiratory syncytial virus entry into different host cells. *Microbiol. Spectr.* **2025**, *13*, e0119225. [CrossRef]
59. Pang, X.; He, X.; Qiu, Z.; Zhang, H.; Xie, R.; Liu, Z.; Gu, Y.; Zhao, N.; Xiang, Q.; Cui, Y. Targeting integrin pathways: Mechanisms and advances in therapy. *Signal Transduct. Target Ther.* **2023**, *8*, 1. [CrossRef]
60. Hanna, S.J.; McCoy-Simandle, K.; Miskolci, V.; Guo, P.; Cammaer, M.; Hodgson, L.; Cox, D. The role of Rho-GTPases and actin polymerization during macrophage tunneling nanotube biogenesis. *Sci. Rep.* **2017**, *7*, 8547. [CrossRef]

membranes

MDPI

Article

Location and Dynamics of Nymphaeol A in a Complex Membrane

José Villalaín

Institute of Research, Development, and Innovation in Healthcare Biotechnology (IDiBE), Universidad "Miguel Hernández", E-03202 Elche-Alicante, Spain; jvillalain@umh.es; Tel.: +34-648891404

Abstract: Nymphaeol A (NYA) is a tetrahydroxyflavanone anchored to a hydrophobic geranyl group, isolated from different sources and a component of propolis, a complex mixture produced by honeybees and used since ancient times as a healthy drug. This complex exhibits significant antioxidant, antifungal, antibacterial, antiviral, anticancer and antimicrobial properties and NYA is one of its main components. NYA is a lipophilic molecule with two domains, one polar and one hydrophobic. NYA can be inserted into membranes, and its membrane properties depend not only on its location but also on the membrane's lipid composition. This work uses molecular dynamics to obtain the dynamics, orientation, location and interactions of NYA in a complex biomembrane. This work shows that in an aqueous solution, NYA forms high-order aggregates where the molecules are joined together by the hydrophobic chain. In the presence of a membrane but initially located in the aqueous media, NYA is capable of inserting itself spontaneously into the membrane. Inside the membrane, NYA can be found in the monomeric form, as well as forming aggregates, tending to remain in its most extended conformation. NYA moves along the x-, y- and z-axes, with the movement along the z-axis larger than that of the membrane's lipids. NYA forms an approximate angle of 35° perpendicular with respect to the membrane and is inserted between the phospholipid hydrocarbon chains, slightly increasing membrane fluidity. Furthermore, NYA prefers POPC and PSM but not POPE or CHOL. NYA's location and movement within the membrane should be well-suited for its potent bioactivity.

Keywords: Nymphaeol; antioxidant; molecular dynamics; membrane

1. Introduction

Nymphaeol A (6-geranyl-3′,4′,5,7-tetrahydroxyflavanone) is a tetrahydroxyflavanone substituted by hydroxy groups at positions 5, 7, 3′ and 4′ with a geranyl group at position 6 (Figure 1A). It has been isolated from different sources, such as *Macaranga tanarius*, and is one of the main components of propolis, a natural product produced by honeybees [1–17]. Propolis is composed of a complex mixture of resins, waxes, essential oils and pollen and has been used since ancient times as a potent drug. It exhibits significant antioxidant, antifungal, antibacterial, antiviral, anticancer and antimicrobial properties. Its phytochemical content varies depending on its botanical and geographical origin, which influences its bioactivity.

Nymphaeol A consists of two well-differentiated parts: a tetrahydroxyflavanone and a geranyl chain. The tetrahydroxyflavanone is a relatively polar domain, whereas the geranyl chain is hydrophobic; however, the whole molecule is very lipophilic (xLogP3 of 5.8). Interestingly, the geranyl group plays an important role in exhibiting antioxidant activity [16]. It is already known that to counteract the existence of free radicals, organisms have different antioxidant defense mechanisms which can be either from exogenous or endogenous origin [18–23]; however, antioxidants are the preferred mechanism to inhibit the so-called reactive oxygen species (ROS)'s formation and/or propagation [18,19,24]. High ROS levels lead to oxidative stress, which gives place to neurodegeneration, premature aging and many other illnesses [25]. Since NYA has a great hydrophobic propensity, it could be inserted into a membrane, where it would be capable of quenching and/or scavenging reactive oxygen or nitrogen species at both the surface and interior parts of the membrane. Furthermore, NYA's position in the membrane could depend on the membrane's lipid composition and it could have different membrane locations. Because of this, it is important to study its location and interactions in a complex biomembrane.

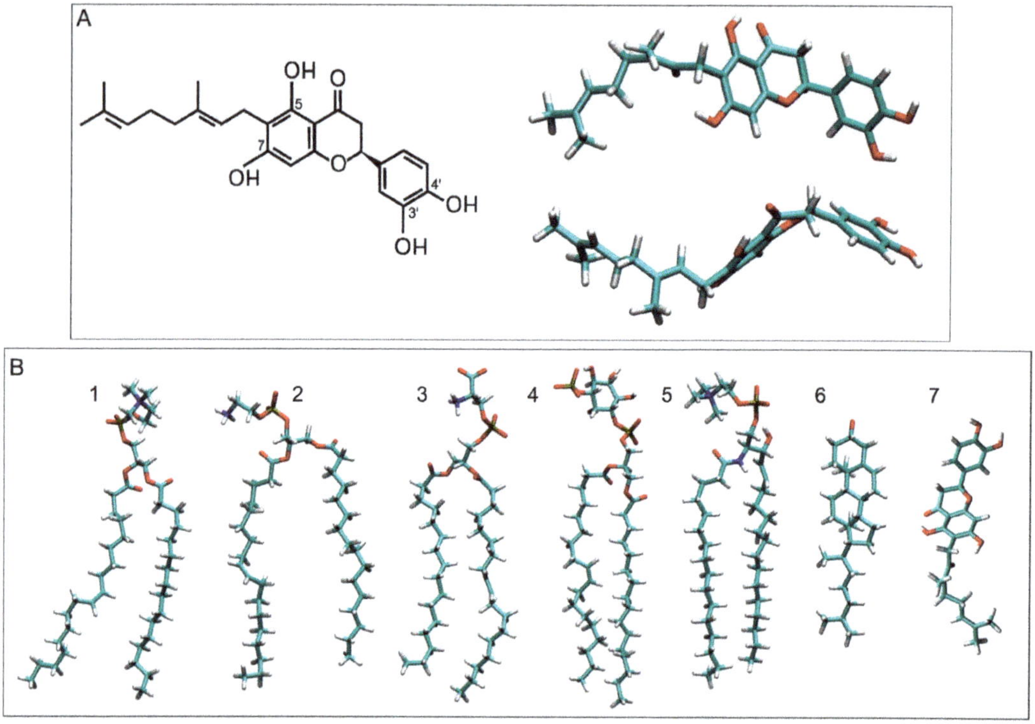

Figure 1. (**A**) Chemical and molecular structures of Nymphaeol A (NYA) and (**B**) the molecular structures of the lipids studied in this work, i.e., POPC (1), POPE (2), POPS (3), PI-3P (4), PSM (5), CHOL (6) and NYA (7), to compare molecular sizes. The molecules are shown in licorice representation.

Understanding how NYA acts is imperative to determine its location, orientation and dynamics inside a membrane and at the same time know its possible effects on the membrane, which could be accountable for its valuable properties on cellular health. Molecular dynamics is suitable to obtain useful data on the structure, dynamics, location and interaction of bioactive molecules within biomembranes [26,27] and it is used in this work to obtain the dynamics, orientation, location and interactions of NYA in a complex biomembrane [28,29] (Table 1). This work shows that NYA molecules in aqueous solution form high-order aggregates where the molecules are joined together by the hydrophobic chain. However, NYA molecules spontaneously insert into the membrane and are finely inserted between the phospholipid hydrocarbon chains, slightly increasing their fluidity. NYA's location and movement within the membrane is therefore very well-suited for its potent antioxidant activity.

Table 1. Membrane systems and number of components. The NaCl concentration was 0.15 M. The time to obtain the production trajectories for each one of the systems is also indicated. The total number of lipid molecules was 500, with 250 per monolayer.

Systems		1	2	3	4
Nymphaeol A		1	-	-	-
Nymphaeol A		-	4	-	-
Nymphaeol A		-	-	8	-
Nymphaeol A		-	-	-	8
MD time (ns)		1000	1000	1000	600
POPC	28.8%	144	144	144	-
POPE	17.2%	86	86	86	-
PI-3P	5.6%	28	28	28	-
POPS	6.4%	32	32	32	-
PSM	12%	60	60	60	-
CHOL	30%	150	150	150	-
ATOMS		166,370	166,547	166,234	167,176
H_2O		36,413	36,413	36,230	55,464
H_2O/LIPID		72.8	72.8	72.5	-
Na^+		219	219	219	156
Cl^-		103	103	103	156
DIMENSIONS x/y/z (Å)		128/129/120	128/129/120	128/129/120	128/129/120

2. Materials and Methods

2.1. Molecular Dynamics Simulation

Unrestrained all-atom molecular dynamics was performed using NAMD 2.14 [30] and the CHARMM36 protein and lipid force fields [31–33]. All molecular dynamics parameters have been previously described [28,29]. The TIP3P model was used for water [34]. The membrane systems were minimized for 150,000 steps to remove bad atomic contacts and then equilibrated for 10 ns. The temperature was 310 K (Table 1).

2.2. Molecular Dynamics Specifications

The membranes were obtained using Charmm-Gui (http://www.charmm-gui.org (accessed on 3 January 2025) [35]). All systems contained NaCl at physiological conditions, i.e., a concentration of 150 mM and an excess of water [36] in a neutral environment (Table 1) [28,37].

The composition of the membrane resembles a general plasma membrane (Table 1) [38,39]. It contained 144 molecules of 1-palmitoyl-2-oleoyl-sn-glycero-3-phosphocholine (POPC), 86 molecules of 1-palmitoyl-2-oleoyl-sn-glycero-3-phosphoethanolamine (POPE), 32 molecules of 1-palmitoyl-2-oleoyl-sn-glycero-3-phosphoserine (POPS), 28 of 1-palmitoyl-2-oleoyl-sn-glycero-3-phosphoinositol-3-phosphorous (PI-3P), 60 molecules of N-stearoyl-D-erythro-sphingosylphosphorylcholine (PSM) and 150 of cholesterol (CHOL) (Table 1) [38,39]. The relative molar percentage of the lipids was 28.8%, 17.2%, 6.4%, 5.6%, 12% and 30% for POPC, POPE, POPS, PI-3P, PSM and CHOL, respectively (Table 1) [38,39]. Three different membrane systems were studied: a membrane system containing only one NYA molecule, system 1 (Figure 2A), a membrane system containing four NYA molecules, system 2 (Figure 2B) and a membrane system containing eight NYA molecules, system 3 (Figure 2C). Additionally, system 4 contained eight NYA molecules inside a water box (Figure 2D). These systems are appropriate to study the interaction of small bioactive molecules with lipids in a membrane [28,37]. The number of lipids in the membrane was 500, with 250 in each layer. The lipid chemical structures are shown in Figure 1A,B. Membrane fluidity was amplified using one oleoyl hydrocarbon chain in the phospholipids [28,37]. PSM contained a palmitoyl hydrocarbon chain apart from the sphingosyl one. The original proportions of the complete systems are shown in Table 1. The molecular structure of NYA was obtained from PubChem (https://pubchem.ncbi.nlm.nih.gov/compound/Nymphaeol-A (accessed on 3 January 2025)) and was reviewed, adjusted and minimized using Discovery Studio 4.0 (Accelrys Inc., San Diego, CA, USA). The CHARMM General Force Field stream files were obtained using Charmm-Gui (http://www.charmm-gui.org [35]).

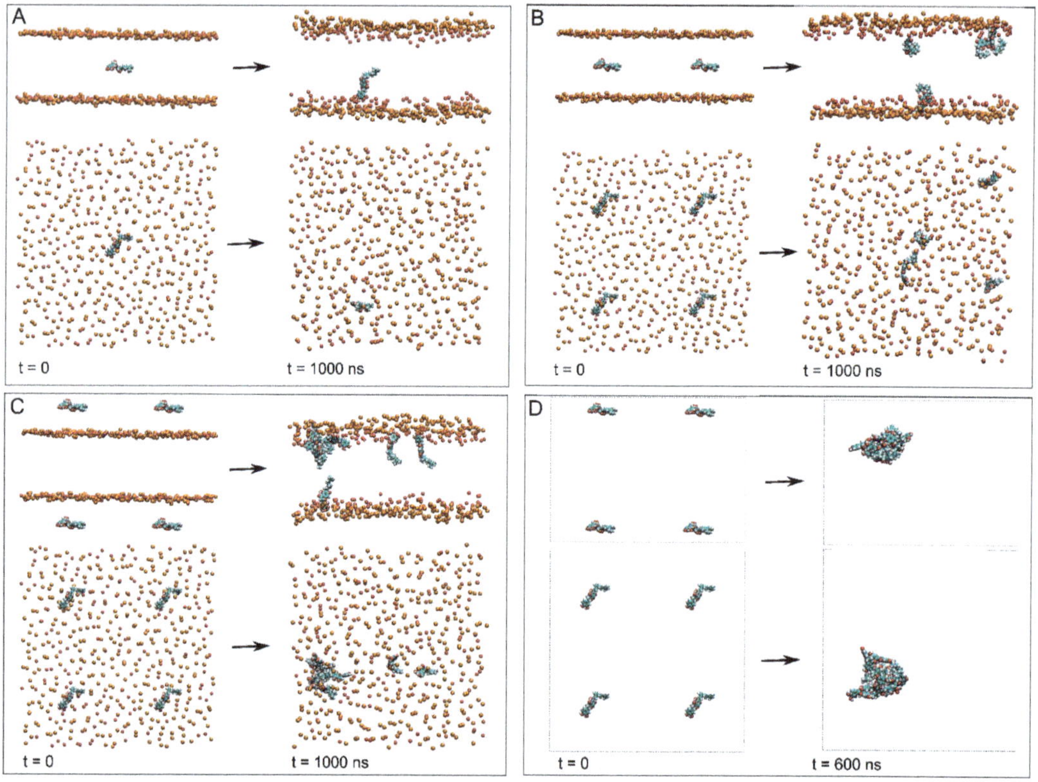

Figure 2. Apical and lateral views of the initial, t = 0 ns, and final, t = 1000/600 ns, dispositions of (**A**) system 1, (**B**) system 2, (**C**) system 3 and (**D**) system 4. The NYA molecules are presented in VDW representation, whereas the phosphate atoms of the phospholipids, defining the upper and lower boundaries of the membrane, and the oxygen atoms of CHOL are represented in VDW and orange and red colors, respectively. The water and lipid molecules and the chloride and sodium ions have been removed for clarity.

2.3. Molecular Dynamics Analysis

The analysis and visualization were conducted using VMD (Visual Molecular Dynamics, Theoretical and Computational Biophysics Group, University of Illinois at Urbana-Champaign) software and plugins [40–49]. Data were obtained using the VMD "Membplugin" and "Density Profile Tool" plugins [41,44–47,49]. The order parameter, S_{CD}, was defined as $S_{CD} = \frac{1}{2}\langle 3\cos^2\theta - 1\rangle$, where θ is the angle between the C-D vector and the bilayer normal and the brackets symbolize an ensemble average. Hydrogen bonds were as defined previously [45,46]. Data uncertainties represent standard deviation unless otherwise stated.

3. Results

3.1. NYA in Water

Before studying NYA molecules in the membrane and in order to compare their structures, their behavior in an aqueous solution was studied (system 4, Figure 2D, for the initial and final systems of NYA in water, and Table 1). The system contained eight NYA molecules inside a water box, separated at the beginning by about 60 Å in the x–y plane and by about 70 Å in the z plane. The molecules were free to wander in the water solution, so that at the end of the simulation, i.e., 600 ns, all eight molecules formed an aggregate (Figure 2D and Supplementary Figure S1A). The average distances between the carbon atoms at position 4′ and 13 (the last carbon of the geranyl chain) for the last 30 ns are shown in Supplementary Figure S1B. All of them are around 13–14 Å, the global average of the eight NYA molecules being 13.7 ± 1.0 Å (rounded to the first decimal). Since the distance between these two atoms in the NYA's most extended conformation is about 15.8 Å, it is possible to say that NYA in an aqueous solution remains nearly in its most extended conformation. As was commented above, the NYA molecule could be divided into two distinct parts: a relatively polar part (tetrahydroxyflavanone group) and a hydrophobic part (geranyl group) (Figure 1A). The oligomer structure is shown in Supplementary Figure S1A, where it is possible to see that the molecules are joined together by the hydrophobic part, i.e., the geranyl chain, forming a kind of sphere, where the polar groups tend to be at the surface. It seems that hydrophobic forces are accountable for attraction of different molecules, whereas the polar terminal groups might stabilize the aggregate (Supplementary Figure S1A).

3.2. Membrane Systems

Three different membrane systems were studied: a membrane system containing only one NYA molecule, system 1 (Figure 2A), a membrane system containing four NYA molecules, system 2 (Figure 2B), and a membrane system containing eight NYA molecules, system 3 (Figure 2C). The NYA molecules in systems 1 and 2 were located in the center of the membrane at time 0, whereas the NYA molecules in system 3 were located in the center of each one of the water layers outside the membrane [37]. It should be taken into account that they were very diluted systems, since the membrane contained 500 lipids, with 250 in each monolayer. The membrane thickness and lipid areas were checked to assess the equilibration of the systems during the molecular dynamics [50,51]. After ~70 ns, the mean membrane thickness was almost the same for all the systems, being for the last 30 ns nearly indistinguishable, i.e., between 45.8 and 46 Å, as shown in Supplementary Figure S2. These data are comparable to those described previously [52]. The mean molecular areas for all systems and for the last 30 ns of molecular dynamics for POPC, POPE, POPS and PI-3P ranged between 56 and 57 Å^2, for PSM ranged between 50 and 51 Å^2, and for CHOL ranged between 28.6 and 28.9 Å^2. These areas were similar to those formerly reported [44,52,53]. Accordingly, all membranes were equilibrated very early and were in a steady state after ~70 ns of molecular dynamics.

3.3. NYA Inside the Membrane

System 1 originally had one molecule of NYA located at the membrane's central part (Figure 2A). At $t = \infty$ of the molecular dynamics simulation, i.e., 1000 ns, the NYA molecule proceeded to a location adjacent to the membrane interphase and inserted in between the phospholipid hydrocarbon layer, with the hydrophobic chain pointed towards the middle of the membrane (Figure 2A). Its behavior throughout the molecular dynamics simulation can be observed by studying the whole-molecule z-axis center-of-mass (z direction normal to the bilayer plane, z-COM) (Figure 3A). This can be compared to the phosphate atoms' z-COMs, which define the membrane surface, and the CHOL oxygen atoms, which define the interphase (Figure 3A). As commented above, at the beginning, the NYA molecule was located in the middle of the membrane; however, with time, it spontaneously positioned itself at the middle of one of the monolayers, this being its average position for the last 30 ns of −7.4 Å ± 1.8 Å (Figure 3A). This location was maintained for nearly all molecular dynamics simulations. It is also possible to observe that its displacement along the z-axis was larger than those of the phosphate atoms of the phospholipids and the oxygen atom of CHOL. However, NYA is a relatively long molecule, its maximum extension being about 15.8 Å (Figure 1B). Because of that, this is interesting, not only knowing its global location but also the location of its extremes, namely the z-COM values of the carbon C4′ atom (−14.1 Å ± 1.6 Å) and C13 atom (−0.7 Å ± 2.4 Å) (Figure 4A). This would imply that the molecule moves in a perpendicular way to the membrane's surface along the z-axis, with the tetrahydroxyflavanone domain pointing towards the membrane interphase, whereas the geranyl domain points towards the middle of the membrane. This work also analyzed the distance between carbons C4′ and C13 and the data are shown in Supplementary Figure S3A. As shown in the figure, the average distance is 15.0 ± 1.7 Å for the last 30 ns of the simulation. Since the distance between these two atoms in the most extended conformation is about 15.8 Å, the NYA molecule inside the membrane is near its most extended conformation.

System 2 originally had four molecules of NYA located at the membrane's central part (Figure 2B). At $t = \infty$ of the molecular dynamics simulation, i.e., 1000 ns, the NYA molecules proceeded to a location close to the membrane interphase and inserted in between the phospholipid hydrocarbon layer, with the hydrophobic chain pointed towards the middle of the membrane in a similar way to the NYA molecule in system 1 (Figure 2A). Although the NYA molecules were located in the middle of the membrane, with time, they spontaneously positioned themselves at the middle of the monolayers, their average positions for the last 30 ns being −15.1 Å ± 2.8 Å, 12.0 Å ± 1.6 Å, 12.0 Å ± 2.2 Å and 11.8 Å ± 1.8 Å (Figure 3A). As observed in the Figure, of the four molecules, three arranged themselves rapidly in the middle of each of the monolayers; the remainder remained in the middle of the membrane until about 800 ns, and from that time onwards it was located in the middle of one of the monolayers (Figure 3A). Similarly to the NYA molecule in system 1, their displacement along the z-axis was larger than those of the phosphate atoms of the phospholipids and the oxygen atom of CHOL. The locations of their extremes, namely the z-COM values of the carbon C4′ and C13, were −20.7 Å ± 3.3 Å and −8.8 Å ± 3.4 Å, 17.6 Å ± 1.8 Å and 5.3 Å ± 2.3 Å, 17.9 Å ± 1.9 Å and 8.0 Å ± 4.1 Å and finally 14.4 Å ± 1.3 Å and 8.6 Å ± 4.3 Å, respectively (Figure 4B). This would imply that the molecules are capable of moving in a perpendicular way to the membrane's surface along the z-axis, with the tetrahydroxyflavanone domain pointed towards the membrane interphase, whereas the geranyl domain points towards the middle of the membrane. This work also analyzed the distance between the carbon C4′ and C13 atoms and the terminal carbon of the geranyl chain, and the data are shown in Supplementary Figure S3B. As shown in the figure, the average distances were 14.6 Å ± 2.1 Å, 13.9 Å ± 2.4 Å, 12.6 Å ± 2.3 Å and 14.3 Å ± 1.0 Å, respectively, for the last 30 ns of the simulation. Similarly to the NYA molecule in system 1, the NYA molecules in system 2 are close to their most extended conformation.

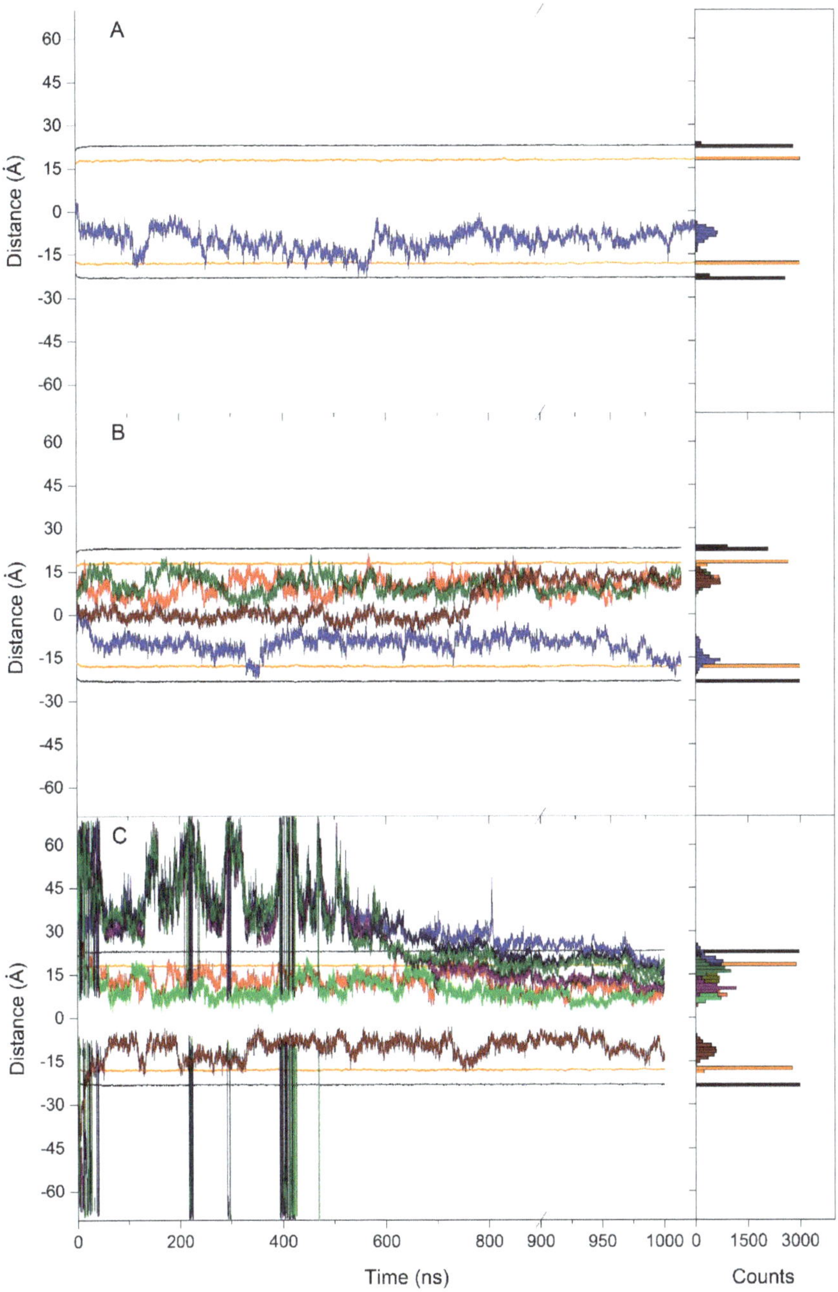

Figure 3. Time variation of the NYA whole-molecule z-axis COM distance pertaining to (**A**) system 1 (one NYA molecule), (**B**) system 2 (four NYA molecules) and (**C**) system 3 (eight NYA molecules). Each NYA molecule is represented by a different color. The corresponding histograms are shown at the right margin (the data are the averages of the last 30 ns of molecular dynamics simulation). The z-axis COM distance of the phosphate atoms of the phospholipids, defining the membrane surface, is depicted in black, whereas the z-axis distance of the oxygen atoms of cholesterol is depicted in orange (upper and lower boundaries).

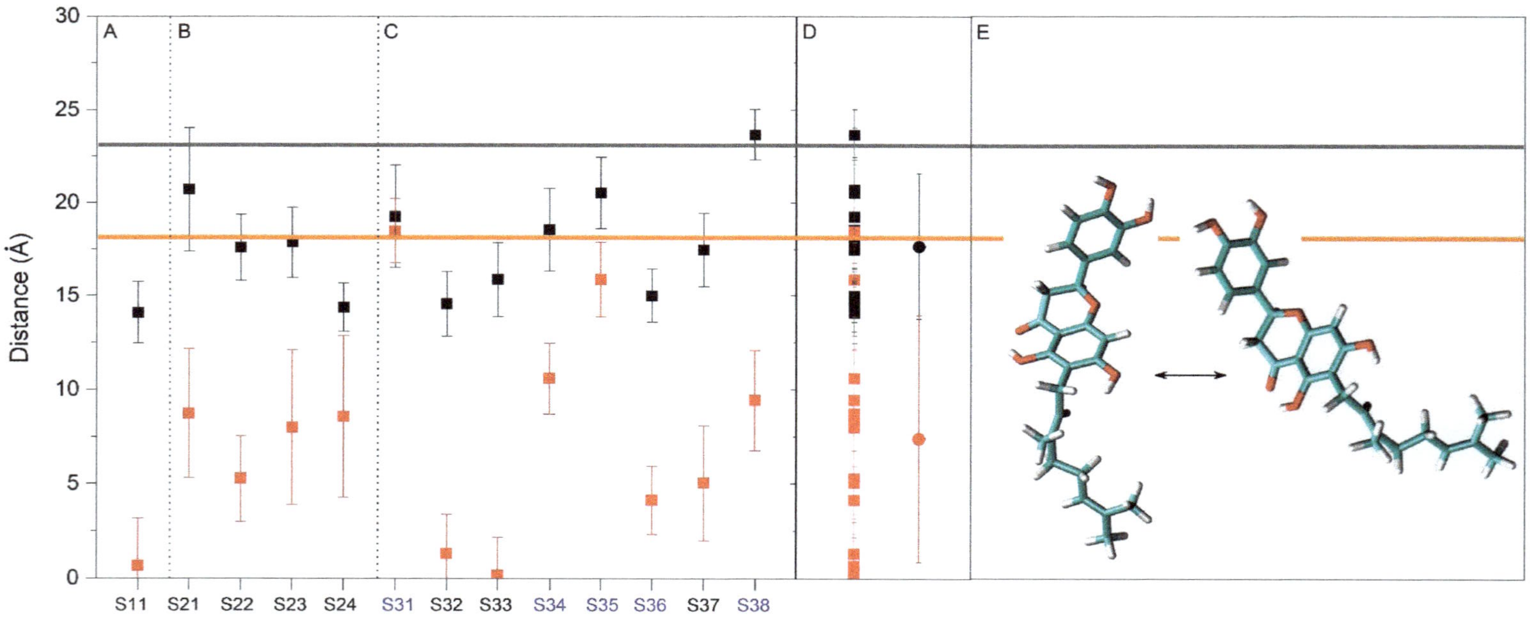

Figure 4. Average z-axis COM distance of the last 30 ns of molecular dynamics simulation of the carbon C4′ atom (■) and the terminal carbon of the geranyl chain (■) pertaining to (**A**) system 1 (one NYA molecule), (**B**) system 2 (four NYA molecules) and (**C**) system 3 (eight NYA molecules). Blue marks in (**C**) denote those NYA molecules in aggregated form inside the membrane. In (**D**), all data are represented at the same time, as well as their average ± SD, whereas in (**E**) the possible location of NYA in the membrane is represented. The z-axis COM distance of the phosphate atoms of the phospholipids, defining the membrane surface, is depicted as a dark grey line, whereas the z-axis distance of the oxygen atoms of cholesterol is depicted as an orange line.

System 3 originally had eight molecules of NYA positioned in the middle of each water layer (Figure 2C). At t = ∞ of the molecular dynamics simulation, i.e., 1000 ns, the NYA molecules proceeded to a position close to the membrane interphase and inserted in between the phospholipid hydrocarbon layer, with the hydrophobic chains pointed towards the middle of the membrane in a similar way to the NYA molecules in system 1 and system 2 (Figure 2A,B). Although the NYA molecules were located in the water layers, at different times they spontaneously positioned themselves at the middle of the monolayers, remaining there until the end of the simulation. Their average positions for the last 30 ns were 19.7 Å ± 2.2 Å, 8.0 Å ± 1.6 Å, 8.4 Å ± 2.0 Å, 13.3 Å ± 1.9 Å, 17.8 Å ± 2.1 Å, 10.4 Å ± 1.4 Å, −11.4 Å ± 2.1 Å and 16.9 Å ± 1.5 Å, respectively (Figure 3A). Three of the NYA molecules did insert into the membrane very quickly (at 16, 18 and 20 ns, respectively), while the remaining five did so later and at different times (Figure 3A). These NYA molecules interacted with and inserted into the membrane at about 610, 620, 630, 714 and 825 ns (Figure 3A). Similarly to the NYA molecules in the other systems, once inserted into the membrane, their displacement along the z-axis was larger than those of the phosphate atoms of the phospholipids and the oxygen atom of CHOL. The locations of their extremes, namely the z-COM values of the carbon C4′ and the terminal carbon of the geranyl chain, were 19.3 Å ± 2.8 Å and 18.5 Å ± 1.7 Å, 14.6 Å ± 1.7 Å and 1.3 Å ± 2.1 Å, 15.9 Å ± 2.0 Å and 0.2 Å ± 2.0 Å, 18.5 Å ± 2.2 Å and 10.6 Å ± 1.9 Å, 20.5 Å ± 1.9 Å and 15.9 Å ± 2.0 Å, 15.0 Å ± 1.4 Å and 4.1 Å ± 1.8 Å, −17.5 Å ± 2.0 Å and −5.0 Å ± 3.0 Å and finally 23.6 Å ± 1.4 Å and 9.4 Å ± 2.7 Å, respectively (Figure 4C). It is highlighted that the three NYA molecules that were rapidly inserted into the membrane remained in a monomeric state for the entire simulation, while the NYA molecules that were inserted later joined together, forming a 5-unit oligomer at the membrane interface. The five monomers formed an oligomer in solution beforehand, and this oligomer subsequently inserted into the membrane through the geranyl chains of two NYA molecules (about 620 ns, Figure 3C). We have also analyzed the distance between carbon C4′ and C13 atoms and the data are shown in Supplementary Figure S3C. As shown in the figure, the average distances were 15.5 Å ± 0.7 Å, 14.7 Å ± 1.7 Å, 16.2 Å ± 0.9 Å, 14.1 Å ± 2.0 Å, 15.3 Å ± 0.9 Å, 14.7 Å ± 1.4 Å, 15.6 Å ± 1.1 Å and 15.0 Å ± 1.9 Å, respectively, for the last 30 ns of the simulation. Similarly to the NYA molecules in systems 1 and 2, the NYA molecules are close to their most extended conformation. In a similar way to the oligomer formed in water, the molecules in the oligomer inside the membrane were joined together by the hydrophobic part, i.e., the geranyl chain.

Summarizing the results, it is possible to say that NYA inserts spontaneously inside the membrane and its most probable location is in the middle of the membrane monolayer (the global average location of NYA inside the membrane is 14.7 Å ± 1.4 Å, Supplementary Figure S3D). The NYA molecules were capable of moving in a perpendicular way to the membrane's surface along the z-axis, with the tetrahydroxyflavanone domain pointed towards the membrane interphase, whereas the geranyl domain pointed towards the middle of the membrane (Supplementary Figure S3D,E). Interestingly, NYA is always close to its most extended conformation and it is possible for the molecule to be a monomer or in an aggregated form inside the membrane.

3.4. Angles Formed Between NYA and the Membrane

As commented above, the NYA molecules insert well in between the hydrocarbon chains of the phospholipids, remaining in their nearly most extended conformation, 14–15 Å. Using the z-COM data of the extremes of the molecule, it is possible to infer the relative angle of NYA in the membrane (Figure 4E). Nevertheless, this work obtained the average of the angle formed by the whole molecules in the membrane for the last

30 ns of molecular dynamics, and the data can be observed in Supplementary Figure S4 (average angles formed by carbons C4′ and C13, the last one of the geranyl chain, for the NYA molecules in the monomeric form with respect to the membrane's surface). The average angles were relatively similar for all of the NYA molecules in the monomeric form, i.e., 24.2° ± 9.3°, 43.6° ± 17.1°, 46.3° ± 12.7°, 27.5° ± 14.3°, 75.0° ± 6.9°, 24.9° ± 13.9°, 38.3° ± 12.6° and 11.3° ± 6.7° (Supplementary Figure S4). The global average angle was 36.4° ± 19.4° (Supplementary Figure S4). These data are in agreement with the data inferred by the z-COM data commented above. Although the NYA molecules present different conformations, their preferred disposition is approximately the same for all of them and forms an approximate angle of 36° with respect to the membrane.

3.5. *Lipid Density Profiles*

The lipid mass density profiles are shown in Supplementary Figure S6A,B for system 1, Supplementary Figure S6C,D for system 2 and Supplementary Figure S6E,F for system 3. The lipid mass density profiles were essentially symmetric between the two membrane layers, representing the comparable behavior of the lipids. This is the expected comportment, since all the systems were very diluted. It can be also observed in the figures the positions of the NYA molecules with respect to the lipids in the membrane, where all of them reside relatively at the same depth in the membrane. We should remember that the NYA molecules are relatively narrow and long molecules with two distinct moieties; the hydrophobic tail and the tetrahydroxyflavanone moiety. As observed in Supplementary Figure S5, the tetrahydroxyflavanone moiety tends to be between the oxygen atom of CHOL and the first carbons of the phospholipid hydrocarbon chain, whereas the geranyl chains tends to be located between the middle of the phospholipid hydrocarbon chain and the membrane middle. These data also show that the locations of the NYA molecules in the membrane are very similar between the different molecules and relatively well defined. This should be related to the geometry and structure of the molecules and their location should favor their bioactivity.

3.6. *Phospholipid Hydrocarbon Chain Order in the Presence of NYA*

Bioactive molecules inside a membrane can alter the phospholipid hydrocarbon chain order. As such, this work explored the consequence of the NYA molecules obtaining the deuterium order parameter, S_{CD}, for the phospholipids in systems 1, 2 and 3 (NYA molecules in the monomeric form). The S_{CD} values are shown in Supplementary Figure S6. The S_{CD} data of the hydrocarbon chains of the bulk phospholipids were comparable with the profiles previously observed for biomembranes for both experimental and simulated data [33,54–56]. For the phospholipid hydrocarbon chains near the NYA molecules, globally speaking, there were some differences in the S_{CD} profiles: the S_{CD} values were slightly higher in the first half of the chains in the presence of the NYA molecules, but in the second half the S_{CD} values were slightly lower. This is in accordance with the data shown above: the tetrahydroxyflavanone moiety tends to be between the oxygen atom of CHOL and the first carbons of the phospholipid hydrocarbon chain, whereas the geranyl chain tends to be located between the middle of the phospholipid hydrocarbon chain and the membrane's middle. To know the global effect these bioactive molecules have on lipid fluidity, this work detracts the S_{CD} values of the bulk lipids minus the S_{CD} values of the surrounding ones: if the result is negative, the molecule has an ordering effect; if positive, it has a disordering effect (Supplementary Figure S6K). These data show that NYA molecules are ordered in the first half of the phospholipid hydrocarbon chains, where the tetrahydroxyflavanone tends to be located (decrease of membrane fluidity), but disordered in the second half, where the geranyl chain tends to be located (increase of membrane fluidity). However, they do not affect dramatically the hydrocarbon chains' anisotropy.

3.7. Lipid Molecules near NYA

This work obtained the type and number of lipid molecules at a distance of 6 Å from NYA and compared them with the bulk quantities, i.e., analyzed if any lipid type was overrepresented or diminished near the bioactive molecules (average over the last 30 ns of the molecular dynamics and for systems 1, 2 and 3, Figure 5). The percentages of POPC and PSM were relatively higher than the bulk percentages (about 38% vs. 28.8% and about 16% vs. 12%, respectively), whereas the percentages of POPE and CHOL were lower than the bulk ones (about 13% vs. 17.2% and about 22% vs. 30%, respectively) (Figure 5). The percentages of POPS and PI-P3 were relatively similar to the corresponding bulk percentages (6% vs. 6.4% and 4% vs. 5.6%, respectively). In general, it can be said that NYA would prefer POPC and PSM but not POPE or CHOL.

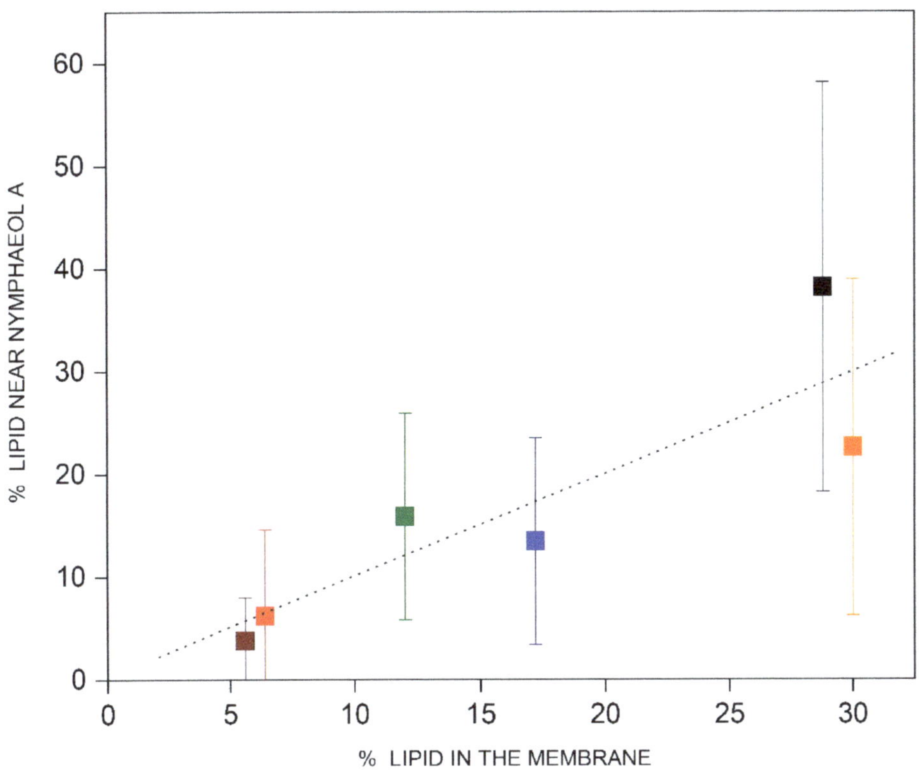

Figure 5. Percentage of the number of lipid molecules in the membrane versus observed number of lipid molecules at a distance of 6 Å from the NYA molecules: POPC (-■-), POPE (-■-), POPS (-■-), PI-3P (-■-), PSM (-■-) and CHOL (-■-). The dotted line represents an identical observed versus expected number of lipid molecules. The analysis was carried out for the last 30 ns of simulation.

3.8. Diffusion Coefficients

This work measured the diffusion coefficients of all the bioactive and lipid molecules in all the systems and the data are shown in Supplementary Figure S7. As observed in Supplementary Figure S7, the lipid diffusion coefficients on the x- and y-axes range from about 3.4 $Å^2/ns$ (POPC) to 8.8 $Å^2/ns$ (PI-P3), with the values of the other lipids in between.

The average x–y diffusion coefficient of the lipids is 5.7 ± 2.8 Å^2/ns. These values can be compared with the x–y diffusion coefficient of NYA, which is 0.5 ± 0.3 Å^2/ns. Therefore, the x–y diffusion coefficient of NYA is much lower than the x–y diffusion coefficient of the lipids. The lipid diffusion coefficients on the z-axis range from about 0.08 Å^2/ns (POPE) to 0.05 Å^2/ns (CHOL), with the values of the other lipids in between. The average z diffusion coefficient of the lipids is 0.06 ± 0.02 Å^2/ns. These values can be compared with the z-axis diffusion coefficient of NYA which is 0.2 ± 0.1 Å^2/ns, i.e., about three times larger than the lipids. Therefore, the diffusion of NYA in the membrane x–y plane is lower than the diffusion of the lipids and the diffusion of NYA in the z direction is larger than the diffusion of the lipids (Supplementary Figure S7).

3.9. Hydrogen Bonds

This work also obtained the mean number of hydrogen bonds formed between the NYA molecules and the membrane lipids for the last 30 ns of the simulation (Supplementary Figure S8). POPC and PSM were the lipids which showed a relatively major proportion of hydrogen bonds (about 1 and 2, respectively); the other lipids did not show the presence of any hydrogen bonds at all (POPE, POPS, PI-3P and CHOL). The data shown are in accordance with the above-commented data that NYA would prefer POPC and PSM rather than the other lipids. Since the relative molar percentage of POPC and PSM in the membrane is about 40% of the total lipids, NYA should be relatively free to move within the membrane.

4. Discussion

Nymphaeol A is a tetrahydroxyflavanone anchored to a hydrophobic geranyl group, isolated from different sources and a component of propolis, a complex mixture produced by honeybees and used since ancient times as a healthy drug. Nymphaeol A is a relatively lipophilic molecule, having two domains, one polar and another hydrophobic, and therefore it could insert into a membrane, where it would be capable of quenching and/or scavenging reactive oxygen or nitrogen species. I have used molecular dynamics to obtain the location, orientation, dynamics and interactions of NYA in a complex biomembrane (Figure 6). This work shows that NYA molecules in an aqueous solution form high-order aggregates where the molecules are joined together by the hydrophobic chain. In the presence of a membrane but initially located in aqueous media, NYA molecules were capable of inserting themselves into the membrane; inside the membrane, NYA can be found in its monomeric form, as well as forming aggregates, tending to remain in its most extended conformation. Interestingly, NYA can form oligomers in aqueous solution, but its formation does not preclude the oligomer from inserting spontaneously into the membrane. NYA moves along the x-, y- and z-axes, with the movement along the z-axis larger than that of the membrane lipids, but lower in the x–y plane. In this way, NYA moves much better perpendicular to the membrane's surface. Although NYA molecules in the membrane can present different conformations, they tend to form an approximate angle of 36°, perpendicular with respect to the membrane, and be inserted between the phospholipid hydrocarbon chains. Therefore, NYA tends to be in its most extended conformation. In doing so, NYA increases membrane fluidity, but without dramatically affecting its anisotropy. It is interesting to comment that NYA would prefer POPC and PSM but not POPE or CHOL. Regardless of this preference, which is not exclusive, the fact that, on the one hand, fluidity is little affected by NYA and, on the other hand, that it had a certain preference for PSM but not for CHOL indicates there is no evidence that the incorporation of NYA into the membrane can lead to the formation of specific domains. The average location of NYA in the membrane lies in the upper part of the membrane's hydrocarbon structure, with the tetrahydroxyflavanone domain pointed towards the membrane interphase, whereas the geranyl domain points towards the middle of the membrane. In this way, NYA can reach the membrane interface where the carbonyl groups of phospholipids are located. Since one of the most important bioactive properties of NYA is that it is an antioxidant molecule, its location in the membrane is essential to protect lipids from the harmful effects of reactive oxygen species. NYA's location and movement within the membrane is therefore well-suited for its bioactivity.

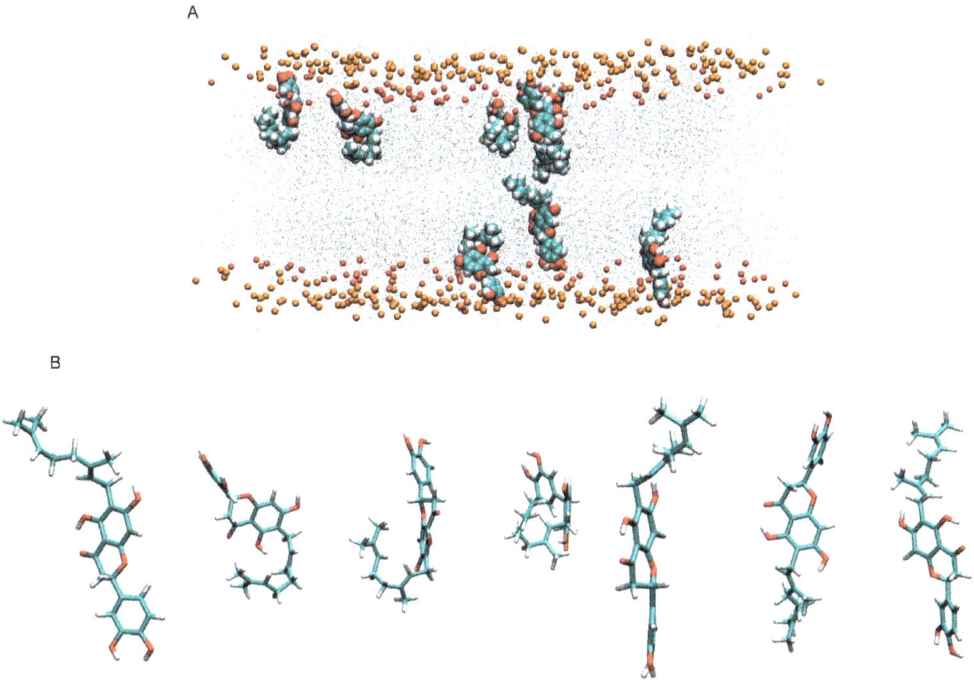

Figure 6. (**A**) Representation of the location of NYA molecules in the membrane (monomeric form, final frame of the molecular dynamics simulation). The NYA molecules are presented in VDW representation, as well as the phosphate atoms of the phospholipids (orange, size reduced by half), defining the upper and lower boundaries of the membrane, and the oxygen atoms of cholesterol (red, size reduced by half). For clarity, lipid atoms are represented as grey dots, while water and chloride and sodium ions have been removed. (**B**) NYA molecules in (**A**) highlighting the different conformations observed.

5. Conclusions

NYA, a bioactive molecule composed of two characteristic domains, is relatively lipophilic and spontaneously inserts into the biological membrane. Using molecular dynamics, this work has been able to determine that it forms oligomers in an aqueous solution without preventing its insertion into the membrane, it is capable of moving much faster along the z-axis than membrane lipids, it tends to be in its most extended conformation, it slightly increases membrane fluidity, it has a preference for POPC and PSM but not for POPE and CHOL and it is also capable of locating itself at the upper part of each monolayer, reaching the carbonyl groups of phospholipids. The location and movement of NYA within the membrane are well-suited to its bioactivity and essential for protecting lipids from the damaging effects of reactive oxygen species.

Supplementary Materials: The following supporting information can be downloaded at: https://www.mdpi.com/article/10.3390/membranes15060163/s1, Figure S1: (A) Structure of the aggregate of NYA molecules at the last frame for system 4 (t = 600 ns). NYA molecules (eight) are shown in licorize representation, whereas the oxygens are shown in VDW representation (hydroxyl, carbonyl and ether oxygens in red, green and orange colours, respectively). (B) Average distance between the carbons C4′ and the last one of the geranyl chain in the aggregate for each one of the NYA molecules as well as their average. Data correspond to the last 30 ns of molecular dynamics simulation; Figure S2: Time variation of membrane thickness for the whole simulation time and the corresponding histograms for the last 30 ns of molecular dynamics simulation. Thickness corresponds to the phosphate atoms of the phospholipids (upper lines) and the oxygen atoms of cholesterol (lower lines) for system 1 (black), system 2 (blue) and system 3 (red); Figure S3: Average distance between carbons C4′ and the last one of the geranyl chain for the NYA molecules pertaining to (A) system 1 (one NYA molecule), (B) system 2 (four NYA molecules), and (C) system 3 (eight NYA molecules). In (D) the average $\pm$ SD is shown. Data correspond to the last 30 ns of molecular dynamics simulation; Figure S4: Histograms and average angles ($\pm$ SD) formed by carbons C4′ and the last one of the geranyl chain for the NYA molecules in monomeric form with respect to the membrane surface. The full symbols correspond to the average angle of each one of the NYA molecules, whereas the empty symbol corresponds to the global average. The data is the average for the last 30 ns of molecular dynamics simulation time; Figure S5: Mass density profiles for the last 30 ns of the molecular dynamics for (A,B) system 1, (C,D) system 2 and (E,F) system 3. (A,C,E) Whole lipid molecules and (B,D,F) phosphorous atoms of phospholipids and the oxygen atoms of CHOL. POPC is shown in black, POPE in blue, POPS in red, PI-3P in olive, PSM in wine and CHOL in orange. The monomeric and aggregated NYA molecules are shown in dotted black and blue colours; Figure S6: Average deuterium order parameter, $-S_{CD}$, calculated for the hydrocarbon chains of the phospholipids. (A,C,E,G) Oleoyl and (B,D,F,H) palmitoyl acyl chains of (A, B) POPC, (C, D) POPE, (E, F) POPS, (G,H) PI-3P, as well as the palmitoyl (I) and sphyngosyl (J) acyl chains of PSM. The data correspond to the bulk phospholipid acyl chains (-■-) and the phospholipid acyl chains within 6 Å of the NYA molecule (-•-). The analysis was carried out for the last 30 ns of simulation. (K) Sum of differences between the $-S_{CD}$ values of the bulk lipids minus the S_{CD} values of the surrounding ones; Figure S7: (A) x-y and (B) z diffusion coefficients ($Å^2$/ns) for the NYA molecules in systems 1, 2 and 3 in the monomeric form: POPC (■), POPE (■), POPS (■), PI-3P (■), PSM (■), CHOL (■) and NYA (■). The data is the average for the last 30 ns of molecular dynamics simulation time; Figure S8: Average number of hydrogen bonds between the membrane lipids and NYA: POPC (-■-), POPE (-■-), POPS (-■-), PI-3P (-■-), PSM (-■-) and CHOL (-■-). The analysis was carried out for the last 30 ns of simulation.

Funding: This research was partially funded by ICAR_2023.

Institutional Review Board Statement: Not applicable.

Data Availability Statement: The original contributions presented in this study are included in the article/Supplementary Materials. Further inquiries can be directed to the corresponding author.

Acknowledgments: NAMD was developed by the Theoretical and Computational Biophysics Group in the Beckman Institute for Advanced Science and Technology at the University of Illinois at Urbana-Champaign. I am very grateful to SIPT, Universitas Miguel Hernández (UMH), for the generous use of the UMH Computer Cluster. This work was partially funded by "Programa de Ayudas para la Investigación del Envejecimiento del Centro Internacional para la Investigación del Envejecimiento de la Comunidad Valenciana, ICAR, Convocatoria 2023".

Conflicts of Interest: The author declares that there are no conflicts of interest.

Abbreviations

NYA	6-geranyl-3′,4′,5,7-tetrahydroxyflavanone
CHOL	cholesterol
PI-3P	1-palmitoyl-2-oleoyl-sn-glycero-3-phosphoinositol-3-phosphate

POPC	1-palmitoyl-2-oleoyl-sn-glycero-3-phosphocholine
POPE	1-palmitoyl-2-oleoyl-sn-glycero-3-phosphoethanolamine
POPS	1-palmitoyl-2-oleoyl-sn-glycero-3-phospho-L-serine
PSM	N-palmitoyl-D-erythro-sphingosylphosphorylcholine
z-COM	z-axis center-of-mass (z direction normal to the bilayer plane)

References

1. Arung, E.T.; Syafrizal; Kusuma, I.W.; Paramita, S.; Amen, Y.; Kim, Y.U.; Naibaho, N.M.; Ramadhan, R.; Ariyanta, H.A.; Fatriasari, W.; et al. Antioxidant, anti-inflammatory and anti-acne activities of stingless bee (*Tetragonula biroi*) propolis. *Fitoterapia* **2023**, *164*, 105375. [CrossRef] [PubMed]
2. Nam, S.H.; Yamano, A.; Kim, J.A.; Lim, J.; Baek, S.H.; Kim, J.E.; Kwon, T.G.; Saito, Y.; Teruya, T.; Choi, S.Y.; et al. Prenylflavonoids isolated from *Macaranga tanarius* stimulate odontoblast differentiation of human dental pulp stem cells and tooth root formation via the mitogen-activated protein kinase and protein kinase B pathways. *Int. Endod. J.* **2021**, *54*, 1142–1154. [CrossRef] [PubMed]
3. Shahinozzaman, M.; Taira, N.; Ishii, T.; Halim, M.A.; Hossain, M.A.; Tawata, S. Anti-Inflammatory, Anti-Diabetic, and Anti-Alzheimer's Effects of Prenylated Flavonoids from Okinawa Propolis: An Investigation by Experimental and Computational Studies. *Molecules* **2018**, *23*, 2479. [CrossRef] [PubMed]
4. Massaro, C.F.; Simpson, J.B.; Powell, D.; Brooks, P. Chemical composition and antimicrobial activity of honeybee (*Apis mellifera ligustica*) propolis from subtropical eastern Australia. *Naturwissenschaften* **2015**, *102*, 68. [CrossRef]
5. Tsuchiya, I.; Hosoya, T.; Ushida, M.; Kunimasa, K.; Ohta, T.; Kumazawa, S. Nymphaeol-A Isolated from Okinawan Propolis Suppresses Angiogenesis and Induces Caspase-Dependent Apoptosis via Inactivation of Survival Signals. *Evid. Based Complement. Altern. Med.* **2013**, *2013*, 826245. [CrossRef]
6. Zakaria, I.; Ahmat, N.; Jaafar, F.M.; Widyawaruyanti, A. Flavonoids with antiplasmodial and cytotoxic activities of *Macaranga triloba*. *Fitoterapia* **2012**, *83*, 968–972. [CrossRef]
7. Chen, C.N.; Hsiao, C.J.; Lee, S.S.; Guh, J.H.; Chiang, P.C.; Huang, C.C.; Huang, W.J. Chemical modification and anticancer effect of prenylated flavanones from Taiwanese propolis. *Nat. Prod. Res.* **2012**, *26*, 116–124. [CrossRef]
8. Ebner, D.W.; Woods, D.C.; Tran, T.D. Bioactive Polyphenolic Compounds from Propolis of *Tetragonula carbonaria* in the Gibberagee Region, New South Wales, Australia. *Foods* **2025**, *14*, 965. [CrossRef] [PubMed]
9. Amalia, N.R.; Budhy, T.I.; Ridwan, R.D.; Rianti, D.; Bramantoro, T.; Luthfi, M.; Ramadhani, N.F.; Pramusita, A.; Putranti, N.A.R.; Nugraha, A.P.; et al. Propolis nanoemulsion extract from celebes stingless bee (*Tetragonula biroi*) phytochemistry and antibacterial analysis to periodontopathogen bacteria. *J. Oral Biol. Craniofacial Res.* **2025**, *15*, 576–584. [CrossRef]
10. Balasubramaniam, A.K.; Elangovan, A.; Rahman, M.A.; Nayak, S.; Swain, D.; Babu, H.P.; Narasimhan, A.; Monga, V. Propolis: A comprehensive review on the nature's polyphenolic wonder. *Fitoterapia* **2025**, *183*, 106526. [CrossRef]
11. Deniz, F.S.S.; Orhan, I.E.; Filipek, P.A.; Ertas, A.; Gstir, R.; Jakschitz, T.; Bonn, G.K. Evaluation of the Anti-Aging Properties of Ethanolic Extracts from Selected Plant Species and Propolis by Enzyme Inhibition Assays and 2D/3D Cell Culture Methods. *Pharmaceuticals* **2025**, *18*, 439. [CrossRef] [PubMed]
12. Martinotti, S.; Bonsignore, G.; Ranzato, E. Propolis: A Natural Substance with Multifaceted Properties and Activities. *Int. J. Mol. Sci.* **2025**, *26*, 1519. [CrossRef]
13. Nguyen, B.C.Q.; Yoshimura, K.; Kumazawa, S.; Tawata, S.; Maruta, H. Frondoside A from sea cucumber and nymphaeols from Okinawa propolis: Natural anti-cancer agents that selectively inhibit PAK1 in vitro. *Drug Discov. Ther.* **2017**, *11*, 110–114. [CrossRef]
14. Kumazawa, S.; Murase, M.; Momose, N.; Fukumoto, S. Analysis of antioxidant prenylflavonoids in different parts of *Macaranga tanarius*, the plant origin of Okinawan propolis. *Asian Pac. J. Trop. Med.* **2014**, *7*, 16–20. [CrossRef]
15. Kumazawa, S.; Nakamura, J.; Murase, M.; Miyagawa, M.; Ahn, M.R.; Fukumoto, S. Plant origin of Okinawan propolis: Honeybee behavior observation and phytochemical analysis. *Naturwissenschaften* **2008**, *95*, 781–786. [CrossRef] [PubMed]
16. Kumazawa, S.; Ueda, R.; Hamasaka, T.; Fukumoto, S.; Fujimoto, T.; Nakayama, T. Antioxidant prenylated flavonoids from propolis collected in Okinawa, Japan. *J. Agric. Food Chem.* **2007**, *55*, 7722–7725. [CrossRef] [PubMed]
17. Kumazawa, S.; Goto, H.; Hamasaka, T.; Fukumoto, S.; Fujimoto, T.; Nakayama, T. A new prenylated flavonoid from propolis collected in Okinawa, Japan. *Biosci. Biotechnol. Biochem.* **2004**, *68*, 260–262. [CrossRef]
18. Sood, B.; Patel, P.; Keenaghan, M. Coenzyme Q10. In *StatPearls*; Disclosure: Preeti Patel declares no relevant financial relationships with ineligible companies. Disclosure: Michael Keenaghan declares no relevant financial relationships with ineligible companies; Treasure Island (FL) Ineligible Companies: Treasure Island, FL, USA, 2024.
19. Cores, A.; Carmona-Zafra, N.; Clerigue, J.; Villacampa, M.; Menendez, J.C. Quinones as Neuroprotective Agents. *Antioxidants* **2023**, *12*, 1464. [CrossRef]
20. Gasmi, A.; Bjorklund, G.; Mujawdiya, P.K.; Semenova, Y.; Piscopo, S.; Peana, M. Coenzyme Q_{10} in aging and disease. *Crit. Rev. Food Sci. Nutr.* **2024**, *64*, 3907–3919. [CrossRef]

21. Jomova, K.; Raptova, R.; Alomar, S.Y.; Alwasel, S.H.; Nepovimova, E.; Kuca, K.; Valko, M. Reactive oxygen species, toxicity, oxidative stress, and antioxidants: Chronic diseases and aging. *Arch. Toxicol.* **2023**, *97*, 2499–2574.
22. Rizzo, A.M.; Berselli, P.; Zava, S.; Montorfano, G.; Negroni, M.; Corsetto, P.; Berra, B. Endogenous antioxidants and radical scavengers. *Adv. Exp. Med. Biol.* **2010**, *698*, 52–67. [PubMed]
23. Orsucci, D.; Mancuso, M.; Ienco, E.C.; LoGerfo, A.; Siciliano, G. Targeting mitochondrial dysfunction and neurodegeneration by means of coenzyme Q10 and its analogues. *Curr. Med. Chem.* **2011**, *18*, 4053–4064. [CrossRef]
24. Bentinger, M.; Brismar, K.; Dallner, G. The antioxidant role of coenzyme Q. *Mitochondrion* **2007**, *7*, S41–S50. [CrossRef] [PubMed]
25. Wilson, D.M., III; Cookson, M.R.; Van Den Bosch, L.; Zetterberg, H.; Holtzman, D.M.; Dewachter, I. Hallmarks of neurodegenerative diseases. *Cell* **2023**, *186*, 693–714. [CrossRef]
26. Ingolfsson, H.I.; Melo, M.N.; van Eerden, F.J.; Arnarez, C.; Lopez, C.A.; Wassenaar, T.A.; Periole, X.; de Vries, A.H.; Tieleman, D.P.; Marrink, S.J. Lipid organization of the plasma membrane. *J. Am. Chem. Soc.* **2014**, *136*, 14554–14559. [CrossRef] [PubMed]
27. Perricone, U.; Gulotta, M.R.; Lombino, J.; Parrino, B.; Cascioferro, S.; Diana, P.; Cirrincione, G.; Padova, A. An overview of recent molecular dynamics applications as medicinal chemistry tools for the undruggable site challenge. *Medchemcomm* **2018**, *9*, 920–936. [CrossRef]
28. Villalain, J. Location and interaction of idebenone and mitoquinone in a membrane similar to the inner mitochondrial membrane. Comparison with ubiquinone 10. *Free Radic. Biol. Med.* **2024**, *222*, 211–222. [CrossRef]
29. Villalain, J. Bisphenol F and Bisphenol S in a Complex Biomembrane: Comparison with Bisphenol A. *J. Xenobiot.* **2024**, *14*, 1201–1220. [CrossRef]
30. Phillips, J.C.; Hardy, D.J.; Maia, J.D.C.; Stone, J.E.; Ribeiro, J.V.; Bernardi, R.C.; Buch, R.; Fiorin, G.; Henin, J.; Jiang, W.; et al. Scalable molecular dynamics on CPU and GPU architectures with NAMD. *J. Chem. Phys.* **2020**, *153*, 044130. [CrossRef]
31. Best, R.B.; Zhu, X.; Shim, J.; Lopes, P.E.; Mittal, J.; Feig, M.; Mackerell, A.D., Jr. Optimization of the additive CHARMM all-atom protein force field targeting improved sampling of the backbone ϕ, ψ and side-chain χ_1 and χ_2 dihedral angles. *J. Chem. Theory Comput.* **2012**, *8*, 3257–3273. [CrossRef]
32. Vanommeslaeghe, K.; Hatcher, E.; Acharya, C.; Kundu, S.; Zhong, S.; Shim, J.; Darian, E.; Guvench, O.; Lopes, P.; Vorobyov, I.; et al. CHARMM general force field: A force field for drug-like molecules compatible with the CHARMM all-atom additive biological force fields. *J. Comput. Chem.* **2010**, *31*, 671–690. [CrossRef] [PubMed]
33. Klauda, J.B.; Venable, R.M.; Freites, J.A.; O'Connor, J.W.; Tobias, D.J.; Mondragon-Ramirez, C.; Vorobyov, I.; MacKerell, A.D., Jr.; Pastor, R.W. Update of the CHARMM all-atom additive force field for lipids: Validation on six lipid types. *J. Phys. Chem. B* **2010**, *114*, 7830–7843. [CrossRef] [PubMed]
34. Ozu, M.; Alvarez, H.A.; McCarthy, A.N.; Grigera, J.R.; Chara, O. Molecular dynamics of water in the neighborhood of aquaporins. *Eur. Biophys. J.* **2013**, *42*, 223–239. [CrossRef]
35. Wu, E.L.; Cheng, X.; Jo, S.; Rui, H.; Song, K.C.; Davila-Contreras, E.M.; Qi, Y.; Lee, J.; Monje-Galvan, V.; Venable, R.M.; et al. CHARMM-GUI Membrane Builder toward realistic biological membrane simulations. *J. Comput. Chem.* **2014**, *35*, 1997–2004. [CrossRef]
36. Murzyn, K.; Rog, T.; Jezierski, G.; Takaoka, Y.; Pasenkiewicz-Gierula, M. Effects of phospholipid unsaturation on the membrane/water interface: A molecular simulation study. *Biophys. J.* **2001**, *81*, 170–183. [CrossRef] [PubMed]
37. Villalain, J. Localization and Aggregation of Honokiol in the Lipid Membrane. *Antioxidants* **2024**, *13*, 1025. [CrossRef] [PubMed]
38. van Meer, G.; Voelker, D.R.; Feigenson, G.W. Membrane lipids: Where they are and how they behave. *Nat. Rev. Mol. Cell Biol.* **2008**, *9*, 112–124. [CrossRef]
39. Esposito, D.L.; Nguyen, J.B.; DeWitt, D.C.; Rhoades, E.; Modis, Y. Physico-chemical requirements and kinetics of membrane fusion of flavivirus-like particles. *J. Gen. Virol.* **2015**, *96*, 1702–1711. [CrossRef]
40. Humphrey, W.; Dalke, A.; Schulten, K. VMD: Visual molecular dynamics. *J. Mol. Graph.* **1996**, *14*, 33–38. [CrossRef]
41. Guixa-Gonzalez, R.; Rodriguez-Espigares, I.; Ramirez-Anguita, J.M.; Carrio-Gaspar, P.; Martinez-Seara, H.; Giorgino, T.; Selent, J. MEMBPLUGIN: Studying membrane complexity in VMD. *Bioinformatics* **2014**, *30*, 1478–1480. [CrossRef]
42. Galiano, V.; Villalain, J. Aggregation of 25-hydroxycholesterol in a complex biomembrane. Differences with cholesterol. *Biochim. Biophys. Acta Biomembr.* **2020**, *1862*, 183413. [CrossRef] [PubMed]
43. Villalain, J. Envelope E protein of dengue virus and phospholipid binding to the late endosomal membrane. *Biochim. Biophys. Acta Biomembr.* **2022**, *1864*, 183889. [CrossRef]
44. Villalain, J. Epigallocatechin-3-gallate location and interaction with late endosomal and plasma membrane model membranes by molecular dynamics. *J. Biomol. Struct. Dyn.* **2019**, *37*, 3122–3134. [CrossRef]
45. Villalain, J. Bergamottin: Location, aggregation and interaction with the plasma membrane. *J. Biomol. Struct. Dyn.* **2023**, *41*, 12026–12037. [CrossRef] [PubMed]
46. Villalain, J. Procyanidin C1 Location; Interaction, and Aggregation in Two Complex Biomembranes. *Membranes* **2022**, *12*, 692. [CrossRef]

47. Giorgino, T. Computing 1-D atomic densities in macromolecular simulations: The Density Profile Tool for VMD. *Comput. Phys. Commun.* **2014**, *185*, 317–322. [CrossRef]
48. Baylon, J.L.; Tajkhorshid, E. Capturing Spontaneous Membrane Insertion of the Influenza Virus Hemagglutinin Fusion Peptide. *J. Phys. Chem. B* **2015**, *119*, 7882–7893. [CrossRef]
49. Giorgino, T. Computing diffusion coefficients in macromolecular simulations: The Diffusion Coefficient Tool for VMD. *J. Open Source Softw.* **2019**, *4*, 1698. [CrossRef]
50. Kandt, C.; Ash, W.L.; Tieleman, D.P. Setting up and running molecular dynamics simulations of membrane proteins. *Methods* **2007**, *41*, 475–488. [CrossRef] [PubMed]
51. Anezo, C.; de Vries, A.H.; Höltje, H.D.; Tieleman, D.P.; Marrink, S.J. Methodological issues in lipid bilayers simulations. *J. Phys. Chem. B* **2003**, *107*, 9424–9433. [CrossRef]
52. Bera, I.; Klauda, J.B. Molecular Simulations of Mixed Lipid Bilayers with Sphingomyelin, Glycerophospholipids, and Cholesterol. *J. Phys. Chem. B* **2017**, *121*, 5197–5208. [CrossRef]
53. Mukhopadhyay, P.; Monticelli, L.; Tieleman, D.P. Molecular dynamics simulation of a palmitoyl-oleoyl phosphatidylserine bilayer with Na+ counterions and NaCl. *Biophys. J.* **2004**, *86*, 1601–1609. [CrossRef]
54. Tsai, H.H.; Lee, J.B.; Li, H.S.; Hou, T.Y.; Chu, W.Y.; Shen, P.C.; Chen, Y.Y.; Tan, C.J.; Hu, J.C.; Chiu, C.C. Geometrical effects of phospholipid olefinic bonds on the structure and dynamics of membranes: A molecular dynamics study. *Biochim. Biophys. Acta* **2015**, *1848*, 1234–1247. [CrossRef]
55. Bockmann, R.A.; Hac, A.; Heimburg, T.; Grubmuller, H. Effect of sodium chloride on a lipid bilayer. *Biophys. J.* **2003**, *85*, 1647–1655. [CrossRef]
56. Tieleman, D.P.; Marrink, S.J.; Berendsen, H.J. A computer perspective of membranes: Molecular dynamics studies of lipid bilayer systems. *Biochim. Biophys. Acta* **1997**, *1331*, 235–270. [CrossRef]

membranes

MDPI

Article

Anti-Inflammatory Properties of Novel 1,2-Benzothiazine Derivatives and Their Interaction with Phospholipid Model Membranes

Berenika M. Szczęśniak-Sięga [1,*], Jadwiga Maniewska [1], Benita Wiatrak [2], Tomasz Janek [3], Paulina Nowotarska [4] and Żaneta Czyżnikowska [5,*]

1 Department of Medicinal Chemistry, Faculty of Pharmacy, Wroclaw Medical University, Borowska 211, 50-556 Wroclaw, Poland; jadwiga.maniewska@umw.edu.pl
2 Department of Pharmacology, Faculty of Medicine, Wroclaw Medical University, J. Mikulicza-Radeckiego 2, 50-345 Wroclaw, Poland; benita.wiatrak@umw.edu.pl
3 Department of Biotechnology and Food Microbiology, Faculty of Biotechnology and Food Science, Wroclaw University of Environmental and Life Sciences, Chełmońskiego 37, 51-630 Wroclaw, Poland; tomasz.janek@upwr.edu.pl
4 Department of Biostructure and Animal Physiology, Wroclaw University of Environmental and Life Sciences, Norwida 25/27, 50-375 Wroclaw, Poland; paulina.nowotarska@upwr.edu.pl
5 Department of Basic Chemical Sciences, Faculty of Pharmacy, Wroclaw Medical University, Borowska 211a, 50-556 Wroclaw, Poland
* Correspondence: berenika.szczesniak-siega@umw.edu.pl (B.M.S.-S.); zaneta.czyznikowska@umw.edu.pl (Ż.C.)

Abstract: The design of novel anti-inflammatory drugs remains a critical area of research in the development of effective treatments for inflammatory diseases. In this study, a series of 1,2-benzothiazine was evaluated through a multifaceted approach. In particular, we investigated the potential interactions of the potential drugs with lipid bilayers, an important consideration for membrane permeability and overall pharmacokinetics. In addition, we evaluated their ability to inhibit cyclooxygenase 1 and cyclooxygenase 2 activity and selectivity using both a cyclooxygenase inhibition assay and molecular docking simulations. To evaluate their therapeutic potential, we performed in vitro assays to measure cytokine mRNA expression in inflamed cells. The antioxidant activity was evaluated using both in vitro assays, such as 2,2-diphenyl-1-picrylhydrazyl and 2,2-azino-*bis*-3-ethylbenzothiazoline-6-sulphonic acid scavenging, to determine the compounds' capacity to neutralize free radicals and reduce oxidative stress. Theoretical calculations, including density functional theory, were used to predict the reactivity profiles of the compounds.

Keywords: 1,2-benzothiazine derivatives; drug–membrane interaction; anti-inflammatory activity; antioxidant properties

1. Introduction

Cyclooxygenase (E.C. 1.14.99.1., prostaglandin cyclic peroxide synthase, PGHS, COX) is an enzyme belonging to the metalloproteinase family, located in the lumen of the endoplasmic reticulum, the Golgi apparatus, and the nuclear membrane of the cell, embedded with hydrophobic fragments in the lipid bilayer of the membrane. COX catalyzes the process of prostanoid synthesis, i.e., prostaglandins, prostacyclin, and thromboxane from arachidonic acid. Three isoforms of this enzyme have been identified so far: COX-1, COX-2, and COX-3 [1]. Since COX-2 is the isoform induced by the inflammation process, great efforts to find selective COX-2 inhibitors have been made between the last decade of 1900 and the beginning of 2000, resulting in several compounds being approved worldwide (e.g., celecoxib). Some of these were subsequently withdrawn from the market due to cardiovascular toxicity (e.g., rofecoxib and valdecoxib) [2]. However, since inflammation has been linked to cancer development, the search for new, safer anti-inflammatory drugs has become even more important [3–6]. It has been established that in the process of carcinogenesis, overexpression, increased secretion, or inappropriate activation of proinflammatory mediators, such as COX-2, prostaglandins, cytokines, inducible nitric oxide synthase and nitric oxide, and many other intracellular signaling molecules, facilitate tumor promotion and progression [7,8]. In addition, inflammation induces the formation of reactive oxygen species (ROS) and nitrogen species (RNS), which cause damage to very important cellular elements such as DNA, proteins, or lipids, which may directly or indirectly contribute to neoplastic transformation [9,10]. Taking these data into account, finding new COX-2 inhibitors that also have free radical scavenging properties would be beneficial both in inflammatory diseases and in cancer treatment.

In our previous studies, we obtained a series of 1,2-benzothiazine derivatives with anti-inflammatory and analgesic activity confirmed in in vivo tests on mice [11]. In addition, these compounds were devoid of ulcerogenic activity characteristic of most non-steroidal anti-inflammatory drugs (NSAIDs). These compounds were modified in relation to the model drug—meloxicam (Figure 1), in positions 2 and 3 of the thiazine ring. The modification in position 2 seems to be more important for the activity of the new compounds because it consisted of replacing the small methyl substituent in meloxicam with a large, extended arylpiperazine group. According to Hatnapure, the high electron-donating capacity of the piperazine moiety can be clearly correlated with the high anti-inflammatory activity of the compounds, which was consistent with many other scientific studies [12–17].

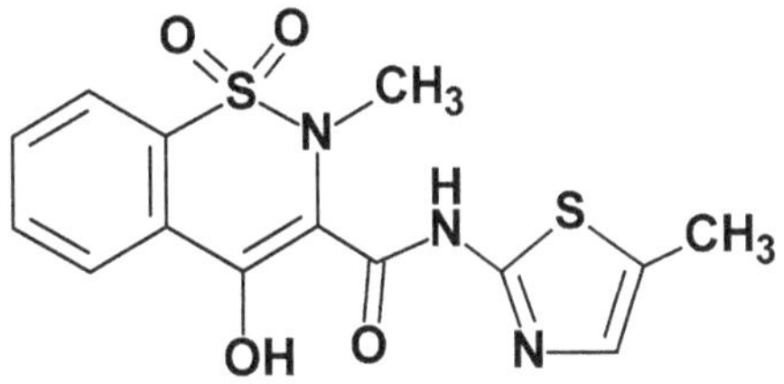

Figure 1. Structure of the model drug—meloxicam.

One of these tested and non-toxic compounds was a 1,2-benzothiazine derivative containing a heterocyclic pyrimidine ring and a bromine atom in its structure (Figure 2a). We decided to check whether these substituents significantly influenced the anti-inflammatory activity of this compound, and therefore, we planned to obtain a new derivative in which the pyrimidine ring was replaced with a benzene ring (Figure 2b). Another planned modification was to replace the acetyl linker between the thiazine and piperazine nitrogen atoms with a propyl linker because compounds with such a linker also showed strong anti-inflammatory activity and did not cause ulcer formation, as shown by our previous studies (Figure 2c) [11]. Subsequent modifications included replacing the bromine atom with other electron-withdrawing substituents, such as a fluorine atom, or electron-donating substituents such as a methyl or methoxy group. On the other hand, the pyrimidine ring was replaced with a pyridine ring to assess the effect of the type of ring on the anti-inflammatory activity of the compound (Figure 2d).

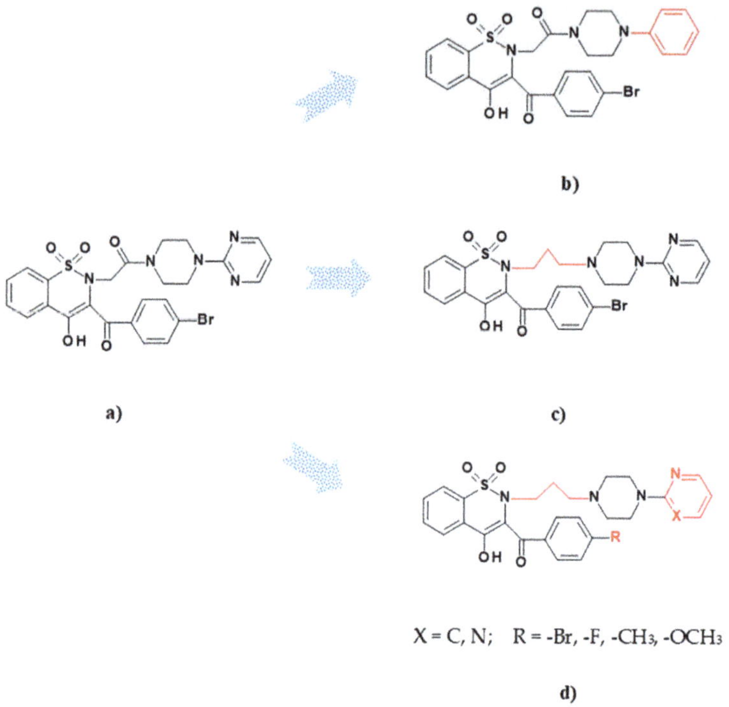

Figure 2. Structures of a non-ulcerogenic 1,2-benzothiazine (**a**) and new designed arylpiperazine derivatives of 1,2-benzothiazine (**b–d**).

The primary objective of this study was to determine whether compounds based on an arylpiperazine-1,2-benzothiazine core could serve as a lead structure for new, safer anti-inflammatory and at the same time antioxidant compounds. Hence, we performed an in vitro evaluation of the impact of the tested compounds on the activity of COX-2 and COX-1 to assess the selectivity ratio. The anti-inflammatory effect of the new compounds was also confirmed in the MTT and DCF-DA assays on normal human dermal fibroblast (NHDF) cells exposed to lipopolysaccharide (LPS). The radical scavenging activity was evaluated with two different in vitro measurements, using 2-diphenyl-1-picrylhydrazyl (DPPH) and 2,2′-azino-bis(3-ethylbenzothiazoline-6-sulphonic acid (ABTS). An additional goal was to assess the ability of designed compounds to interact with phospholipid bilayers because cyclooxygenase is associated with biological membranes. Experimental results were supported by theoretical calculations. The global reactivity parameters were obtained based on the density level of theory. The binding mode of designed compounds to both cyclooxygenase isoforms was established by using molecular docking.

2. Materials and Methods

2.1. Chemistry

All commercial chemicals were used as supplied unless otherwise indicated. ^{1}H and ^{13}C NMR spectra were recorded on a Bruker 300 MHz spectrometer using $CDCl_3$ as a solvent. Chemical shifts for proton nuclear magnetic resonance (^{1}H NMR) spectra are reported in parts per million (ppm) relative to the signal of tetramethylsilane at 0 ppm (internal standard). Splitting patterns are designated as follows: s, singlet; brs, broad singlet; d, doublet; t, triplet; q, quartet; m, multiplet. Chemical shifts for carbon nuclear magnetic resonance (^{13}C NMR) spectra are reported in parts per million (ppm) relative to the center line of the $CDCl_3$ triplet at 77.02 ppm. FT-IR spectra were recorded on a Perkin-

Elmer Spectrum Two UATR FT-IR spectrometer. Mass data were acquired on a Bruker Daltonics micrOTOF-Q Mass Spectrometer in a positive ion mode with flow injection electron spray ionization (ESI). The elemental analyses were carried out on a Carlo Erba NA 1500 analyzer and were within ±0.4% of the theoretical value. Melting points were determined in open glass capillaries using a MEL-TEMP melting-point apparatus and were uncorrected. Reaction progress was monitored using thin-layer chromatography (TLC) performed on silica gel 60 F254 pre-coated aluminum sheets. Spots were visualized under a 254 nm UV lamp. A rotary evaporator under reduced pressure conditions was used to concentrate the reaction solutions. Solvents were of reagent grade and purified with standard methods when necessary.

Synthesis and experimental data of compounds **3a–d** and **4a–d** were previously reported [18,19], and compounds **5**, **6**, and **7** were reported also [20,21].

Synthesis and Experimental Data of Compounds **BS23–BS30**

A solution of 5 mL of sodium ethoxide (2.3%) was added to the stirred mixture of 5 mmol of corresponding 1,2-benzothiazines **4a–d** in 20 mL of anhydrous ethanol. Then, 5 mmol of compound **5**, **6**, or **7** was added and refluxed with stirring for 10 h. When the reaction ended, which was controlled on TLC plates, ethanol was distilled off, and the residue was treated with 50 mL of chloroform and insoluble materials (e.g., sodium chloride) were filtered off. The filtrate was then evaporated, and the residue was purified by crystallization from ethanol to give desirable products **BS23–BS30** with a yield of 25–50%.

3-(4-Methoxybenzoyl)-4-hydroxy-2-{3-[4-(2-pyridyl)-1-piperazinyl)propyl]}-2H-1,2-benzothiazine 1,1-dioxide **BS23**

Yellow powder, 33.05% yield, mp 107–111 °C; FT-IR (cm^{-1}): 1605 (C=O), 1345, 1170 (SO_2). ^1H NMR (300 MHz, $CDCl_3$) δ (ppm): 1.41 (brs, 2H, $CH_2CH_2CH_2$), 2.11–2.30 (m, 6H, $CH_2N(CH_2)_2$), 3.25–3.48 (m, 6H, $CH_2N(CH_2)_2$), 3.88 (s, 3H, OCH_3), 6.57–8.21 (m, 12H, ArH), 15.75 (s, 1H, OH_{enolic}). ^{13}C NMR (300 MHz, $CDCl_3$) δ (ppm): 190.01, 163.46, 158.98, 147.86, 138.08, 137.64, 133.46, 132.83, 131.98, 128.97, 127.98, 127.83, 123.38, 116.72, 113.92, 107.12, 55.53, 55.06, 52.26, 51.71, 44.41, 23.11. HRMS (ESI) calculated for $C_{28}H_{30}N_4O_5S$ $[M+H]^+$ 535.2010; found the following: 535.1957. Analysis calculated for $C_{28}H_{30}N_4O_5S$ (534.63); C, 62.90; H, 5.66; N, 10.48; found the following: C, 62.88; H, 5.80; N, 10.42.

3-(4-Methoxybenzoyl)-4-hydroxy-2-{3-[4-(2-pyrimidyl)-1-piperazinyl)propyl]}-2H-1,2-benzothiazine 1,1-dioxide **BS24**

Yellow powder, 36.10% yield, mp 60–65 °C; FT-IR (cm^{-1}): 1605 (C=O), 1345, 1175 (SO_2). ^1H NMR (300 MHz, $CDCl_3$) δ (ppm): 1.44 (brs, 2H, $CH_2CH_2CH_2$), 2.12–2.29 (m, 6H, $CH_2N(CH_2)_2$), 3.23 (brs, 2H, $-CH_2-$), 3.70–3.83 (brs, 4H, $N(CH_2)_2$), 3.87 (s, 3H, OCH_3), 6.49–8.29 (m, 11H, ArH). ^{13}C NMR (300 MHz, $CDCl_3$) δ (ppm): 189.86, 169.20, 163.45, 161.33, 157.74, 137.96, 133.47, 132.87, 131.99, 128.97, 127.94, 127.87, 123.47, 116.85, 113.91, 110.32, 55.53, 55.10, 52.31, 51.83, 42.61, 23.08. HRMS (ESI) calculated for $C_{27}H_{29}N_5O_5S$ $[M+H]^+$ 536.1962; found the following: 536.1906. Analysis calculated for $C_{27}H_{29}N_5O_5S$ (535,61); C, 60.55; H, 5.46; N, 13.08; found the following: C, 60.80; H, 5.70; N, 13.35.

3-(4-Methylbenzoyl)-4-hydroxy-2-{3-[4-(2-pyridyl)-1-piperazinyl)propyl]}-2H-1,2-benzothiazine 1,1-dioxide **BS25**

Yellow crystals, 34.14% yield, mp 139–141 °C; FT-IR (cm^{-1}): 1600 (C=O), 1350, 1180 (SO_2). ^1H NMR (300 MHz, $CDCl_3$) δ (ppm): 1.42 (brs, 2H, $CH_2CH_2CH_2$), 2.10–2.32 (m, 6H, $CH_2N(CH_2)_2$), 2.42 (s, 3H, CH_3), 3.20 (brs, 2H, $-CH_2-$), 3.52 (brs, 4H, $N(CH_2)_2$), 6.58–8.21 (m, 12H, ArH). ^{13}C NMR (300 MHz, $CDCl_3$) δ (ppm): 190.50, 169.98, 158.85, 147.91, 143.71, 138.07, 137.68, 133.58, 132.88, 129.64, 129.25, 128.95, 127.92, 123.52, 117.10, 113.78, 107.16, 55.06, 52.14, 44.23, 23.14, 21.74. HRMS (ESI) calculated for $C_{28}H_{30}N_4O_4S$ $[M+H]^+$ 519.2061; found the following: 519.2008. Analysis calculated for $C_{28}H_{30}N_4O_4S$ (518,63); C, 64.84; H, 5.83; N, 10.80; found the following: C, 65.15; H, 5.92; N, 10.66.

3-(4-Fluorobenzoyl)-4-hydroxy-2-{3-[4-(2-pyridyl)-1-piperazinyl)propyl]}-2H-1,2-benzothiazine 1,1-dioxide **BS26**

Beige powder, 50.15% yield, mp 155–158 °C; FT-IR (cm^{-1}): 1600 (C=O), 1345, 1175 (SO_2). ^1H NMR (300 MHz, $CDCl_3$) δ (ppm): 1.44 (brs, 2H, $CH_2CH_2CH_2$), 2.14–2.39 (m, 6H, $CH_2N(CH_2)_2$), 3.21 (brs, 2H, -CH_2-), 3.54 (brs, 4H, $N(CH_2)_2$), 6.59–8.21 (m, 12H, ArH). ^{13}C NMR (300 MHz, $CDCl_3$) δ (ppm): 190.00, 170.00, 158.77, 147.85, 138.03, 137.74, 133.72, 132.98, 132.24, 132.12, 128.01, 123.58, 116.94, 115.91, 115.62, 113.84, 107.24, 55.04, 52.20, 44.25, 23.13. HRMS (ESI) calculated for $C_{27}H_{27}FN_4O_4S$ $[M+H]^+$ 523.1810; found the following: 523.1759. Analysis calculated for $C_{27}H_{27}FN_4O_4S$ (522.59); C, 62.05; H, 5.21; N, 10.72; found the following: C, 62.19; H, 5.27; N, 10.51.

3-(4-Fluorobenzoyl)-4-hydroxy-2-{3-[4-(2-pyrimidyl)-1-piperazinyl)propyl]}-2H-1,2-benzothiazine 1,1-dioxide **BS27**

Beige powder, 40.40% yield, mp 151–153 °C; FT-IR (cm^{-1}): 1600 (C=O), 1340, 1170 (SO_2). ^1H NMR (300 MHz, $CDCl_3$) δ (ppm): 1.42 (brs, 2H, $CH_2CH_2CH_2$), 2.12–2.31 (m, 6H, $CH_2N(CH_2)_2$), 3.21 (brs, 2H, -CH_2-), 3.69–3.79 (m, 4H, $N(CH_2)_2$), 6.49–8.29 (m, 11H, ArH). ^{13}C NMR (300 MHz, $CDCl_3$) δ (ppm): 189.79, 169.78, 163.66, 161.35, 157.75, 138.04, 133.72, 132.97, 132.25, 132.13, 127.99, 123.58, 116.92, 115.92, 115.63, 110.36, 55.04, 52.37, 51.96, 42.66, 23.16. HRMS (ESI) calculated for $C_{26}H_{26}FN_5O_4S$ $[M+H]^+$ 524.1762; found the following: 524.1705. Analysis calculated for $C_{26}H_{26}FN_5O_4S$ (523,58); C, 59.64; H, 5.01; N, 13.38; found the following: C, 59,60; H, 5.30; N, 13.31.

3-(4-Bromobenzoyl)-2-[2-(4-phenylpiperazin-1-yl)-2-oxoethyl]-4-hydroxy-2H-1,2-benzothiazine 1,1-dioxide **BS28**

Yellow powder, 36.22% yield, mp 134–138 °C; FT-IR (cm^{-1}): 1600, 1660 (C=O), 1340, 1180 (SO_2). ^1H NMR (300 MHz, $CDCl_3$) δ (ppm): 2.98–3.07 (m, 4H, $N(CH_2)_2$), 3.41 (brs, 4H, $N(CH_2)_2$), 4.37–4.39 (m, 2H, CH_2CO), 7.00–8.26 (m, 13H, ArH), 15.46 (s, 1H, OH_{enolic}).^{13}C NMR (300 MHz, $CDCl_3$) δ (ppm): 187.09, 171.86, 164.35, 138.63, 134.04, 133.39, 132.65, 132.04, 130.64, 129.53, 129.39, 127.77, 122.36, 117.50, 116.15, 58.42, 51.94, 50.32, 44.52, 41.06, 18.43. HRMS (ESI) calculated for $C_{27}H_{24}BrN_3O_5S$ $[M+H]^+$ 584.0676; found the following: 584.0616. Analysis calculated for $C_{27}H_{24}BrN_3O_5S$ (582,46); C, 55.68; H, 4.15; N, 7.21; found the following: C, 55.57; H, 4.47; N, 6.88.

3-(4-Bromobenzoyl)-4-hydroxy-2-{3-[4-(2-pyridyl)-1-piperazinyl)propyl]}-2H-1,2-benzothiazine 1,1-dioxide **BS29**

Yellow powder, 25.20% yield, mp 87–90 °C; FT-IR (cm^{-1}): 1602 (C=O), 1350, 1175 (SO_2). ^1H NMR (300 MHz, $CDCl_3$) δ (ppm): 1.48 (brs, 2H, $CH_2CH_2CH_2$), 2.21–2.45 (m, 6H, $CH_2N(CH_2)_2$), 3.20 (brs, 2H, -CH_2-), 3.60 (brs, 4H, $N(CH_2)_2$), 6.60–8.20 (m, 12H, ArH). ^{13}C NMR (300 MHz, $CDCl_3$) δ (ppm): 158.66, 147.86, 137.78, 133.77, 133.05, 131.82, 131.01, 128.09, 123.73, 117.31, 113.96, 107.30, 55.13, 52.19, 44.11, 23.18. HRMS (ESI) calculated for $C_{27}H_{27}BrN_4O_4S$ $[M+H]^+$ 585.0992; found the following: 585.0924. Analysis calculated for $C_{27}H_{27}BrN_4O_4S$ (583.50); C, 55.58; H, 4.66; N, 9.60; found the following: C, 55.62; H, 4.74; N, 9.33.

3-(4-Bromobenzoyl)-4-hydroxy-2-{3-[4-(2-pyrimidyl)-1-piperazinyl)propyl]}-2H-1,2-benzothiazine 1,1-dioxide **BS30**

Yellow powder, 37.30% yield, mp 152–156 °C; FT-IR (cm^{-1}): 1600 (C=O), 1340, 1170 (SO_2). ^1H NMR (300 MHz, $CDCl_3$) δ (ppm): 1.52 (brs, 2H, $CH_2CH_2CH_2$), 2.20–2.44 (m, 6H, $CH_2N(CH_2)_2$), 3.20 (brs, 2H, -CH_2-), 3.90 (brs, 4H, $N(CH_2)_2$), 6.52–8.31 (m, 11H, ArH). ^{13}C NMR (300 MHz, $CDCl_3$) δ (ppm): 170.56, 161.20, 157.82, 137.76, 133.87, 133.12, 131.84, 131.08, 128.14, 123.81, 117.36, 110.65, 55.13, 52.27, 42.15, 23.01. HRMS (ESI) calculated for $C_{26}H_{26}BrN_5O_4S$ $[M+H]^+$ 586.0944; found the following: 586.0867. Analysis calculated for $C_{26}H_{26}BrN_5O_4S$ (584.48); C, 53.43; H, 4.48; N, 11.98; found the following: C, 53.72; H, 4.56; N, 11.86.

2.2. *In Vitro Study*

2.2.1. Cyclooxygenase Inhibition Assay

To conduct the cyclooxygenase inhibition assay, we utilized a kit from the manufacturer that contained 150 μL of assay buffer, 10 μL of heme, and 10 μL of either COX-1 or COX-2 enzyme. Each sample was prepared in triplicate, with the addition of 10–500 μM of the test compounds and 10 μM of solvents such as methanol, ethanol, and dimethyl sulfoxide (DMSO). We then added 20 μL of N,N,N cent,N cent-tetramethyl-p-phenylenediamine (TMPD) to each well. Arachidonic acid was introduced to initiate the reaction, which was allowed to proceed for 2 min. Following the incubation period, the extent of TMPD oxidation was measured using the multiscan Go microplate reader at a wavelength of 590 nm (Thermo Fisher Scientific, Waltham, MA, USA).

2.2.2. Biological Evaluation

Cell Line and Culture Conditions

NHDF (normal human dermal fibroblast) cells, sourced from the American Type Culture Collection (ATCC), were maintained in DMEM without phenol red, enhanced with 10% fetal bovine serum (FBS), 2 mM L-glutamine, and 25 μg/mL gentamicin. The cells were incubated under conditions of 5% CO_2, at a temperature of 37 °C, and 95% humidity. We monitored the cells bi-weekly, utilizing or subculturing them once they reached over 70% confluence. For all experimental setups, cells were plated at a density of 10,000 cells per well and allowed to recover in a CO_2 incubator for 24 h.

Tested Compounds

The compounds under investigation were dissolved in DMSO to achieve a final concentration of 10 mM and then stored at −20 °C. Lipopolysaccharide (LPS) from *Escherichia coli* (Sigma-Aldrich, Saint Louis, MO, USA; catalog no. L2630) was dissolved in distilled water to a concentration of 1 mM. The stock solution of LPS was stored at −20 °C for up to 6 months. Final concentrations of LPS used in this study were 50 μg/mL. Prior to use, the solutions were thawed at room temperature. For each compound, 8 different concentrations in the range of 500–10 μM were prepared in the culture medium.

MTT Assay

The cytotoxicity of the test compound was assessed using the MTT assay. After cell seeding and overnight incubation, the supernatant was removed, and the compounds were added at concentrations ranging from 500 to 10 μM. Control cells were treated with complete medium only, while cells treated with medium containing DMSO at the highest compound concentration were included as the solvent control. After the incubation period, the supernatant was removed, and 50 μL of a 1.0 mg/mL MTT solution was added to each well and incubated for 2 h at 37 °C. The medium was then removed, and 100 μL of DMSO was added to each well. Finally, absorbance was measured at 570 nm using a Multiskan GO spectrophotometer.

Cell viability was calculated according to Equation (1):

$$(Ai - Am)/(Ac - Am) \times 100(Ai - Am)/(Ac - Am) \times 100 \quad (1)$$

where Ai is the absorbance of the tested concentration, Am is the absorbance of the medium, and Ac is the absorbance of the control. Based on the cytotoxicity test results, the IC_{50} was calculated using a 4-point lognormal distribution with a Hill coefficient. Experiments were performed in triplicate.

Although the data showed a normal distribution, due to the lack of homogeneity of variance, Welch's ANOVA was used for analysis. Post-hoc analysis was performed using Dunnett's multiple comparison test to identify specific group differences. A significance level of $p < 0.05$ was used in all statistical tests. All statistical analyses of the MTT assay were performed using GraphPad Prism 8.0.2 software (GraphPad Software Inc., San Diego, CA, USA).

Inflammation Model

Cells seeded in 96-well plates were treated with 50 ug/mL LPS for 24 h for the MTT assay or 1 h for the DCF-DA assay. Then, the culture was washed and treated with the compounds with the most favorable COX-2/COX-1 ratio—**BS23** and **BS28** at concentrations of 10, 50, and 100 μM for the next 24 or 1 h. After this time, MTT and DCF-DA assays were performed according to the procedure described in Wiatrak et al. [22]

Gene Expression of TNF-α and IL-6 Detected by Real-Time RT-PCR

The NHDF cells were treated according to the previously outlined procedure. The control and LPS groups were incubated separately in culture dishes containing DMEM, while the test compounds groups were placed in dishes with 10 μM **BS23** and **BS28**, respectively. After a 24 h incubation at 37 °C in a 5% CO_2 atmosphere, the NHDF cells were scraped from the bottom of the microplate using a rubber, and then they were centrifuged at 9000× *g* for 5 min. The pellet was washed with ice-cold PBS twice. RNA was isolated using a Total RNA Mini Plus kit (A&A Biotechnology, Gdańsk, Poland) followed by DNase I (Thermo Fisher Scientific, Waltham, MA, USA) treatment according to the producer's instructions. cDNA synthesis was performed using a Fast Gene Scriptase II cDNA (Nippon Genetics Europe, Duren, Germany). RT-qPCR analyses were performed using the DyNAmo Flash SYBR Green qPCR Kit (Thermo Fisher Scientific, Waltham, MA, USA) and the CFX Connect Real-Time PCR Detection System (Bio-Rad, Hercules, CA, USA) . Relative gene expression was normalized to the reference gene, β-actin, using the $2^{-\Delta\Delta CT}$ method. The primer sequences utilized in this study were as follows: *IL-6*: F, 5′-TGTCCCATGCCACTCAGAGA-3′; R, 5′-AGCAGGTGCTCCGGTTGTAT-3′; *TNF-α*: F, 5′-AGCCCATGTTGTAGCAAACC-3′; R, 5′-TGAGGTACAGGCCCTCTGAT-3′; *β-actin*: F, 5′-CTGTCTGGCGGCACCACCAT-3′; R, 5′-GCAACTAAGTCATAGTCCG-3′. All experiments were performed at least three times.

In Vitro Antioxidant Assays

2,2-Diphenyl-1-Picrylhydrazyl (DPPH) as an Antioxidant Capacity

The DPPH assay was conducted to assess the radical scavenging capabilities of the synthesized compounds. Stock solutions of the compounds were formulated in DMSO and subsequently diluted serially to achieve test concentrations ranging from 500 to 10 μM in ethanol. Each sample's ethanolic DPPH solution was treated with the respective test concentrations and incubated at 37 °C. After a 30 min incubation period, the absorbance was measured at 517 nm, using ethanol as the blank control. Ascorbic acid served as the reference standard, and the IC_{50} values for the synthesized compounds were determined using the following Equation (2):

$$\%\text{DPPH scavenging activity} = \left(\frac{\text{Abs(control)} - \text{Abs (sample)}}{\text{Abs (control)}}\right) \times 100 \quad (2)$$

ABTS Assay

The activity of ABTS (2,2′-azino-bis(3-ethylbenzothiazoline-6-sulphonic acid)) was assessed to evaluate the radical scavenging potential of the synthesized compounds. The ABTS reagent was generated by combining equal volumes of a 7 mM aqueous solution of ABTS and a 2.45 mM potassium persulfate solution. This mixture was then incubated in the dark at room temperature for 16 h. Following incubation, the solution was diluted with ethanol until the absorbance reached approximately 0.7 ± 0.05 at a wavelength of 734 nm. The stock solutions of the synthesized compounds in DMSO were serially diluted to obtain concentrations ranging from 500 μM to 10 μM in ethanol. The ABTS reagent was then treated with various concentrations of each compound and incubated in the dark at room temperature. After 30 min of incubation, the absorbance was measured at 734 nm, using ethanol as the blank control. Ascorbic acid served as the standard for comparison, and the IC_{50} values for the synthesized compounds were calculated using the following Equation (3):

$$\%\text{ABTS scavenging activity} = \left(\frac{\text{Abs(control)} - \text{Abs (sample)}}{\text{Abs (control)}}\right) \times 100 \quad (3)$$

2.3. Differential Scanning Calorimetry (DSC)

Phospholipids: 1,2-dipalmitoyl-n-glycero-3-phosphatidylcholine (DPPC) and 1,2-dimyristoyl-*sn*-glycero-3-phosphatidylcholine (DMPC) were obtained from Merck KGaA (Darmstadt, Germany). Phospholipids were used as delivered without any purification. Tris-EDTA buffer solution was purchased from Merck KGaA (Darmstadt, Germany). Meloxicam was purchased in Alfa Aesar (Karlsruhe, Germany).For each calorimetric sample, 2 mg of phospholipid was dissolved in the appropriate amount of chloroform stock solution (5 mM) of the compound studied. The compound/lipid molar ratios in the samples were 0.06, 0.08, 0.10, and 0.12. Then, the solvent was evaporated by a stream of nitrogen, and the residual solvent was removed under a vacuum for 2 h (Rotary evaporator, Büchy Poland, Warsaw, Poland). Samples were hydrated by 20 µL of Tris–EDTA buffer (pH 7.4). Hydrated mixtures of phospholipid, compounds studied, and buffer closed in Eppendorf tubes were heated (Labnet Dry Bath, Labnet International Inc., Edison, NJ, USA) to a temperature higher by 10 °C than the main phase transition temperature of the phospholipid used and vortexed (neoVortex, neoLab) until homogeneous dispersion was obtained. Samples were sealed in aluminum pans type Concavus® (Netzsch GmbH & Co., Selb, Germany). A pan of the same type, filled with 20 µL of Tris–EDTA buffer (pH 7.4), was employed as a reference. Calorimetric measurements were performed using a differential scanning calorimeter DSC 214 Polyma (Netzsch GmbH & Co., Selb, Germany) equipped with an Intracooler IC70 (Netzsch GmbH & Co., Selb, Germany) in the Laboratory of Elemental Analysis and Structural Research, Faculty of Pharmacy, Wroclaw Medical University. Measurements of the phospholipid's main phase transition were performed using the heat-flow measurement method at a heating rate of 1 °C per minute over a temperature range of 30–50 °C in a nitrogen dynamic atmosphere (25 mL/min). Data were analyzed offline using Netzsch Proteus® 7.1.0 (Netzsch GmbH & Co., Selb, Germany) analysis software. The transition enthalpies were stated in [J/g]. The measured heat was normalized per gram of lipid. The calorimeter was calibrated using standard samples from calibration set 6.239.2–91.3.00 supplied by Netzsch (Netzsch GmbH & Co., Selb, Germany). All samples were weighed on a Sartorius CPA225D-0CE analytical balance (Sartorius AG, Gottingen, Germany) with a resolution of 0.01 mg.

2.4. In Silico Study

2.4.1. The Reactivity of Designed Compounds

The structures of arylpiperazine-1,2-benzothiazine derivatives were optimized at the B3LYP ++6-31G** level of theory in the Gaussian 2016 C.01 software package [23]. The solvent effects were included using the polarization continuum model (PCM) [24]. The reactivity parameters of compounds were calculated by the analogous method. The Avogadro molecular visualization program was used to visualize the calculated energies of frontier orbitals [25].

The reactivity of the molecules is related to the calculated energy values of the highest occupied molecular orbital (HOMO) and lowest unoccupied molecular orbital (LUMO) and their energy gap ($\Delta\varepsilon = \varepsilon\text{LUMO} - \varepsilon\text{HOMO}$) [26,27]. Another parameter, the ionization potential (I) can be expressed as the negative energy of HOMO energy according to Koopman's theory [28]. The electron affinity (A) was calculated as the negative value of LUMO energy. The methodology of calculation of the remaining global reactivity parameters is described in detail in the results section.

2.4.2. Molecular Docking

In the present study, we used the standard docking protocol implemented in the AutoDock4.2 package to predict the binding mode of ligands to both cyclooxygenase isoforms (COX-1 and COX-2) [29]. The molecular targets downloaded from the Protein Data Bank were the structures of COX-1 (PDB ID:4O1Z) and COX-2 (PDB ID:4M11) co-crystalized with meloxicam [30]. The validation was performed by re-docking of reference drug into the protein crystal and comparing its position with the original structures. The root mean square deviation (RMSD) was estimated on the LigRMSD web server to evaluate the accuracy of docking prediction [31]. Due to the RMSD being found to be less than 2 Å, we assumed that the binding mode of compounds was correctly predicted. It should be underlined here that the scoring functions used in the molecular docking give approximate values of free energy of binding, so the computational results were validated by in vitro measurements. The preparation method of molecular target and inhibitor structures was presented in detail in our previous studies preparation [32,33]. According to previous research, the Lamarckian genetic algorithm is the most efficient and reliable approach of AutoDock4.2, so it was used with a total of 500 runs for each binding site [34]. The visualization of obtained results was performed in BIOVIA Discovery Studio visualizer and a Chimera package [35,36].

3. Results

3.1. Chemistry

The first stage of synthesis of final compounds **BS23–BS30** was the condensation of saccharin **1** with 2-bromoacetophenones substituted in the para position with a fluorine **2c** or bromine **2d** atom or a methyl **2b** or methoxy group **2a** (Scheme 1). The next step was the *Gabriel–Colman* rearrangement reaction consisting of opening the 1,2-thiazole ring under the influence of sodium ethoxide at a temperature of 40–55 °C. In the second stage of this reaction, excess sodium ethoxide leads to ring closure by incorporating a carbon atom and forming a six-membered 1,2-thiazine ring. Subsequently, alkyl (**5** and **7**) and acyl (**6**) derivatives of arylpiperazine were obtained. Compounds **5** and **7** were obtained by alkylation of 1-(2-pyrido)piperazine or 1-(2-pyrimido)piperazine with 1-chloro-3-bromopropane, whereas compound **6** was obtained by reaction of 1-phenylpiperazine with chloroacetyl chloride. The last step towards obtaining the planned compounds **BS23–BS30** was the condensation of compounds **4a–d** with **5**, **6**, or **7** in anhydrous ethanol with the addition of sodium ethoxide (2.3%) at the boiling point of ethanol. The progress of the reaction was monitored by thin-layer chromatography (TLC). The obtained products were crystallized from ethanol and then submitted to spectroscopic analyses to confirm their structure and purity.

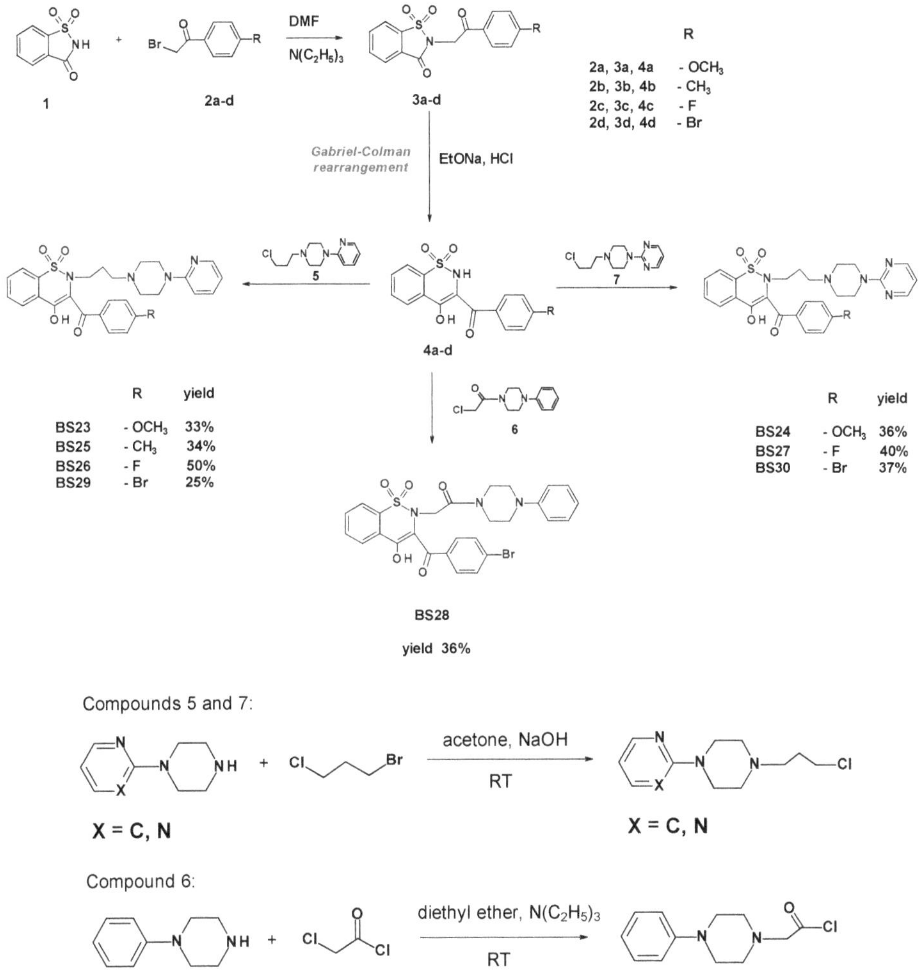

Scheme 1. Synthesis of new 1,2-benzothiazine derivatives **BS23–BS30**.

3.2. Evaluation of COX-1/COX-2 Selectivity and Potency

3.2.1. In Vitro Cyclooxygenase Inhibition Assay

Cyclooxygenase (COX) peroxidase activity was evaluated using a colorimetric assay. The impact of meloxicam and the analyzed compounds on the activities of both cyclooxygenase isoforms (COX-1 and COX-2) was measured. The incubation period was maintained at 2 min, following the manufacturer's guidelines. Considering the well-established side effects of nonsteroidal anti-inflammatory drugs (NSAIDs), such as aspirin, naproxen, and ketoprofen, which typically inhibit COX-1, our goal was to compare the effects of newly synthesized compounds with meloxicam. Meloxicam inhibits both isoforms but has a higher affinity for COX-2. We determined the concentrations of all compounds required for 50% inhibition of COX-1 and COX-2. IC_{50} values were then calculated, and the ratio of IC_{50} values for both cyclooxygenase isoforms allowed us to assess the selectivity of the tested structures for COX-1 and COX-2. The full set of IC_{50} values for both enzymes and the selectivity coefficient are shown in Table 1.

Table 1. IC_{50} values (mean ± SD) calculated for COX-1 and COX-2 enzymes after incubation for 2 min with the tested compounds, COX selectivity ratio, and IC_{50} values (mean ± SD) calculated for NHDF cell cultures after 24 h of incubation with the tested compounds.

Compound	Cyclooxygenase Inhibition Assay				Selectivity Ratio: (COX-2/COX-1)	MTT Assay IC_{50} (μM)]
	COX-1		COX-2			
	IC_{50} [μM]	SD	IC_{50} [μM]	SD		
BS23	241.64 ± 4.2		13.19 ± 2.1		0.05	228.40 ± 28.1
BS24	126.54 ± 3.6		26.34 ± 3.4		0.21	163.46 ± 15.7
BS25	130.96 ± 7.5		62.63 ± 1.7		0.48	200.10 ± 20.9
BS26	128.73 ± 9.1		18.20 ± 4.3		0.14	203.28 ± 22.9
BS27	116.89 ± 4.3		25.55 ± 2.0		0.22	190.60 ± 12.5
BS28	95.88 ± 5.7		12.46 ± 1.9		0.13	179.74 ± 13.6
BS29	124.81 ± 6.7		17.80 ± 2.1		0.14	232.51 ± 11.8
BS30	121.51 ± 3.4		59.22 ± 3.7		0.49	168.50 ± 30.4
meloxicam	267.71 ± 8.1		112.67 ± 3.3		0.42	174.23 ± 20.3

In the cyclooxygenase inhibition assay, **BS** compounds exhibited varying degrees of efficacy against COX-1 and COX-2 isoforms, with selectivity ratios (COX-2/COX-1) indicating a relative preference for COX-2 over COX-1.

3.2.2. Cytotoxicity Assay

Meloxicam demonstrated an IC_{50} value of 267.71 ± 8.1 μM for COX-1 and 112.67 ± 3.3 μM for COX-2, resulting in a selectivity ratio of 0.42. This indicates that meloxicam has a moderate preference for inhibiting COX-2 over COX-1. The IC_{50} value for meloxicam in the MTT assay was 174.23 ± 20.3 μM.

Among the **BS** compounds, **BS23** exhibited an IC_{50} value of 241.64 ± 4.2 μM for COX-1 and 13.19 ± 2.1 μM for COX-2, with a selectivity ratio of 0.05. This compound showed a significantly higher selectivity for COX-2 compared to meloxicam. Additionally, the IC_{50} value in the MTT assay was 228.40 ± 28.1 μM, indicating lower cytotoxicity compared to meloxicam. **BS24** had an IC_{50} value of 126.54 ± 3.6 μM for COX-1 and 26.34 ± 3.4 μM for COX-2, giving a selectivity ratio of 0.21. This selectivity is still more pronounced for COX-2 than for meloxicam, with an IC_{50} value in the MTT assay of 163.46 ± 15.7 μM, indicating slightly higher cytotoxicity. **BS25** had IC_{50} values of 130.96 ± 7.5 μM for COX-1 and 62.63 ± 1.7 μM for COX-2, with a selectivity ratio of 0.56. This compound showed a lower selectivity for COX-2 compared to meloxicam, with an IC_{50} value in the MTT assay of 200.10 ± 20.9 μM. **BS26** showed IC_{50} values of 128.73 ± 9.1 μM for COX-1 and 18.20 ± 4.3 μM for COX-2, with a selectivity ratio of 0.14, indicating greater selectivity for COX-2 than meloxicam. Additionally, this compound showed lower cytotoxicity compared to meloxicam, with an IC_{50} value of 203.28 ± 22.9 μM. **BS27** had IC_{50} values of 116.89 ± 4.3 μM for COX-1 and 25.55 ± 2.0 μM for COX-2, with a selectivity ratio of 0.22, also indicating greater selectivity for COX-2 than meloxicam. Its MTT assay IC_{50} was 190.60 ± 12.5 μM. **BS28** showed IC_{50} values of 95.88 ± 5.7 μM for COX-1 and 12.46 ± 1.9 μM for COX-2, with a selectivity ratio of 0.13, indicating a strong preference for COX-2 over COX-1 and better selectivity than meloxicam. Its MTT assay IC_{50} was 179.74 ± 13.6 μM. **BS29** presented IC_{50} values of 124.81 ± 6.7 μM for COX-1 and 17.80 ± 2.1 μM for COX-2, with a selectivity ratio of 0.14, indicating higher selectivity for COX-2 compared to meloxicam. Its MTT assay IC_{50} was 232.51 ± 11.8 μM. **BS30** had IC_{50} values of 121.51 ± 3.4 μM for COX-1 and 59.22 ± 3.7 μM for COX-2, with a selectivity ratio of 0.49, which is comparable to meloxicam. Its MTT assay IC_{50} was 168.50 ± 30.4 μM, indicating slightly higher cytotoxicity.

In summary, many **BS** compounds showed higher selectivity for COX-2 over COX-1 compared to meloxicam, with **BS23, BS26, BS28**, and **BS29** demonstrating particularly strong selectivity for COX-2. This suggests that these compounds may have potential advantages as anti-inflammatory agents with fewer side effects associated with COX-1 inhibition.

The simplification by removing the nitrogen atom from the aromatic rings appears to be beneficial. These modifications significantly increase the selectivity for COX-2. For example, **BS23**, in which piperazine has a pyridine ring instead of pyrimidine, shows a strong preference for COX-2 with a selectivity coefficient of 0.05 compared to 0.42 for meloxicam. This suggests that removing the nitrogen atom may reduce interactions with COX-1, thereby increasing the selectivity for COX-2.

Additionally, introducing halogen atoms (e.g., fluorine and bromine) into the structure may affect the binding affinity for COX enzymes. For instance, **BS26**, **BS28**, and **BS29**, which contain a fluorine or bromine atom, show selectivity coefficients of 0.14, 0.13, and 0.14, respectively, indicating an increased selectivity for COX-2.

The structural modifications in **BS** compounds highlight the delicate balance between achieving high COX-2 selectivity and maintaining low cytotoxicity. Removing specific atoms, simplifying ring structures, and introducing or removing functional groups can significantly impact the binding affinity and selectivity toward COX enzymes.

3.3. The Interactions with Phospholipid Bilayers

Cyclooxygenase is a membrane protein, associated with the phospholipid bilayer surrounding the endoplasmic reticulum, as well as the cell nucleus, which is why it is so important to investigate the interaction of new compounds with model membranes.

In the present work, we have used differential scanning calorimetry (DSC) to determine the interactions of studied compounds with phospholipid bilayers as artificial models of biological membranes.

The impact of two studied compounds on the lipid thermal behavior is presented in Figure 3, showing the example thermograms of DPPC mixed with **BS23** and **BS24** at different molar ratios. The addition of the compounds caused the disappearance of the DPPC pretransition and concentration-dependent shift of the main transition temperature towards lower values, accompanied by a decrease in the transition peaks area and the broadening of the peaks. Despite the slight difference in the chemical structure of both compounds (pyrimidine instead of pyridine), the effects are more pronounced for **BS24** (Figure 3B) than for **BS23** (Figure 3A).

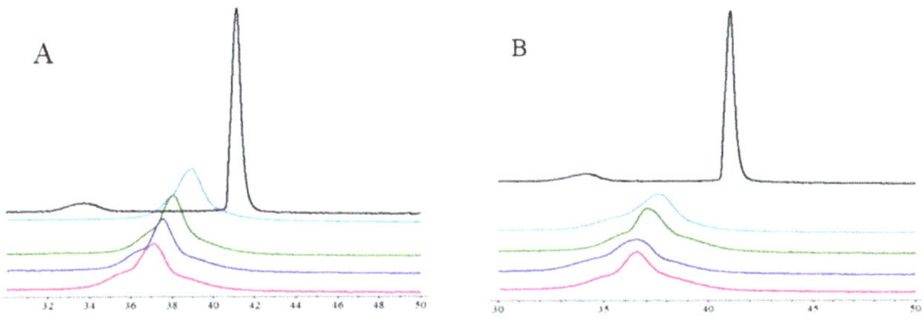

Figure 3. The exemplary thermograms obtained for DPPC mixed with compounds: (**A**) **BS23** and (**B**) **BS24** as well as for pure lipid (the first curve from the top—black color): DPPC molar ratios from the top to the bottom: 0 (pure lipid), 0.06, 0.08, 0.10, and 0.12; the direction of the endothermic reaction is upwards.

All examined compounds decreased the main transition temperature of DPPC in a concentration-dependent manner (Figure 4A). The addition of studied compounds to DPPC also resulted in the broadening of the transition peaks. Moreover, all examined compounds decreased the enthalpy of the DPPC main phase transition. In the case of all DPPC gel–liquid crystalline phase transition parameters, the most pronounced effects were found for compounds **BS24**, **BS23**, and **BS30**.

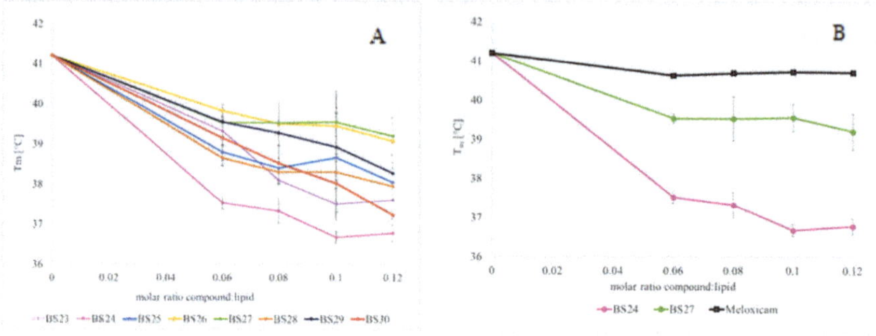

Figure 4. Influence of all studied compounds on the DPPC main transition temperature (**A**); influence of two studied compounds (**BS24** and **BS27**) and meloxicam on the DPPC main transition temperature (**B**).

Compounds **BS26** and **BS27** had the smallest impact on the transition temperature. The **BS27** compound with the highest molar ratio (0.12) lowered the temperature of the main phase transition to 39.2 °C, which is a reduction in the transition temperature of the DPPC phospholipid by only 2 °C. The **BS27** compound at the same concentration reduced the temperature to 39.075 °C, i.e., by 2.13 °C compared to the pure DPPC phospholipid. The greatest impact on the main transition temperature change was observed for the compounds **BS23, BS24**, and **BS30**. For example, for the **BS24** compound in the highest test compound: the lipid ratio, the change in the temperature of the main phase transformation in relation to the pure DPPC phospholipid, was 4.425 °C. All tested compounds had a more pronounced effect on model phospholipid membranes than meloxicam (Figure 4).

Figure 5 shows the influence of studied compounds on the DPPC peak half-width of the DPPC main phase transition. Each compound tested caused an increase in the half-width of the main phase transition peak in relation to the pure DPPC. The following compounds had the smallest impact on this parameter: **BS25, BS26**, and **BS27**. For example, **BS25** and **BS27** at the highest molar ratios broadened the main phase transition peak by only 1.25. The greatest increase in the half-width of the DPPC main phase transition was caused by **BS24** and **BS30**. (The changes in the DPPC peak half-width in relation to the pure phospholipid for these derivatives were 2 and 2.125).

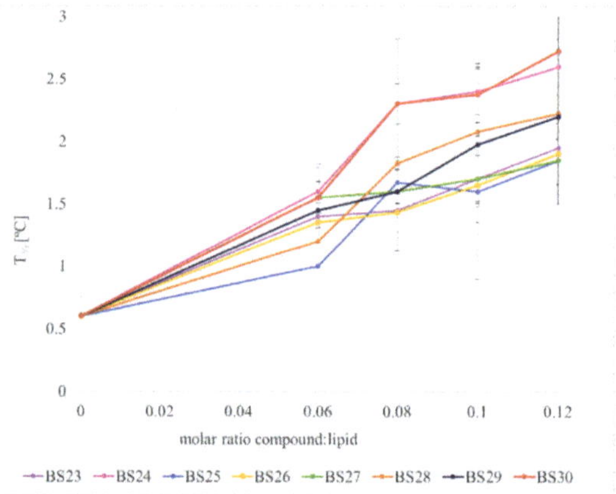

Figure 5. Influence of studied compounds on the DPPC peak half-width.

All examined compounds decreased the main transition temperature of DPPC and DMPC, so they perturbed the phospholipid multi-bilayer structure. The stronger change in this parameter was observed in the presence of DMPC (Figure 6 open symbols) in comparison to DPPC (Figure 6 full symbols). The greatest reduction in Tm was demonstrated by the following compounds: **BS23**, **BS25**, and **BS30** in comparison to DMPC. The smallest effect on the main transition temperature was observed for compounds **BS26** and **BS27** for DPPC. DMPC is a phospholipid with a shorter acyl chain than DPPC. The weaker interactions between hydrocarbon chains and the more loosely packed structure of DMPC in comparison to DPPC, could provoke easier compound incorporation into bilayers formed from DMPC.

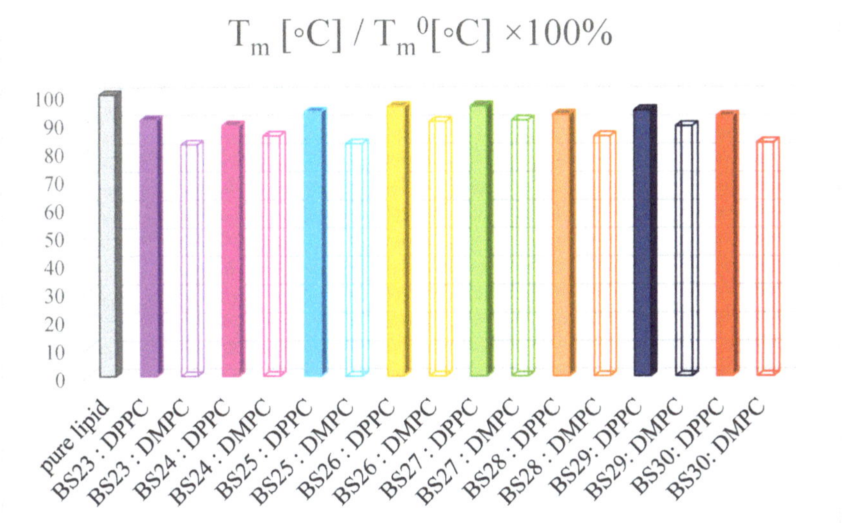

Figure 6. Relative main transition temperature (T_m/T_m^0) for DPPC and DMPC in the presence of studied compounds at compound: phospholipid molar ratio 1:10; T_m^0—the main transition temperature for pure lipid; T_m—the main transition temperature in the presence of studied compounds at compound: phospholipid molar ratio 1:10; full symbols—DPPC, open symbols—DMPC.

3.4. *Anti-Inflammatory Properties of Designed Compounds*

3.4.1. Inflammation Assay

NHDF cell cultures were treated with LPS at a concentration of 50 µg/mL to induce pro-inflammatory cytokines such as IL-1β, TNF-α, and IL-6 through the activation of the TLR4 pathway [37]. Subsequently, the cells were treated with selected compounds **BS23** and **BS28** due to their favorable COX2/COX1 ratio. The results are presented in Figure 7A. Within the tested concentration range, the cell viability after treatment with **BS23** was around 70% of the value for untreated (control) cells. For **BS28** at the same concentrations, the absorbance values ranged between 0.64 and 0.68. Additionally, the absorbance value for LPS was 0.56. Compared to the LPS control, both compound samples showed higher cell viabilities, suggesting that these compounds may reduce LPS-induced cytotoxicity. The **BS23** compound exhibited a slightly stronger protective effect than **BS28**.

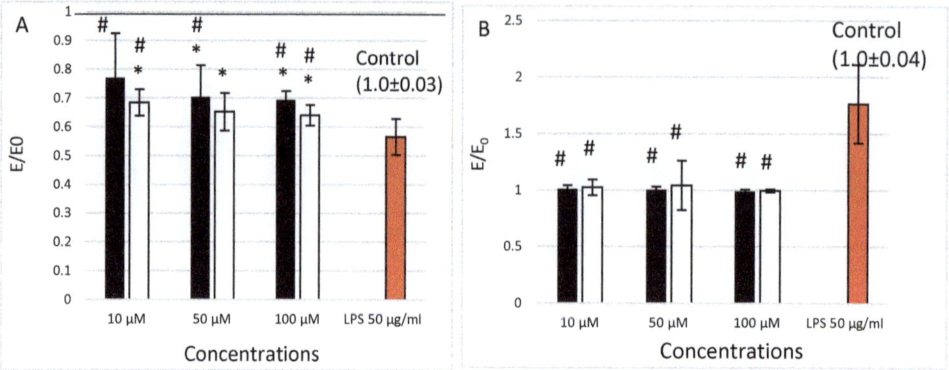

Figure 7. Anti-inflammatory effect after incubation of NHDF cells with lipopolysaccharide (LPS) and then the tested compounds. The black column presents the **BS23** compound and the white column the **BS28** compound. (**A**) MTT assay; (**B**) DCF-DA assay. Data presented as a mean and SD; * $p < 0.05$—significant difference compared to the control (cells with only medium); # $p < 0.05$ significant difference compared to the positive control (cells treated with only LPS).

The results presented in Figure 7B pertain to the ROS (reactive oxygen species) test using DCF-DA, conducted on NHDF cells incubated with LPS at a concentration of 50 μg/mL for 1 h, followed by treatment with compounds **BS23** and **BS28** for an additional hour. For both tested compounds, **BS23** and **BS28**, within the examined concentration range, the ROS values were similar to the ROS levels in the control culture (cells incubated in medium only). These values were significantly lower than those for the LPS control, where the ROS level was nearly twice as high compared to the control (medium culture), with a value of 1.76.

3.4.2. Effect of New Meloxicam Derivatives on the Cytokine mRNAs Level in Inflamed Cells

After a 24 h incubation period of inflamed normal human dermal fibroblast (NHDF) cells with meloxicam derivatives, specifically **BS23** and **BS28**, the expression of pro-inflammatory cytokines TNF-α and IL-6 was assessed through real-time RT-PCR analysis (Figure 8). Compared to the control group, the lipopolysaccharide (LPS)-treated cells exhibited a significant upregulation of TNF-α (Figure 8 left panel) and IL-6 (Figure 8 right panel), demonstrating that LPS stimulation effectively induces the expression of genes associated with inflammatory responses. Conversely, treatment with meloxicam derivatives (**BS23** and **BS28**) markedly reduced the expression levels of TNF-α and IL-6 in comparison to the LPS group. These findings suggest that **BS23** and **BS28** may suppress inflammation by inhibiting the gene expression of key pro-inflammatory cytokines in inflamed NHDF cells.

The observed anti-inflammatory effects of **BS23** and **BS28** align with previous research on nonsteroidal anti-inflammatory drug (NSAID) derivatives, which have been shown to mitigate the inflammatory response by targeting pathways that regulate cytokine production. Studies suggest that TNF-α and IL-6 play pivotal roles in mediating chronic inflammation and are often used as biomarkers to assess the efficacy of anti-inflammatory treatments [38]. The significant downregulation of these cytokines following treatment with meloxicam derivatives highlights their potential therapeutic value. Similar outcomes have been reported with other NSAID analogs, which also demonstrate the capacity to inhibit pro-inflammatory gene expression, thereby reducing the overall inflammatory response [39]. Further exploration into the molecular mechanisms underlying the action of **BS23** and **BS28** could provide valuable insights into their clinical applications.

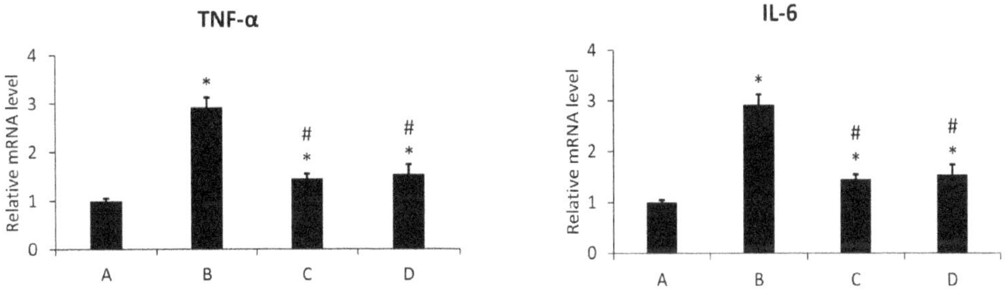

Figure 8. Relative expression of TNF-α (**left panel**) and IL-6 (**right panel**) in NHDF cells. (A) Control—only NHDF cells, (B) NHDF cells incubated with lipopolysaccharide (LPS), (C) NHDF cells incubated with lipopolysaccharide (LPS) treated with 10 μM of **BS23,** and (D) NHDF cells incubated with lipopolysaccharide (LPS) treated with 10 μM of **BS28**. The values represent mean ± SD from three experiments. * $p < 0.05$—significant difference compared to the control (cells with only medium); # $p < 0.05$ significant difference compared to the positive control (cells treated with only LPS).

3.4.3. Antioxidant Assays

The antioxidant properties of meloxicam and its synthesized derivatives (**BS23–BS30**) were evaluated using two different in vitro assays, reflecting the common practice of applying methods to assess the antioxidant capacity of a sample [40]. The + radical scavenging of these compounds was summarized using IC_{50} values (Table 2).

Table 2. IC_{50} values of test compounds in relation to ABTS and DPPH inhibition. The results are presented as mean values ± SD. n = 3.

Compound	ABTS Assay IC_{50} [μM]	DPPH Assay IC_{50} [μM]
BS23	54.21 ± 2.2	>500
BS24	63.22 ± 2.8	>500
BS25	75.26 ± 1.4	>500
BS26	83.72 ± 3.2	>500
BS27	83.34 ± 2.5	>500
BS28	88.44 ± 1.9	>500
BS29	53.74 ± 1.7	>500
BS30	88.28 ± 1.6	>500
meloxicam	89.45 ± 1.5	>500
ascorbic acid	124.27 ± 2.3	218.31 ± 1.8

Commonly employed assays like DPPH and ABTS are widely recognized for assessing the radical scavenging potential of compounds by measuring their capacity to neutralize free radicals or donate hydrogen atoms, thereby determining antioxidant activity [41]. The ABTS assay is considered superior to DPPH due to its solubility in both water and organic solvents and its quicker reaction rate with the test compounds [42]. The interaction of the tested compounds with the stable DPPH radical demonstrated their free radical scavenging ability, with this interaction being concentration dependent. However, the IC_{50} values for meloxicam and its derivatives in the DPPH assay exceeded 500 μM. In contrast, in the ABTS radical scavenging assay, the meloxicam derivatives displayed a notably higher scavenging potential in a concentration-dependent manner compared to both meloxicam and ascorbic acid (standard antioxidant). While meloxicam is known to possess inherent antioxidant properties and has been reported to reduce reactive oxidant formation [43], its derivatives like **BS23** and **BS28** significantly improved its ABTS radical scavenging capacity.

3.5. The Reactivity of Designed Compounds

It is well known that frontier molecular orbital energies can be used to determine various reactivity parameters of compounds. The ability of molecules to donate electrons is indicated by the high value of HOMO energy. The compounds with a low value of LUMO energy are capable of accepting electrons. Therefore, the visualization of these orbital features enables the identification of molecular regions responsible for interactions in the molecular target pocket. As presented in Table 3 and in the Supplementary Materials, the regions of high electron density in the case of pyrimidinylpiperazine derivatives (**BS24**, **BS30**) are localized on 1,2-benzotiazine fragment of molecules. Such a distribution of electron density may indicate the region of compounds capable of scavenging free radicals as well as the area participating in charge transfer, hydrophobic, and π-stacking interactions. In the cases of compounds **BS23**, **BS25**, **BS26**, **BS28**, and **BS29**, the region of high electron density is localized on phenylpiperazine and pyridinylpiperazine fragments. In all considered compounds, LUMO orbitals existed over a phenyl ring with various alkyl substituents.

Table 3. The distribution map of HOMO/LUMO densities and the HOMO-LUMO gap ($\Delta\varepsilon$) of compounds **BS23**, **BS24**, and **BS28**. The negative electron density is presented in blue color while the red color indicates the positive electron density.

	HOMO	Δε	LUMO
BS23	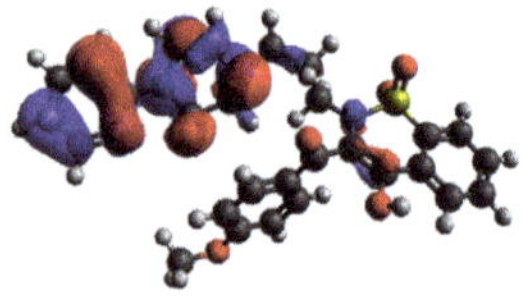	3.523 eV →	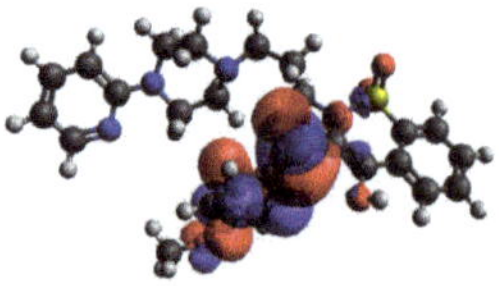
BS24	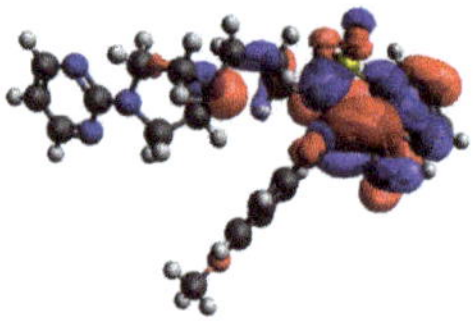	3.600 eV →	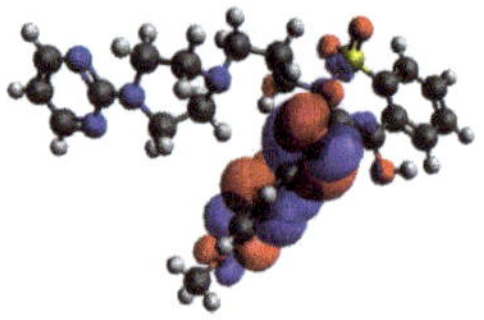
BS28	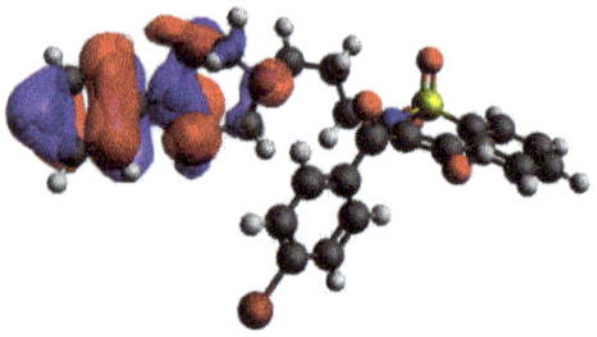	2.997 eV →	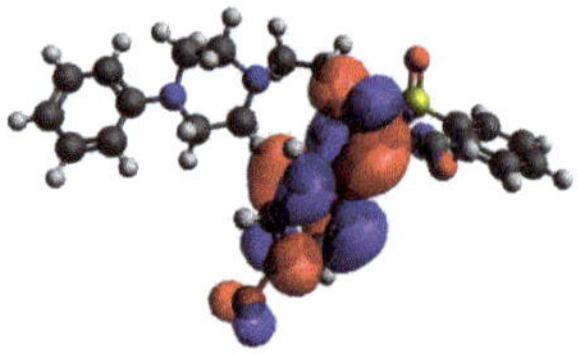

It should be also mentioned that the HOMO-LUMO energy gap determines the stability and reactivity of the molecules. Namely, the larger energy gap indicates the lower reactivity and the greater stability of molecular systems. According to the data presented in Table 3, compound **BS24** has the largest value of $\Delta\varepsilon$ (3.600 eV), which suggests its lower reactivity and greater stability. Compound **BS29** is characterized by the lower value of the gap. It indicates higher reactivity of the system. The energy of HOMO orbitals of the molecules corresponds with their ionization potential (I). Due to the lowest value of the I compound, **BS28** shows better donor electron properties than other derivatives. On the other hand, the largest values of electron affinity of **BS29** and **BS30** suggest their ability to accept electrons.

$$A = -E_{LUMO};\ I = -E_{HOMO};\ \eta = \frac{I - A}{2};\ s = \frac{1}{2\eta};\ \mu = \frac{-(I + A)}{2};\ \chi = \frac{(I + A)}{2};\ \omega = \frac{\mu^2}{2\eta}$$

The next parameter, the hardness of molecules (η), can be calculated following Expression (4):

$$\eta = \frac{I - A}{2} \tag{4}$$

based on Janak's theorem [44]. The second related parameter, namely softness (s), is calculated from Equation (5).

$$s = \frac{1}{2\eta} \tag{5}$$

Both could be useful in determining the intermolecular charge transfer phenomenon of molecular systems. It is known that the hardness increases while the polarizability of molecules decreases. Additionally, η and s determine the stability and reactivity of compounds.

Compound **BS24** is characterized by the highest value of hardness (4.726), and compound **BS28** is the softest molecule (0.116).

Another parameter based on the thermodynamic properties of the molecules is the global electrophilicity index (ω) expressed by Equation (6) [45].

$$\omega = \frac{\mu^2}{2\eta} \tag{6}$$

The electrophilicity index is defined as the change of energy due to the flow of electrons between the donor and acceptor groups of molecules. Studies have shown an existing correlation between ω and the biological activity of compounds [46]. It is clear from Table 4 that compounds **BS28, BS29,** and **BS30** are characterized by the highest values of electrophilicity indexes and can therefore be the most biologically active.

Table 4. Energy of HOMO/LUMO orbitals and other reactivity parameters [eV].

	HOMO	LUMO	GAP	I	A	η	s	μ	χ	ω
BS23	−5.764	−2.241	3.523	5.764	2.241	4.643	0.107	−3.993	3.993	1.717
BS24	−5.853	−2.253	3.600	5.853	2.253	4.726	0.105	−4.053	4.053	1.737
BS25	−5.762	−2.370	3.392	5.762	2.370	4.570	0.109	−4.066	4.066	1.806
BS26	−5.769	−2.437	3.332	5.769	2.437	4.550	0.109	−4.103	4.103	1.849
BS27	−5.886	−2.449	3.437	5.886	2.449	4.661	0.107	−4.167	4.167	1.862
BS28	−5.576	−2.579	2.997	5.576	2.579	4.286	0.116	−4.077	4.077	1.939
BS29	−5.764	−2.591	3.173	5.764	2.591	4.468	0.111	−4.177	4.177	1.952
BS30	−5.908	−2.591	3.317	5.908	2.591	4.612	0.108	−4.249	4.249	1.957

Very important parameters are the chemical potential of the molecule (μ) and its electronegativity (χ). The first one describes the charge transfer within a system in the ground state and can be expressed as follows (see Equation (7)).

$$\mu = \frac{-(I + A)}{2} \tag{7}$$

Compound **BS30** has the greatest value of chemical potential, so it is the most reactive of all derivatives. Electronegativity measures the attraction of an atom for electrons in a covalent bond numerically; it can be expressed with Equation (8).

$$\chi = \frac{(I + A)}{2} \tag{8}$$

According to the calculations, compound **BS30** had a higher value χ and better electron-attracting capability than others.

3.6. Molecular Docking Studies

According to the scoring function including free enthalpy of binding, all designed compounds show a slightly stronger binding affinity for COX-2 (see Table in the Supplementary Materials). The data are in agreement with this experimental study. The value of $\Delta G_{binding}$, which can characterize the ability of compounds to bind to the active site of protein, varies from −15.1 kcal/mol to −15.6 kcal/mol (Table 5).

Table 5. Intermolecular interactions in the active center of COX-2.

	$\Delta G_{binding}$ [kcal/mol]	Van Der Waals	Hydrogen Bonds	π-Type Interactions
BS23	−15.6	Leu93, Met113, Leu117, Ser119, Ser353, Tyr355, Leu359, Phe381, Tyr385, Trp387, Leu534	Ser530	Val116, Arg120, Ile345, Leu384, Gly526, Leu531, Met535
BS24	−15.3	Leu93, Met113, Leu117, Ser119, Phe205, Phe209, Tyr348, Leu359, Phe381, Trp387, Val523, Gly526	Ala527, Ser530, Leu531	Val116, Ile345, Val349, Leu352, Tyr366, Leu534, Met535
BS25	−15.2	Leu93, Met113, Ser119, Ser353, Tyr355, Leu359, Phe381, Leu384, Trp387, Phe518, Gly526, Ser530, Leu534	Ala527	Val116, Arg120, Ile345, Leu352, Val523, Met535
BS26	−15.1	Leu117, Ser353, Phe357, Phe381, Leu384, Tro387, Gly526, Phe518, Ser530, Leu531, Leu534,	Arg120, Tyr355	Val116, Leu359, Val349, Leu352, Met522, Ala527
BS27	−15.4	Met113, Leu117, Ser353, Phe381, Tyr385, Trp387	Arg120, Ser530	Val116, Leu351, Leu384, Leu534, Met535, Met522, Gly526
BS28	−15.4	Val89, Leu93, Met113, Leu117, Ser119, Ser353, Leu534, Tyr355, Leu359, Phe381, Tyr385, Trp387, Phe518, Ser530	Ala527	Val116, Ile345, Gly526, Leu531
BS29	−15.5	Leu93, Met113, Leu117, Ser119, Ser353, Leu534, Tyr355, Leu359, Phe381, Trp387, Phe518, Ser530	Ala527	Val116, Ile345, Leu352, Leu531
BS30	−15.6	Leu117, Ser353, Phe381, Leu384, Tyr385	Arg120, Ser530	Met113, Val116, Ile345, Trp387, Met522, Gly526, Leu531, Met535, Leu543

As presented, the potency of binding to the COX-2 active site in all considered cases is almost the same. The structural modifications of compounds influence insignificantly their binding mode. All designed inhibitors can bind to the specific binding pocket arising in the structure of COX-2 due to the presence of Val523 instead of Ile523 and the change of Tyr355 conformation in the structure of COX-2 [47]. The specific cavity is created by Leu352, Ser353, Tyr355, Phe518, and Val523. As presented in Figure 9, compound **BS23** can form one hydrogen bond with Ser530 as a classic anti-inflammatory agent. The pyridinylpiperazine moiety binds to COX-2 through π-type and van der Waals interactions with Leu384, Met522, Val349, Val523, Gly526, and Ala527 and Phe381, Tyr385, and Trp387, respectively. As determined, the benzothiazine scaffold is responsible mainly for van der Waals interactions with Met113, Leu117, Ser353, and π-type forces with Ile345, Leu531, and Met535 amino acid residues. A similar mode of binding exhibits remaining derivatives (see Supplementary Materials).

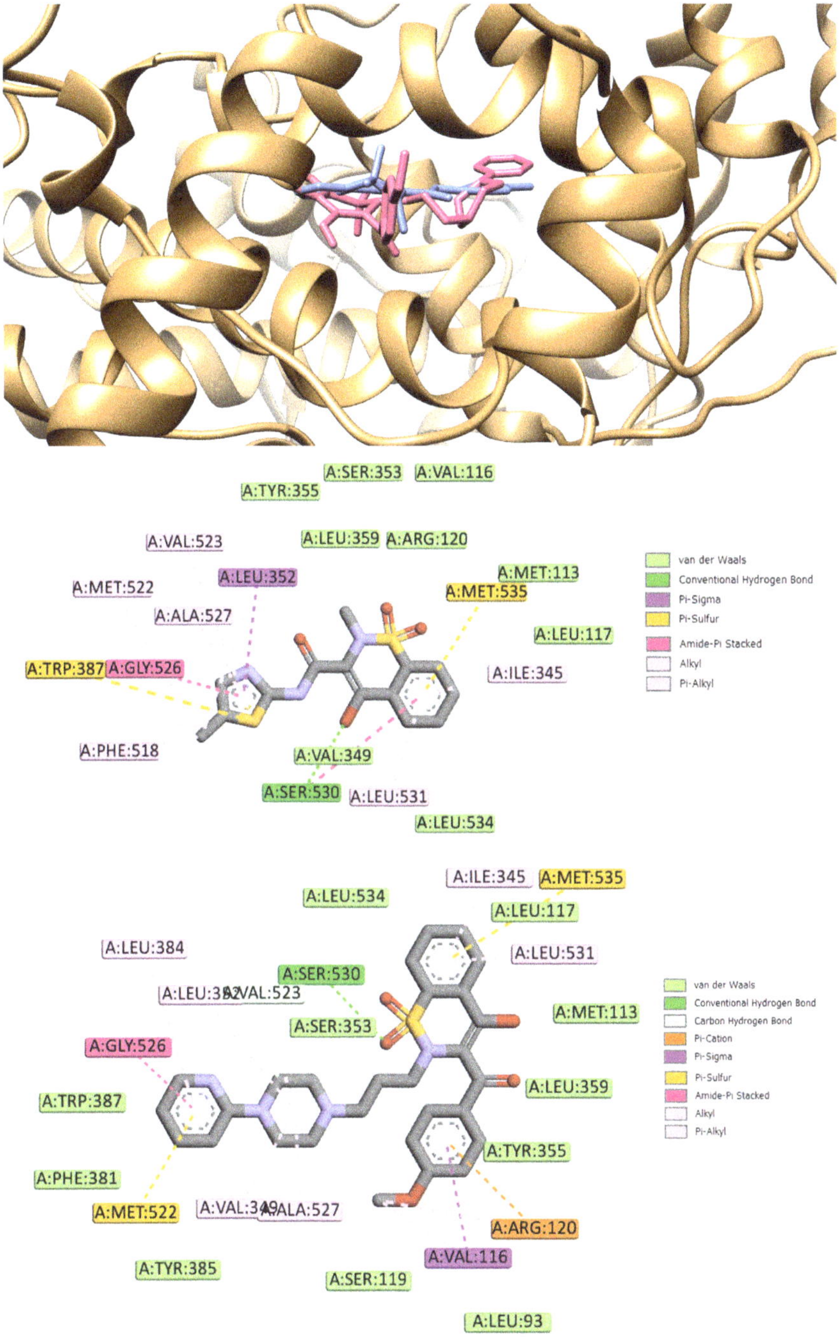

Figure 9. The binding mode of meloxicam (blue and top) and compound **BS23** (pink and bottom) in the active center of COX-2.

The data obtained for the COX-1 in complex with designed inhibitors are presented in detail in the Supplementary Materials.

4. Discussion

Our study presents a comprehensive analysis of synthesized 1,2-benzothiazine derivatives in terms of their potential for anti-inflammatory and antioxidant activity. The results revealed the selective activity of compounds towards COX-2, with some compounds showing significant COX-2 preference compared to meloxicam. The selectivity is advantageous for reducing side effects, aligning with literature on selective COX-2 inhibitors, which suggests their reduced risk of gastrointestinal adverse effects common in COX-1 inhibitors [48–50]. As presented in Table 1, the most selective compound is **BS23**; however, compound **BS24**, **BS25**, **BS27**, **BS28**, and **BS29** show a better selectivity profile than meloxicam. Only two of the proposed derivatives have selectivity close to that of the reference drug (selectivity ratio: (COX-2/COX-1) about 0.40). On the other hand, only compounds **BS24** and **BS30** show greater toxicity than meloxicam. Many studies discussed the structural factors determining for COX-2 selectivity of NSAID derivatives, reinforcing the current study's approach. The binding site of COX-2 is predominantly hydrophobic, facilitating interactions with lipophilic substrates and inhibitors. The presence of polar amino acid residues such as Arg120 and Ser530 allows hydrogen bonding, which is critical for selectivity. In addition, the side pocket of COX-2 is more flexible than that of COX-1, allowing the accommodation of bulkier substituents in selective inhibitors. According to molecular docking proposed compounds similar to classic nonsteroidal anti-inflammatory drugs (diclofenac) can form hydrogen bonds with Arg120, Tyr355, and Ser530 [51]. As presented in Table 5, compounds **BS23** and **BS30** form two hydrogen bonds with arginine and serine amino acid residues of COX-2. On the other hand, similar to meloxicam, all compounds occupy the hydrophobic pocket and are surrounded by Ser353, Leu384, Tyr385, Trp387, Val523, and Met522, and the differences in free energy of binding are negligible [47].

The selectivity of inhibitors with various structural modifications was analyzed by Gautam et al. [47]. They proved that specific substitutions, such as halogens and nitrogen-based groups, enhance COX-2 binding affinity while reducing interactions with COX-1. There are also some studies on benzo[d]thiazol derivatives that indicate a significant role of halogens as electron-withdrawing substituents in determining anti-inflammatory properties providing the best orientation within the active site of COX-2 [52]. This structural approach is reflected in the results of our study, particularly with compounds like **BS26**, **BS28**, and **BS29**, where halogen groups improve COX-2 selectivity. Another study on oxicam analogs revealed that slight changes in molecular structure can significantly impact COX binding [30]. As can be seen from Table 1, removing nitrogen atoms from aromatic substituent appears to be important for selectivity (compound **BS28**). The replacement of the methoxy group with a methyl substituent in the **BS24** structure results in reduced selectivity. As we know, the methoxy group is a strong electron donor substituent. When attached to an aromatic ring, it increases the electron density of the ring, which can improve π–π stacking interactions with the hydrophobic regions of the COX-2 active site. As concluded in the earlier studies on pyrazole analogs with methoxy substituents, these groups contribute additionally hydrogen bonding with residues like Lys68, Tyr108, and Arg106, increasing binding energy and improving COX-2 inhibition compared to the selective inhibitor celecoxib. Figure 9 and Figure S18 clearly show that compounds with the methoxybenzene moiety (**BS23** and **BS24**) promote π-type interactions with amino acids of the protein. As presented, the methoxybenzene ring of compound **BS23** can interact with Val116 and Arg120 via π–σ and π–cation interactions, respectively. In the case of compound **BS24**, π–alkyl and π–π T shape interactions with Val116 and Tyr355 localized specifically for the COX-2 cavity are possible.

Drugs that are COX inhibitors should also interact with cell membranes. The natural substrate for COX is arachidonic acid, which occurs among the phospholipids of the cell membrane. COX is a membrane-bound protein of the endoplasmic reticulum. Potential COX inhibitors should have some ability to interact with the cell membrane to enter the COX channel containing its active site. A potential NSAID candidate must therefore demonstrate an appropriate level of lipophilicity and interaction with the cell membrane. Additionally, interaction with the model membrane will therefore be more informative about whether the compound can penetrate membranes, be absorbed from the gastrointestinal tract, and have an affinity for the membrane. The results of the DSC measurements show that the studied compounds can interact with the phospholipid phase of the membrane. In the case of all phospholipid gel–liquid crystalline phase transition parameters, the most pronounced effects were found for compounds **BS23**, **BS24**, and **BS30**. The vanishing of phospholipid pretransition, broadening of transition peaks, and decrease in the main transition temperature of the phospholipid may suggest that the studied compounds interact with the phospholipid head groups, changing lipid packing, probably by increasing the spacing between lipid heads, which may result in disruption in the rippled phase of the DPPC. Zambrano et al. published a revision of using the differential scanning calorimetry method in investigating drug–membrane interactions. They concluded that the DSC method enables understanding the impact of the studied compounds on lipid packing and stability of the phospholipid bilayer [52]. There are some data concerning selected NSAIDs analyzed by Lichtenberger et al. using a model membrane made of DPPC. The obtained results indicated that these drugs interact with non-charged phospholipids contained in the mucosa of the stomach and intestines by forming complexes with them. This interaction may promote the gastrotoxic effect of those medicines [53]. Lucio et al., on the other hand, showed that acemetacin, indomethacin, and nimesulide influenced the phase transition temperature of the model membrane. It was concluded that this effect may have a negative impact on the gastrointestinal mucosa, which is associated with the occurrence of side effects of these medicines [54]. The thermal effects of selected oxicam derivatives were also evaluated by Kyrikou et al. using the DSC method. Their results were similar to ours. The perturbing effect of the drugs exerted on DPPC model membranes resulted in a decrease in the main phase transition temperature and in the broadening of the transition peaks [55].

As presented in the *Results* section, the anti-inflammatory potential was further supported by in vitro assays, showing that both **BS23** and **BS28** decreased levels of TNF-α and IL-6, cytokines crucial to inflammation pathways in lipopolysaccharide (LPS)-stimulated cells. Additionally, both derivatives visibly reduce the expression levels of pro-inflammatory cytokines in comparison to the LPS group. Hence, compounds can inhibit the gene expression of TNF-α and IL-6 in inflamed cells. This is consistent with reports on other NSAID analogs. Shabbir et al. examined 3-ethoksyhydroxyphenyl derivatives of benzothiazines (EHP). They showed that treatment of EHP significantly reduces pro-inflammatory tissue necrosis factor-α and interleukin-17 [56]. Our results demonstrated a robust correlation between the compounds' COX inhibitory activity and their antioxidant properties. Several compounds exhibited promising anti-inflammatory activity and favorable reactivity profiles including the capacity to neutralize ROS. Additionally, analyzed derivatives (**BS28** and **BS29**) are characterized by the high value of electrophilicity indexes, which is a potential predictor of inflammatory activity. Compound **BS28** demonstrated the lowest value of the HOMO-LUMO gap indicating a stronger tendency to react with free radicals. The antioxidant capacity of proposed compounds also aligns with an earlier study. Erturk et al. proved that NSAID derivatives with strong radical scavenging properties not only reduce inflammation but also mitigate oxidative stress associated with chronic inflammatory conditions [43]. The analysis of structurally similar compounds showed the correlation between the reactivity of thiazine derivatives and their biological properties [57].

In summary, the following structural elements should be considered when designing selective COX-2 inhibitors:

- The substitution of Ile523 in COX-1 with Val523 in COX-2 creates space for selective inhibitors with bulky hydrophobic groups; therefore, compounds with the specific side chains that exploit this pocket should be designed (e.g., 1-(2-pyrido)piperazine fragment of our compounds);
- The sulfonamide group able to interact with polar amino acid residues as arginine or serine via hydrogen bonds (**BS23**, **BS24**, **BS26**, **BS27**, and **BS30**);

- Halogen atoms enhance binding affinity and selectivity by improving hydrophobic and dipole interactions within the COX-2 pocket;
- Aromatic scaffolds enhance π–π stacking interactions with hydrophobic residues in COX-2;
- Methoxy substituent, which contributes to increased hydrophobic interactions within the binding pocket and enhances π–π stacking with aromatic residues in the COX-2 active site (compound **BS23** and **BS24**).

5. Conclusions

Our findings suggest that the 1,2-benzothiazine derivatives could serve as potential lead candidates for the development of novel anti-inflammatory agents. They could induce a dual therapeutic effect, both inhibiting COX-2 and reducing ROS levels, potentially improving therapeutic outcomes in oxidative stress-driven inflammatory diseases. The data revealed that all compounds exhibit high or moderate selectivity towards the cyclooxygenase 2 isoform and low cytotoxicity in comparison to the reference drug. It is certain, however, that further in vivo studies are warranted to validate their therapeutic efficacy and safety.

Supplementary Materials: The following supporting information can be downloaded at https://www.mdpi.com/article/10.3390/membranes14120274/s1, Figure S1: ^{1}H NMR spectrum of **BS23** in $CDCl_3$; Figure S2: ^{13}C NMR spectrum of **BS23** in $CDCl_3$; Figure S3: ^{1}H NMR spectrum of **BS24** in $CDCl_3$; Figure S4: ^{13}C NMR spectrum of **BS24** in $CDCl_3$; Figure S5: ^{1}H NMR spectrum of **BS25** in $CDCl_3$; Figure S6: ^{13}C NMR spectrum of **BS25** in $CDCl_3$; Figure S7. ^{1}H NMR spectrum of **BS26** in $CDCl_3$; Figure S8: ^{13}C NMR spectrum of **BS26** in $CDCl_3$; Figure S9. ^{1}H NMR spectrum of **BS27** in $CDCl_3$; Figure S10: ^{13}C NMR spectrum of **BS27** in $CDCl_3$; Figure S11: ^{1}H NMR spectrum of **BS28** in $CDCl_3$; Figure S12: ^{13}C NMR spectrum of **BS28** in $CDCl_3$; Figure S13. ^{1}H NMR spectrum of **BS29** in $CDCl_3$; Figure S14: ^{13}C NMR spectrum of **BS29** in $CDCl_3$; Figure S15: ^{1}H NMR spectrum of **BS30** in $CDCl_3$; Figure S16: ^{13}C NMR spectrum of **BS30** in $CDCl_3$. Figure S17: The distribution map of HOMO/LUMO densities. The negative electron density is presented in blue color, while the red color indicates the positive electron density; Figure S18: The binding mode of designed compound **BS24**—COX-2 in the active center of COX-2; Figure S19: The binding mode of designed compound **BS25**—COX-2 in the active center of COX-2; Figure S20. The binding mode of designed compound **BS26**—COX-2 in the active center of COX-2; Figure S21: The binding mode of designed compound **BS27**—COX-2 in the active center of COX-2; Figure S22: The binding mode of designed compound **BS28**—COX-2 in the active center of COX-2; Figure S23: The binding mode of designed compound **BS29**—COX-2 in the active center of COX-2; Figure S24: The binding mode of designed compound **BS30**—COX-2 in the active center of COX-2; Table S1. Intermolecular interactions in the active center of COX-1.

Author Contributions: Conceptualization, B.M.S.-S.; methodology, B.M.S.-S., J.M., B.W., T.J., P.N. and Ż.C.; validation, B.M.S.-S., J.M., B.W., T.J., P.N. and Ż.C.; formal analysis, B.M.S.-S., J.M., B.W., T.J. and Ż.C.; investigation, B.M.S.-S., J.M., B.W., T.J., P.N. and Ż.C.; resources, B.M.S.-S., J.M., T.J., P.N. and Ż.C.; data curation, B.M.S.-S., J.M., B.W., T.J., P.N. and Ż.C.; writing—original draft preparation, B.M.S.-S., J.M., B.W., T.J. and Ż.C.; writing—review and editing, B.M.S.-S., J.M. and Ż.C.; visualization, B.M.S.-S., J.M., B.W., T.J. and Ż.C.; supervision, B.M.S.-S. and Ż.C.; project administration, B.M.S.-S.; funding acquisition, B.M.S.-S. All authors have read and agreed to the published version of the manuscript.

Funding: This research was funded by the Ministry of Health subvention according to the number SUBK.D070.24.067 from the IT Simple System of Wroclaw Medical University, Poland.

Institutional Review Board Statement: Not applicable.

Data Availability Statement: Data are contained within the article and Supplementary Materials.

Acknowledgments: Ż.C. gratefully acknowledges the allotment of the CPU time for calculations in the Wroclaw Center of Networking and Supercomputing (WCSS). Molecular graphics and analyses were performed with UCSF ChimeraX, developed by the Resource for Biocomputing, Visualization, and Informatics at the University of California, San Francisco, with support from National Institutes of Health R01-GM129325 and the Office of Cyber Infrastructure and Computational Biology, National Institute of Allergy and Infectious Diseases.

Conflicts of Interest: The authors declare no conflicts of interest.

References

1. Romanelli, M.N. Cyclooxygenase. In *Metalloenzymes*; Elsevier: Amsterdam, The Netherlands, 2024; pp. 431–447.
2. Dogné, J.-M.; Supuran, C.T.; Pratico, D. Adverse Cardiovascular Effects of the Coxibs. *J. Med. Chem.* **2005**, *48*, 2251–2257. [CrossRef] [PubMed]
3. Mantovani, A.; Allavena, P.; Sica, A.; Balkwill, F. Cancer-Related Inflammation. *Nature* **2008**, *454*, 436–444. [CrossRef]
4. Elinav, E.; Nowarski, R.; Thaiss, C.A.; Hu, B.; Jin, C.; Flavell, R.A. Inflammation-Induced Cancer: Crosstalk Between Tumours, Immune Cells and Microorganisms. *Nat. Rev. Cancer* **2013**, *13*, 759–771. [CrossRef]
5. Khandia, R.; Munjal, A. Interplay between Inflammation and Cancer. In *Advances in Protein Chemistry and Structural Biology*; Elsevier: Amsterdam, The Netherlands, 2020; Volume 119, pp. 199–245.
6. Greten, F.R.; Grivennikov, S.I. Inflammation and Cancer: Triggers, Mechanisms, and Consequences. *Immunity* **2019**, *51*, 27–41. [CrossRef]
7. Khatami, M. Chronic Inflammation: Synergistic Interactions of Recruiting Macrophages (TAMs) and Eosinophils (Eos) with Host Mast Cells (MCs) and Tumorigenesis in CALTs. M-CSF, Suitable Biomarker for Cancer Diagnosis! *Cancers* **2014**, *6*, 297–322. [CrossRef] [PubMed]
8. Wen, Y.; Zhu, Y.; Zhang, C.; Yang, X.; Gao, Y.; Li, M.; Yang, H.; Liu, T.; Tang, H. Chronic Inflammation, Cancer Development and Immunotherapy. *Front. Pharmacol.* **2022**, *13*, 1040163. [CrossRef] [PubMed]
9. Wang, Y.; Qi, H.; Liu, Y.; Duan, C.; Liu, X.; Xia, T.; Chen, D.; Piao, H.; Liu, H.-X. The Double-Edged Roles of ROS in Cancer Prevention and Therapy. *Theranostics* **2021**, *11*, 4839–4857. [CrossRef]
10. Cheung, E.C.; Vousden, K.H. The Role of ROS in Tumour Development and Progression. *Nat. Rev. Cancer* **2022**, *22*, 280–297. [CrossRef]
11. Szczęśniak-Sięga, B.M.; Mogilski, S.; Wiglusz, R.J.; Janczak, J.; Maniewska, J.; Malinka, W.; Filipek, B. Synthesis and Pharmacological Evaluation of Novel Arylpiperazine Oxicams Derivatives as Potent Analgesics without Ulcerogenicity. *Bioorg. Med. Chem.* **2019**, *27*, 1619–1628. [CrossRef] [PubMed]
12. Hatnapure, G.D.; Keche, A.P.; Rodge, A.H.; Birajdar, S.S.; Tale, R.H.; Kamble, V.M. Synthesis and Biological Evaluation of Novel Piperazine Derivatives of Flavone as Potent Anti-Inflammatory and Antimicrobial Agent. *Bioorg. Med. Chem. Lett.* **2012**, *22*, 6385–6390. [CrossRef]
13. Boido, A.; Canu Boido, C.; Sparatore, F. Synthesis and Pharmacological Evaluation of Aryl/Heteroaryl Piperazinyl Alkyl Benzotriazoles as Ligands for Some Serotonin and Dopamine Receptor Subtypes. *Il Farm.* **2001**, *56*, 263–275. [CrossRef] [PubMed]
14. Saeed, M.M.; Khalil, N.A.; Ahmed, E.M.; Eissa, K.I. Synthesis and Anti-Inflammatory Activity of Novel Pyridazine and Pyridazinone Derivatives as Non-Ulcerogenic Agents. *Arch. Pharm. Res.* **2012**, *35*, 2077–2092. [CrossRef]
15. Kono, M.; Matsumoto, T.; Imaeda, T.; Kawamura, T.; Fujimoto, S.; Kosugi, Y.; Odani, T.; Shimizu, Y.; Matsui, H.; Shimojo, M.; et al. Design, Synthesis, and Biological Evaluation of a Series of Piperazine Ureas as Fatty Acid Amide Hydrolase Inhibitors. *Bioorg. Med. Chem.* **2014**, *22*, 1468–1478. [CrossRef]
16. Malinka, W.; Redzicka, A.; Jastrzębska-Więsek, M.; Filipek, B.; Dybała, M.; Karczmarzyk, Z.; Urbańczyk-Lipkowska, Z.; Kalicki, P. Derivatives of Pyrrolo[3,4-d]Pyridazinone, a New Class of Analgesic Agents. *Eur. J. Med. Chem.* **2011**, *46*, 4992–4999. [CrossRef]
17. Kono, M.; Matsumoto, T.; Kawamura, T.; Nishimura, A.; Kiyota, Y.; Oki, H.; Miyazaki, J.; Igaki, S.; Behnke, C.A.; Shimojo, M.; et al. Synthesis, SAR Study, and Biological Evaluation of a Series of Piperazine Ureas as Fatty Acid Amide Hydrolase (FAAH) Inhibitors. *Bioorg. Med. Chem.* **2013**, *21*, 28–41. [CrossRef]
18. Krzyżak, E.; Szczęśniak-Sięga, B.; Malinka, W. Synthesis and Thermal Behaviour of New Benzo-1,2-Thiazine Long-Chain Aryl-Piperazine Derivatives. *J. Therm. Anal. Calorim.* **2014**, *115*, 793–802. [CrossRef]
19. Szczęśniak-Sięga, B.M.; Gębczak, K.; Gębarowski, T.; Maniewska, J. Synthesis, COX-1/2 Inhibition and Antioxidant Activities of New Oxicam Analogues Designed as Potential Chemopreventive Agents. *Acta Biochim. Pol.* **2018**, *65*, 199–207. [CrossRef]
20. Maniewska, J.; Gąsiorowska, J.; Szczęśniak-Sięga, B.M.; Michalak, K. The Interaction of New Oxicam Derivatives with Lipid Bilayers as Measured by Calorimetry and Fluorescence Spectroscopy. *Acta Biochim. Pol.* **2018**, *65*, 185–191. [CrossRef]
21. Murty, M.S.R.; Ram, K.R.; Venkateswara Rao, R.; Yadav, J.S.; Venkateswara Rao, J.; R. Velatooru, L. Synthesis of New S-Alkylated-3-Mercapto-1,2,4-Triazole Derivatives Bearing Cyclic Amine Moiety as Potent Anticancer Agents. *Lett. Drug Des. Discov.* **2012**, *9*, 276–281. [CrossRef]
22. Wiatrak, B.; Balon, K. Protective Activity of Aβ on Cell Cultures (PC12 and THP-1 after Differentiation) Preincubated with Lipopolysaccharide (LPS). *Mol. Neurobiol.* **2021**, *58*, 1453–1464. [CrossRef]
23. Gaussian 16, Revision C.01. 2016. Available online: https://gaussian.com/relnotes/ (accessed on 12 November 2024).
24. Tomasi, J.; Mennucci, B.; Cammi, R. Quantum Mechanical Continuum Solvation Models. *Chem. Rev.* **2005**, *105*, 2999–3094. [CrossRef] [PubMed]
25. Hanwell, M.D.; Curtis, D.E.; Lonie, D.C.; Vandermeersch, T.; Zurek, E.; Hutchison, G.R. Avogadro: An Advanced Semantic Chemical Editor, Visualization, and Analysis Platform. *J. Cheminform.* **2012**, *4*, 17. [CrossRef] [PubMed]

26. Fukui, K.; Yonezawa, T.; Shingu, H. A Molecular Orbital Theory of Reactivity in Aromatic Hydrocarbons. *J. Chem. Phys.* **1952**, *20*, 722–725. [CrossRef]
27. Padmaja, L.; Ravikumar, C.; Sajan, D.; Hubert Joe, I.; Jayakumar, V.S.; Pettit, G.R.; Faurskov Nielsen, O. Density Functional Study on the Structural Conformations and Intramolecular Charge Transfer from the Vibrational Spectra of the Anticancer Drug Combretastatin-A2. *J. Raman Spectrosc.* **2009**, *40*, 419–428. [CrossRef]
28. Koopmans, T. Über Die Zuordnung von Wellenfunktionen Und Eigenwerten Zu Den Einzelnen Elektronen Eines Atoms. *Physica* **1934**, *1*, 104–113. [CrossRef]
29. Morris, G.M.; Huey, R.; Lindstrom, W.; Sanner, M.F.; Belew, R.K.; Goodsell, D.S.; Olson, A.J. AutoDock4 and AutoDockTools4: Automated Docking with Selective Receptor Flexibility. *J. Comput. Chem.* **2009**, *30*, 2785–2791. [CrossRef] [PubMed]
30. Xu, S.; Hermanson, D.J.; Banerjee, S.; Ghebreselasie, K.; Clayton, G.M.; Garavito, R.M.; Marnett, L.J. Oxicams Bind in a Novel Mode to the Cyclooxygenase Active Site via a Two-Water-Mediated H-Bonding Network. *J. Biol. Chem.* **2014**, *289*, 6799–6808. [CrossRef]
31. Velázquez-Libera, J.L.; Durán-Verdugo, F.; Valdés-Jiménez, A.; Núñez-Vivanco, G.; Caballero, J. LigRMSD: A Web Server for Automatic Structure Matching and RMSD Calculations among Identical and Similar Compounds in Protein-Ligand Docking. *Bioinformatics* **2020**, *36*, 2912–2914. [CrossRef]
32. Peregrym, K.; Szczukowski, Ł.; Wiatrak, B.; Potyrak, K.; Czyżnikowska, Ż.; Świątek, P. In Vitro and In Silico Evaluation of New 1,3,4-Oxadiazole Derivatives of Pyrrolo[3,4-d]Pyridazinone as Promising Cyclooxygenase Inhibitors. *Int. J. Mol. Sci.* **2021**, *22*, 9130. [CrossRef]
33. Glomb, T.; Wiatrak, B.; Gębczak, K.; Gębarowski, T.; Bodetko, D.; Czyżnikowska, Ż.; Świątek, P. New 1,3,4-Oxadiazole Derivatives of Pyridothiazine-1,1-Dioxide with Anti-Inflammatory Activity. *Int. J. Mol. Sci.* **2020**, *21*, 9122. [CrossRef] [PubMed]
34. Chen, D.; Menche, G.; Power, T.D.; Sower, L.; Peterson, J.W.; Schein, C.H. Accounting for Ligand-bound Metal Ions in Docking Small Molecules on Adenylyl Cyclase Toxins. *Proteins Struct. Funct. Bioinform.* **2007**, *67*, 593–605. [CrossRef] [PubMed]
35. Pettersen, E.F.; Goddard, T.D.; Huang, C.C.; Couch, G.S.; Greenblatt, D.M.; Meng, E.C.; Ferrin, T.E. UCSF Chimera—A Visualization System for Exploratory Research and Analysis. *J. Comput. Chem.* **2004**, *25*, 1605–1612. [CrossRef] [PubMed]
36. BIOVIA. *Dassault Systèmes. Biovia Diccovery Studio Vizualizer*, V21.1.0.20298; Dassault Systèmes: San Diego, CA, USA, 2020.
37. Wiatrak, B.; Krzyżak, E.; Szczęśniak-Sięga, B.; Szandruk-Bender, M.; Szeląg, A.; Nowak, B. Effect of Tricyclic 1,2-Thiazine Derivatives in Neuroinflammation Induced by Preincubation with Lipopolysaccharide or Coculturing with Microglia-like Cells. *Pharmacol. Rep.* **2022**, *74*, 890–908. [CrossRef]
38. Dinarello, C.A. Anti-Inflammatory Agents: Present and Future. *Cell* **2010**, *140*, 935–950. [CrossRef]
39. Vane, J.R.; Botting, R.M. Anti-Inflammatory Drugs and Their Mechanism of Action. *Inflamm. Res.* **1998**, *47*, 78–87. [CrossRef] [PubMed]
40. Ciurko, D.; Łaba, W.; Żarowska, B.; Janek, T. Enzymatic Hydrolysis Using Bacterial Cultures as a Novel Method for Obtaining Antioxidant Peptides from Brewers' Spent Grain. *RSC Adv.* **2021**, *11*, 4688–4700. [CrossRef] [PubMed]
41. Ciurko, D.; Neuvéglise, C.; Szwechłowicz, M.; Lazar, Z.; Janek, T. Comparative Analysis of the Alkaline Proteolytic Enzymes of Yarrowia Clade Species and Their Putative Applications. *Int. J. Mol. Sci.* **2023**, *24*, 6514. [CrossRef] [PubMed]
42. Karadag, A.; Ozcelik, B.; Saner, S. Review of Methods to Determine Antioxidant Capacities. *Food Anal. Methods* **2009**, *2*, 41–60. [CrossRef]
43. Ertürk, Ö.; Değirmenci, A.; Yurdakul Ertürk, E.; Atlı Şekeroğlu, Z.; Çol Ayvaz, M.; Kontaş Yedier, S. Antimicrobial, Antioxidant and Cytotoxic Activities of Some Analgesic or Anti-inflammatory Drugs. *Biologia* **2021**, *76*, 2365–2379. [CrossRef]
44. Janak, J.F. Proof That ∂E∂ni = εin in Density-Functional Theory. *Phys. Rev. B* **1978**, *18*, 7165–7168. [CrossRef]
45. Parr, R.G.; Szentpály, L.; Liu, S. Electrophilicity Index. *J. Am. Chem. Soc.* **1999**, *121*, 1922–1924. [CrossRef]
46. Parthasarathi, R.; Subramanian, V.; Roy, D.R.; Chattaraj, P.K. Electrophilicity Index as a Possible Descriptor of Biological Activity. *Bioorg. Med. Chem.* **2004**, *12*, 5533–5543. [CrossRef]
47. Gautam, R.; Jachak, S.M.; Kumar, V.; Mohan, C.G. Synthesis, Biological Evaluation and Molecular Docking Studies of Stellatin Derivatives as Cyclooxygenase (COX-1, COX-2) Inhibitors and Anti-Inflammatory Agents. *Bioorg. Med. Chem. Lett.* **2011**, *21*, 1612–1616. [CrossRef] [PubMed]
48. Borer, J.S.; Simon, L.S. Cardiovascular and Gastrointestinal Effects of COX-2 Inhibitors and NSAIDs: Achieving a Balance. *Arthritis Res. Ther.* **2005**, *7*, S14. [CrossRef] [PubMed]
49. El-Malah, A.A.; Gineinah, M.M.; Deb, P.K.; Khayyat, A.N.; Bansal, M.; Venugopala, K.N.; Aljahdali, A.S. Selective COX-2 Inhibitors: Road from Success to Controversy and the Quest for Repurposing. *Pharmaceuticals* **2022**, *15*, 827. [CrossRef] [PubMed]
50. Rayar, A.-M.; Lagarde, N.; Ferroud, C.; Zagury, J.-F.; Montes, M.; Sylla-Iyarreta Veitia, M. Update on COX-2 Selective Inhibitors: Chemical Classification, Side Effects and Their Use in Cancers and Neuronal Diseases. *Curr. Top. Med. Chem.* **2017**, *17*, 2935–2956. [CrossRef] [PubMed]
51. Kiefer, J.R.; Pawlitz, J.L.; Moreland, K.T.; Stegeman, R.A.; Hood, W.F.; Gierse, J.K.; Stevens, A.M.; Goodwin, D.C.; Rowlinson, S.W.; Marnett, L.J.; et al. Structural Insights into the Stereochemistry of the Cyclooxygenase Reaction. *Nature* **2000**, *405*, 97–101. [CrossRef]
52. Zambrano, P.; Manrique-Moreno, M.; Petit, K.; Colina, J.R.; Jemiola-Rzeminska, M.; Suwalsky, M.; Strzalka, K. Differential Scanning Calorimetry in Drug-Membrane Interactions. *Biochem. Biophys. Res. Commun.* **2024**, *709*, 149806. [CrossRef] [PubMed]

53. Lichtenberger, L.M.; Wang, Z.-M.; Romero, J.J.; Ulloa, C.; Perez, J.C.; Giraud, M.-N.; Barreto, J.C. Non-Steroidal Anti-Inflammatory Drugs (NSAIDs) Associate with Zwitterionic Phospholipids: Insight into the Mechanism and Reversal of NSAID-Induced Gastrointestinal Injury. *Nat. Med.* **1995**, *1*, 154–158. [CrossRef] [PubMed]
54. Lúcio, M.; Bringezu, F.; Reis, S.; Lima, J.L.F.C.; Brezesinski, G. Binding of Nonsteroidal Anti-Inflammatory Drugs to DPPC: Structure and Thermodynamic Aspects. *Langmuir* **2008**, *24*, 4132–4139. [CrossRef]
55. Kyrikou, I.; Hadjikakou, S.K.; Kovala-Demertzi, D.; Viras, K.; Mavromoustakos, T. Effects of Non-Steroid Anti-Inflammatory Drugs in Membrane Bilayers. *Chem. Phys. Lipids* **2004**, *132*, 157–169. [CrossRef]
56. Shabbir, A.; Shahzad, M.; Ali, A.; Zia-ur-Rehman, M. Discovery of New Benzothiazine Derivative as Modulator of Pro- and Anti-Inflammatory Cytokines in Rheumatoid Arthritis. *Inflammation* **2016**, *39*, 1918–1929. [CrossRef] [PubMed]
57. Al-Fayez, N.; Elsawy, H.; Mansour, M.A.; Akbar Ali, M.; Elghamry, I. Synthesis, Anticancer, Antioxidant, Anti-Inflammatory, Antimicrobial Activities, Molecular Docking, and DFT Studies of Sultams Derived from Saccharin. *Molecules* **2022**, *27*, 7104. [CrossRef] [PubMed]

membranes

Article

"Head-to-Toe" Lipid Properties Govern the Binding and Cargo Transfer of High-Density Lipoprotein

Florian Weber [1,2], Markus Axmann [1], Erdinc Sezgin [2], Mariana Amaro [3], Taras Sych [2], Armin Hochreiner [1], Martin Hof [3], Gerhard J. Schütz [4], Herbert Stangl [5,*] and Birgit Plochberger [1,6,*]

1 Department of Medical Engineering, Upper Austria University of Applied Sciences, 4020 Linz, Austria; markus.axmann@fh-linz.at (M.A.); armin.hochreiner@fh-linz.at (A.H.)
2 Science for Life Laboratory, Department of Women's and Children's Health, Karolinska Institutet, 171 77 Solna, Sweden; erdinc.sezgin@ki.se (E.S.); taras.sych@ki.se (T.S.)
3 J. Heyrovský Institute of Physical Chemistry of the Czech Academy of Science, 182 00 Prague, Czech Republic; mariana.amaro@jh-inst.cas.cz (M.A.); martin.hof@jh-inst.cas.cz (M.H.)
4 Institute of Applied Physics, TU Wien, 1040 Vienna, Austria; gerhard.schuetz@tuwien.ac.at
5 Center for Pathobiochemistry and Genetics, Institute of Medical Chemistry, Medical University of Vienna, 1090 Vienna, Austria
6 Research Group Nanoscopy, Ludwig Boltzmann Institute for Experimental and Clinical Traumatology, 1200 Vienna, Austria
* Correspondence: herbert.stangl@meduniwien.ac.at (H.S.); birgit.plochberger@fh-linz.at (B.P.)

Abstract: The viscoelastic properties of biological membranes are crucial in controlling cellular functions and are determined primarily by the lipids' composition and structure. This work studies these properties by varying the structure of the constituting lipids in order to influence their interaction with high-density lipoprotein (HDL) particles. Various fluorescence-based techniques were applied to study lipid domains, membrane order, and the overall lateral as well as the molecule–internal glycerol region mobility in HDL–membrane interactions (i.e., binding and/or cargo transfer). The analysis of interactions with HDL particles and various lipid phases revealed that both fully fluid and some gel-phase lipids preferentially interact with HDL particles, although differences were observed in protein binding and cargo exchange. Both interactions were reduced with ordered lipid mixtures containing cholesterol. To investigate the mechanism, membranes were prepared from single-lipid components, enabling step-by-step modification of the lipid building blocks. On a biophysical level, the different mixtures displayed varying stiffness, fluidity, and hydrogen bond network changes. Increased glycerol mobility and a strengthened hydrogen bond network enhanced anchoring interactions, while fluid membranes with a reduced water network facilitated cargo transfer. In summary, the data indicate that different lipid classes are involved depending on the type of interaction, whether anchoring or cargo transfer.

Keywords: lipoprotein; membrane order; Laurdan polarity; hydrogen bond network; glycerol region mobility

1. Introduction

The structure and organization of cell membranes are of significant interest, particularly in the context of lipid–protein interactions. In addition to maintaining membrane integrity and acting as signaling molecules, lipids play a structural role in shaping physical membrane properties. Membranes can exhibit multiple possible phase states and their properties are determined by the constituting lipid structure [1]. Moreover, membrane properties such as lateral fluidity strongly depend on temperature. At sufficiently low temperatures, the solid-like gel [2] phase is characterized by low lateral lipid motion [3], high lipid packing/membrane order, and a low level of membrane hydration [4]. Above the phase transition temperature (i.e., at the "melting" temperature T_m), lipids form the so-called liquid-disordered L_d (also known as fluid, liquid–crystalline, or lamellar L_α) phase. The T_m value strongly depends, among other parameters, on the saturation of the fatty acid chains. In the presence of cholesterol (Chol) or similar sterols another phase emerges, known as the liquid-ordered L_o phase [5]. Compared to gel phases, the L_o and L_d phases demonstrate lipid diffusion, though in the case of L_o phases, this diffusion is reduced due to the lipid composition [6–9]. Furthermore, it also exhibits similarities to the gel phase with regard to membrane order and lipid packing. The L_d and L_o phases can coexist, and their size and lifetime are variable and temperature-dependent [1,10]. Their inherent stability (i.e., not merging into a single, larger domain) depends on a high-energy barrier due to domain curvature at the phase boundaries [11].

Glycerophospholipids, where acyl fatty acid chains are attached to a glycerol backbone by ester or ether bonds, are the main lipid constituent of biological membranes. Ether lipids are a substantial structural component in cell membranes, representing around 20% of the total phospholipid content in mammals [12]. Their most significant contribution is to the membranes of the brain, heart, kidneys, lungs, skeletal muscle, and immune system [13]. Plasmalogens, the most abundant form of ether lipids, distinguished by a cis double bond adjacent to the ether linkage, are notably concentrated in the Chol-rich regions of membranes [14,15]. They play a crucial role in facilitating cellular Chol efflux [16]. Notably, the liver contains a very low amount of ether lipids. Comprising Chol esters and triglycerides surrounded by a shell of phospholipids, Chol, and proteins, high-density lipoprotein (HDL) particles facilitate the transport and transfer of non-water-soluble substances to and from cells. HDL particles, which are secreted by the liver and contain approximately 20–30% plasmalogens relative to their total glycerophospholipid pool, transport plasmalogens to other tissues, thus depleting the liver of ether lipids [17]. Cholesterol is an essential building block, comprising up to 40 mol% [18] and, in certain specialized membrane environments, reaching up to 50% [19] in eukaryotic plasma membranes, where it plays a crucial role in regulating membrane fluidity. Notably, Chol enrichment slows down ether lipid uptake [20].

In addition to the diversity of lipids and their composition, the interfaces of lipid membranes create an equally complex environment and play a key role in the dynamics, stability, and function of biological membranes. These interfaces are crucial for important cellular processes such as protein binding [21] or drug interaction [22,23]. The membrane interface is characterized by an abundant network of dynamic hydrogen-bonded water chains bridging the lipid headgroups, some of which form transient lipid clusters [24]. Numerous molecular dynamics simulations [25,26] have been conducted to investigate how water molecules interact with anionic lipids at membrane interfaces and how they are involved in protein binding. Membranes differ in properties such as viscoelastic and compressibility parameters. A wide spectrum of techniques and lipid compositions are applied in an attempt to provide a comprehensive view of these individual segments and how they affect each other and their vicinity [27–29]. The use of environmentally sensitive fluorescence probes like Laurdan is beneficial for studies of biological heterogeneity, in particular within the glycerol region of lipids. Laurdan is sensitive to the polarity and viscosity of its local environment [30]. Thus, Laurdan makes it possible to quantify the hindrance of lipid internal dynamics, so-called glycerol region mobility. Changes in membrane properties, most notably its lipid packing, cause shifts in Laurdan's emission spectrum, which are quantified by the generalized polarization (GP) value. This spectrum change can be influenced by the hydration state, in part, but also by Chol [28] due to an altered local hydrogen bond network. Time-dependent fluorescence shift (TDFS) measurements allow for the mapping of the interfacial region, i.e., the local environment of the lipid's glycerol region, characterizing both the dynamics of the hydrated segments of lipid membranes as well as the level of its hydrogen bond network. Hence, this makes it possible to quantitatively characterize the mobility of hydrated lipid segments that are in close vicinity of the probe fluorophore [31]. Imaging techniques such as single-molecule fluorescence microscopy (SMFM) and fluorescence correlation spectroscopy (FCS) were used to visualize different lipid phases or to determine the motion of biomolecules [32,33]. Phase-specific fluorophores like DiI [34] enable visualizing and characterizing the lipid bilayer's spatial organization. To localize Chol in membrane phases or to monitor sterol uptake and flux in cells [35], Cholesterol-BodipyFL (Chol-BodipyFL) is predominantly adopted. It is a well-established Chol analog with a conjugated boron–dipyrromethene difluoride fluorophore. The partition coefficient (K_P) can be used to describe quantitative partitioning in phase-separated membranes [36]. Previous studies [37] have demonstrated that HDL particles interact directly with the membrane without any receptor involvement. Lipoprotein particles showed strong inhibition in their interaction with artificial lipid membranes as their cholesterol content increased. This study assesses the interaction of HDL particles with the membrane influenced by (i) the glycerol linkage of the fatty acid chains to the head group (i.e., ester vs. ether lipids), (ii) the Chol content of the target membrane, and (iii) the saturation and length of the fatty acid chains using multi-correlative methods, including FCS, TDFS, and SMFM.

2. Materials and Methods

2.1. Materials

Sephadex G-25 fine resin, potassium bromide, sodium chloride, calcium chloride, EthyleneDiamineTetraacetic Acid (EDTA), Tris-(hydroxymethyl)-aminomethan-hydrochlorid (Tris/HCl), 1× Phosphate-Buffered Saline (PBS), 2-(4-(2-HydroxyEthyl)-1-PiperazineEthaneSulfonic acid (HEPES), Nunc® Lab-Tek® Chamber Slides, 1,2-DiOleoyl-sn-glycero-3-PhosphoCholine (DOPC (18:1)), 1,2-DiPalmitoyl-sn-glycero-3-PhosphoCholine (DPPC (16:0)), 1-Palmitoyl-2-Oleoyl-sn-glycero-3-PhosphoCholine (POPC (16:0-18:1)), 1,2-Di-O-hexadecyl-sn-glycero-3-PhosphoCholine (DietherPC (16:0)), 1,2-Di-O-octadecyl-sn-glycero-3-PhosphoCholine (DietherPC (18:0)), 1,2-Di-O-(9Z-octadecenyl)-sn-glycero-3-PhosphoCholine (DietherPC (18:1)), N-(octadecanoyl)-sphing-4-enine-1-phosphocholine (also known as. brain SphingoMyelin (bSM)), Chol, and Chol linked to boron dipyrromethene difluoride at sterol carbon-24 (Chol-BodipyFL) were purchased from Merck, Vienna, Austria. 1,2-DiPalmitoyl-sn-glycero-3-PhosphoEthanolamine linked to abberior STAR RED (DPPE-ASR) was purchased from abberior, Göttingen, Germany. Atto 647 NHS Ester, 1,1′-Dioctadecyl-3,3,3′,3′-tetramethylindocarbocyanine perchlorate (also known as DiI™), Laurdan [38] were purchased from Thermo Fisher Scientific, Vienna, Austria.

2.2. HDL Particle Isolation and Labeling

Lipoprotein particles were isolated as previously described [39]. Lipoprotein isolation was approved by the Ethics Committee, Medical University of Vienna (EK-Nr. 1414/2016). Written informed consent was obtained from all participants. Briefly, plasma was recovered from whole blood obtained from normolipidemic healthy volunteers by centrifugation (twice at 3000× *g*, 4 °C, 20 min) and its density was adjusted to 1.063 g/L using KBr. The samples were centrifuged (52,000× *g*, 4 °C, 20 h) and the upper phase containing VLDL and LDL was discarded. The density of the bottom fraction was adjusted to 1.21 g/L with KBr and the samples were centrifuged as described above. The upper phase containing HDL particles was recovered and was re-adjusted to a density of 1.21 g/L and centrifuged a second time to ensure complete removal of serum albumin. The sample containing HDL particles was dialyzed extensively against 0.9% (*w*/*v*) NaCl and 0.1% (*w*/*v*) EDTA, pH 7.4, to remove KBr. Protein concentration was determined by the Bradford method. The (apo-lipo)protein(s) of HDL particles were labeled with Atto 647 via NHS-based covalent linking according to the manufacturer's protocol. Briefly, the labeling procedure includes conjugation of the dye to amino groups at pH 8.4 at room temperature for 1 h and separation of the labeled particles from the free dye by gel filtration chromatography using Sephadex G-25 fine resin.

2.3. Preparation of Phase-Separated Supported Lipid Bilayers (PSLBs)

For the formation of PSLBs, the following mixtures or pure substances were used: DOPC (18:1), bSM, and Chol at a molar ratio of [2:2:1]; DOPC (18:1) and DPPC (16:0), DOPC (18:1) and DietherPC (16:0), and DOPC (18:1) and DietherPC (18:0). In general, the lipid(s) was/were dissolved in chloroform, mixed in an appropriate glass vial, and the solvent was evaporation using N_2 gas. Before use, the lipid film was rehydrated using PBS (overall lipid concentration of 10 mg/mL) by vigorous vortexing. After ultra-sonicating at 60 °C for 20 min (the turbidity change confirmed the formation of small unilamellar vesicles), an aliquot was diluted in PBS buffer at a ratio of 1:10 and deposited on a freshly cleaved mica surface; 2 mM of $CaCl_2$ was added to the vesicle suspension to improve their spread. The surface was sealed with a home-built chamber and incubated at 55 °C for 10 min. The sample was rinsed at the same temperature at least 10 times with PBS and cooled to 25 °C within 30 min. At this stage, the L_d phase marker DiI was incorporated into the formed lipid bilayer by adding it to the solution. The concentration was between 10^{-2} and 10^{-3} mol%. After 20 min, the sample was rinsed again with PBS to remove non-incorporated DiI. After reaching room temperature, the sample was incubated with an HDL particle solution for approx. 2 min and was washed extensively. For ternary membrane systems, such as Chol/bSM/DOPC, the sample was incubated with 2.8 µg/mL HDL. For all other membrane systems, the sample was incubated with 5 µg/mL HDL. All other preparation steps remained identical. To quantitatively compare the association of protein-bound Atto 647 with the L_o and L_d phases, for DOPC (18:1), bSM, and Chol, we determined the number of single protein-bound Atto 647 signals in the two phases; for all other samples the overall Atto 647 signal was determined. A binary mask obtained from intensity-thresholded DiI signals was used for phase discrimination. The partition coefficient $K_P = I_1/I_2 \cdot A_2/A_1$ where I_i represents the fluorescence intensity in phase i and A_i its area. Statistical differences between samples were assessed using the non-parametric Kruskal–Wallis test (α = 5%). Prior to this, normality was tested using the Shapiro–Wilk test and the homogeneity of variances using Levene's test. Due to significant results ($p < 0.05$) in the Levene's test, indicating unequal variances, the Kruskal–Wallis test was deemed appropriate. Post hoc analyses were performed using Dunn's test with Bonferroni correction. Statistical tests were conducted using JASP (v0.18.3.0).

2.4. Preparation of Giant Unilamellar Vesicles (GUVs)

GUVs were generated by electroformation [40] using custom-built GUV Teflon chambers with two platinum electrodes. This preparation technique produces vesicles with varying sizes from 10 to 100 µm.

For measurements of the diffusion constant and partition coefficient: A volume of 6 µL of lipid (DOPC (18:1) or DOPC (18:1): DietherPC (18:1), molar ratio [3:1]) dissolved in chloroform (overall lipid concentration 1 mg/mL) was homogeneously distributed on the two electrodes, dried with nitrogen, and placed in 300 nM (370 µL) sucrose solution. Electroformation was performed at 2 V and 10 Hz for 1 h, followed by 2 V and 2 Hz for 30 min. After GUV formation, 1 µL DPPE-ASR (10 µg/mL in DMSO) was added to 100 µL GUV solution.

For measurements of the GP value/hydrogen bond network strength: 20 mL of a 10 mM lipid solution (POPC (16:0–18:1), DietherPC (18:1, 18:0, or 16:0), DPPC (16:0), DOPC (18:1), DOPC (18:1) with varying Chol concentration (molar ratio [9:1], [4:1]) in chloroform containing the fluorescent L_d phase lipid marker DiI (0.1 mol%) and/or Laurdan (methanol solution, molar ratio of fluorescent probes to lipids was 1:100) was deposited on Pt electrodes and the solvent was evaporated by a constant N_2 gas flow for 20 min. In total, 300 µL of 300 mM sucrose was added to a headed (60 °C) and modified Lab-Tec® chamber (Fisher Scientific): two holes with a distance of 5 mm were placed on the chamber's cap for the electrodes. The electrodes with dried lipids were placed into the sucrose solution and a voltage of 2 V at 10 Hz was applied for 2 h. The temperature was kept at 60 °C. Afterwards, the solution was allowed to cool down to room temperature. After reaching room temperature, the sample was incubated with a 2.8 μg/mL HDL particle solution for approx. 2 min. In total, 300 μL of a 300 mM glucose solution was added to the solution for imaging (i.e., density change causes GUVs to sink). The images were acquired 20 min after the addition of the HDL particle solution by confocal microscopy. Partitioning was measured with confocal microscopy, and membrane order was quantified by Laurdan microscopy/TDFS.

2.5. *Confocal Microscopy*

GUVs were imaged with a laser scanning confocal microscope (LSM 700 AxioObserver, Zeiss, Oberkochen, Germany). The microscope was equipped with a Plan-Apochromat 63x/1.40 Oil DIC M27 objective (Zeiss). The LSM 700 operates with solid-state lasers (polarization-preserving single-mode fibers) at a wavelength of 639 nm and 488 nm. The laser power was adjusted to 0.1–0.5% of the total laser power, equivalent to 2–10 μW. Signals were detected after appropriate filtering on a photomultiplier tube. Typically, a z-stack with a step size of 500 nm was generated. Detector amplification, laser power, and pinhole opening were kept constant for all measurements.

The images were analyzed by utilizing an algorithm that automatically identifies the interface between the membrane and the inner section of the GUVs. This was carried out by investigating the Atto 647 signal. The information was used to obtain the lateral membrane geometry and yields a binary mask. Fluorescence images (in one image several GUVs were visible) were background-corrected and the GUV surfaces were extracted via a mask. A GUV was selected by defining an ROI of the image. The total amount of fluorescence signal was calculated in both color channels (BodipyFL and Atto 647). Calculating the density was performed by dividing the total amount value by the number of counted masked pixels or by the pure circle length in a pixel. For each experiment, about 50 GUVs were evaluated.

2.6. *Fluorescence Correlation Spectroscopy (FCS)*

Ibidi chambers (#1.5) were coated with BSA (1 mg/mL) and rinsed with PBS after 30 min of incubation. GUVs were allowed to settle in glass-bottomed Ibidi chambers (#1.5) for 15 min. Fluorescence correlation spectroscopy measurements were conducted on the GUVs' top membrane using an LSM 700 AxioObserver (Zeiss). The microscope was equipped with a Plan-Apochromat 40x/1.2 water objective (Zeiss). The LSM 700 operates with a solid-state laser at a wavelength of 639 nm. The laser power was adjusted to 0.1–0.5% of the total laser power, equivalent to 2–10 μW. Signals were detected after appropriate filtering on a photomultiplier tube. Three measurements (10 s each) were performed on each GUV. At least five vesicles were measured for each sample. Three technical replicas of each sample were measured. Before incubation with HDL particles, the initial mobility of the tracer lipid DPPE-ASR was determined and afterward the same GUV chambers were incubated for (at least) 30 min with 14 μL of 4.3 mg/mL HDL particles (the final concentration of HDL in the chamber ~600 μg/mL, of which approximately 20% would be Chol [41] in HDL particles = 120 μg/mL) and measured again. The obtained auto correlation curves were fitted with a 2D diffusion plus triplet model, with Focus point (version: win 1.13.156) [42].

Statistical differences between samples were assessed using the non-parametric Kruskal–Wallis test ($\alpha = 5\%$). Prior to this, normality was tested using the Shapiro–Wilk test and the homogeneity of variances was tested using Levene's test. Due to significant results ($p < 0.05$) in both tests, indicating non-normal distribution and unequal variances, the Kruskal–Wallis test was deemed appropriate. Post hoc analyses were performed using Dunn's test with Bonferroni correction. The statistical tests were conducted using JASP (v0.18.3.0).

2.7. *Single-Molecule Fluorescence Microscopy*

The system is based on a Zeiss Axiovert 200 inverted epi-fluorescence microscope equipped with a 100× NA = 1.45 oil-immersion Plan-Apochromat TIRFM objective (Olympus, Vienna, Austria). Samples were illuminated in an objective-type Total Internal Reflection (TIR) configuration via the epi port using 488 nm light from a solid-state laser (Sapphire 200 mW, Coherent, Dieburg, Germany), 647 nm light from a Kr^+-laser (Innova 301, Coherent), or 532 nm light from a solid-state laser (Millennia X, Spectra Physics, Vienna, Austria), with intensities of 3–10 kW/cm^2. After appropriate filtering, the emitted signals were imaged on a back-illuminated, TE-cooled CCD camera (Andor iXon Du-897 BV, Oxford Instruments, Belfast, UK). For precise control of the illumination timings, acousto-optical modulators (1205C, Isomet, Gröbenzell, Germany) were used. Timing protocols were generated by an in-house program package implemented in LABVIEW (National Instruments, Vienna, Austria). Illumination times were between 1 and 5 ms. Image series were recorded with a delay between two consecutive images of 15 to 300 ms.

2.8. Laurdan Time-Dependent Fluorescence Shift (TDFS)

The temperature in the cuvette holders was maintained using a water-circulating bath at 23 ± 0.5 °C. Steady-state excitation and emission spectra were acquired using a Fluorolog-3 spectrofluorometer (model FL3-11; Jobin Yvon Inc., Edison, NJ, USA) equipped with a xenon arc lamp. The steady-state spectra were recorded in steps of 1 nm (bandwidths of 1.2 nm were chosen for both the excitation and emission monochromators) in triplicate and averaged. Fluorescence decays were recorded on a 5000 U single-photon counting setup using a NanoLED 11 laser diode (375 nm peak wavelength, 1 MHz repetition rate) and a cooled Hamamatsu R3809U-50 microchannel plate photomultiplier (IBH, Glasgow, UK). A 399 nm cut-off filter was used to eliminate scattered light. The signal was kept below 1% of the repetition rate of the light source. Fluorescence emission decays were recorded at a series of wavelengths spanning the steady-state emission spectrum (400–550 nm) in steps of 10 nm. Data were collected until the peak value reached 5000 counts. The full width at half maximum (FWHM) of the instrument response function was 78 ps.

2.9. Laurdan GP

A steady-state emission spectrum (Ex = 378 nm) and two excitation spectra (Em = 440 and 490 nm) were recorded. The excitation spectra were used to calculate excitation generalized polarization spectra (GP_{EX}) [43]:

$$GP_{EX}(\lambda_{EX}) = \frac{I_{440} - I_{490}}{I_{440} + I_{490}}$$

where I_{440} and I_{490} represent the fluorescence intensity emitted at 440 and 490 nm, respectively, at the excitation wavelength λ_{EX}.

2.10. Laurdan TDFS Analysis

The fluorescence decays were fitted to a multi-exponential function via the reconvolution method using IBH DAS6 software (https://www.horiba.com/int/scientific/products/detail/action/show/Product/das6-1377/, accessed on 3 December 2024). The purpose of the fit is to deconvolve the instrumental response from the data and should not be over-parameterized. The fitted decays, together with the steady-state emission spectrum, were used for the reconstruction of time-resolved emission spectra (TRES) by a spectral reconstruction method [44]. The reconstruction routine was implemented in MATLAB 24.2. The position of TRES' maximum $\nu(t)$ and its FWHM(t) were inspected. The two main parameters describing the polarity and mobility of the probed system were derived from $\nu(t)$. The total amount of fluorescence shift $\Delta\nu$ reflects the polarity of the environment of the probe and is calculated as: $\Delta\nu = \nu(0) - \nu(\infty)$ where $\nu(0)$ = 23,800 cm^{-1} is the position of TRES maximum at t = 0, estimated using the method of *Fee and Maroncelli* [45] and $\nu(\infty)$ is the position of the TRES at the fully relaxed state. The TDFS kinetics depend on the dynamics of the polar moieties in the vicinity of the probe and can be expressed as the integrated relaxation time τ_r:

$$\tau_r = \int_0^{\infty} \frac{\nu(t) - \nu(\infty)}{\Delta\nu} dt$$

The intrinsic uncertainties for the TDFS parameters were 50 cm^{-1} and 0.05 ns for $\Delta\nu$ and τ_r, respectively.

2.11. Single Molecule/Particle Tracking

Individual diffraction-limited fluorescence signal spots were selected, fitted with a Gaussian intensity profile, and tracked using in-house algorithms implemented in MATLAB (MathWorks); the single-molecule positions were obtained with an accuracy of $\sigma_{xy} = 20 - 40$ nm. Diffusion constants were determined as described previously [46]. In brief, trajectories are characterized by a sequence of positions $\vec{x}(i)$, with i ranging from 1 to the number of observations of this signal. The mean square displacement $\langle r^2 \rangle$ was calculated as a function of the time lag $t_{lag} = n\left(t_{ill} + t_{delay}\right)$ according to $\langle r^2 \rangle = \left\langle \left(\vec{x}(i) - \vec{x}(n+i)\right)^2 \right\rangle_{i=1;1+n,1+2n;...}$, with n denoting the difference in frame index i. Data were analyzed by fitting with $\langle r^2 \rangle = 4Dt_{lag} + 4\sigma_{xy}^2$ (Equation (1)), yielding the lateral diffusion constant D and the single-molecule localization precision σ_{xy}.

To discriminate between two mobile fractions (see Figure S2), a bimodal model as previously described was used [46]. Briefly, the cumulative density function of the square displacements (sd) was fitted with $cdf = 1 - \alpha \cdot exp\left[-\frac{sd}{msd_1(t_{lag})}\right] - (1-\alpha) \cdot exp\left[-\frac{sd}{msd_2(t_{lag})}\right]$ (Equation (2)), yielding the proportion of the two mobile components, α and $(1 - \alpha)$, and the t_{lag}-dependent mean square displacements msd_1 and msd_2. Confined diffusion (diffusion constant D) within an impermeable circle of radius R was fitted with $msd = R^2\left(1 - exp\left[-\frac{4Dt_{lag}}{R^2}\right]\right)$.

3. Results

Phosphatidylcholine (PC) lipids play an important role in interacting with HDL particles, given their role as the major outer leaflet lipid of mammalian cell membranes [40]. This work studies the interaction of HDL particles with membranes made of different PC lipids, whose structure differs in the linkage (i.e., ester- vs. ether-linked) of the glycerol region at sn-1 and sn-2 to, and the length/saturation of two acyl chains (see Figures 1 and S1). A polar headgroup was attached at glycerol's sn-3 via a phosphodiester bond. Planar membranes were primarily used to determine phase preferences, while spherical membranes were used to analyze diffusion. In both cases, the interaction with HDL particles was examined.

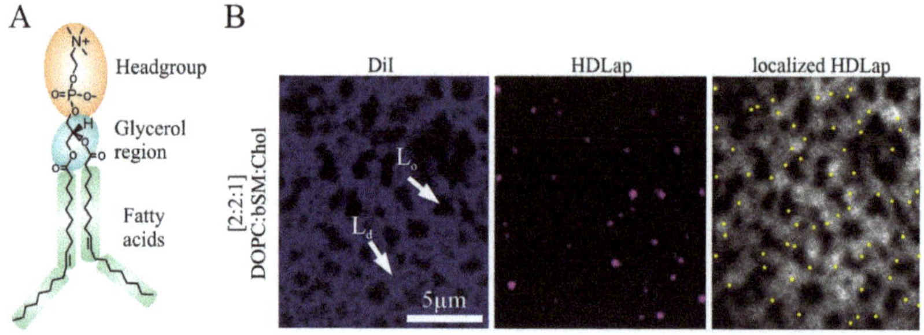

Figure 1. (**A**) The nomenclature used, exemplified by a structural model of a single DOPC (18:1) lipid. (**B**) HDL-associated protein (HDLap, fluorescently labeled with Atto 647, purple) partitions preferential to the L$_d$ phase (blue) in PSLBs ([2:2:1] mixture of DOPC (18:1), bSM, and Chol) supported on a glass surface. The bilayer was treated with DiI as an L$_d$ phase marker (shown in the first color

channel, corresponding to the image on the left) and HDL particles (shown in the second color channel, corresponding to the middle image, and represented as yellow dots in the right image, illustrating the center of mass.) (partition coefficient of K_p = 3 a.u. of HDLap (phase 1: L_d, phase 2: L_o)). The last image displays the positions of single HDLap-Atto 647 signals (indicated as yellow dots) overlaid with the DiI fluorescence image.

3.1. HDL Particles Transfer Cargo Molecules via L_d Phase Interaction

It is speculated that the plasma membrane is divided into relatively L_o and L_d lipid environments on nanometer length scales, where L_o phases represent the hypothesized "lipid rafts", mimicking low-fluidity membranes [47]. Membrane fluidity is a dynamic parameter that varies depending on the cell type and state; therefore, it is important to understand how it affects HDL particle interaction. This speculation finds support through domain formation occurring in synthetic lipid vesicles and bilayers containing saturated glycerophospholipids or sphingolipids and Chol [48]. Such biomimetic bilayers are frequently utilized as minimal systems that maintain biological relevance such as the ternary model membrane system composed of DOPC (18:1), Chol, and brain Sphingomyelin (bSM) [1]. Experiments conducted on phase-separated supported lipid bilayers (PSLBs) using a 2:2:1 mixture of DOPC (18:1), bSM, and Chol yielded two distinct phases (Figure 1B). The L_d phase is mostly enriched in DOPC (18:1), while the L_o phase is solely composed of bSM and Chol [7]. To visually distinguish the phases, specific markers such as the fluorescent lipid DiI can be used, which predominantly populate the L_d phase [34] (Figure 1B, left image).

After the addition of HDL particles, the fluorescently labeled HDL-associated proteins (HDLap) yielded a clear preference for the Ld phase (Figure 1B, middle image). The calculated partition coefficient for HDLap is K_p = 3 a.u. ± 1.2 a.u. (standard deviation) for phase 1 (L_d) and phase 2 (L_o), as illustrated in Figure 1B (right-hand image). Approximately 70% of all HDLap fluorescence signals were immobile; the remaining 30% showed confined diffusion (Figure S2A; Movie S1). Drawing from previous studies [37], these diffusion data suggest that most HDL particles were immobilized due to fusion, while the remaining fraction is primarily attached to the membrane and thus shows confined diffusion. A predominant localization of HDLap was detected at the phase interfaces (~30% of all signals), indicating interaction with phase boundaries. Line tension at phase boundaries may offer anchor points for fusion processes through the energy generated by the hydrophobic mismatch between the L_o and L_d phases being comparable to that required to form a lipid stalk intermediate [49]. The Chol-BodipyFL transferred by the HDL particle into the L_d shows free diffusion (Figure S2B). Notably, after maintaining contact with the HDL particles, Chol-BodipyFL is transferred into the L_d phase and triggers a rearrangement within the L_o phase, while the HDLap signal is retained in its initial environment (Figure S2C). In summary, these results suggest that HDL particles interact with fluid membranes predominately to facilitate cargo exchange. The next step will be to determine which specific biophysical properties of the target membrane, particularly the lipids, are responsible for enabling protein interaction or inducing cargo transfer.

3.2. Phospholipid Backbone Influences HDL Interaction: Ester vs. Ether

We compared two lipids with identical fatty acid chains, but where the ester linkage is replaced by an ether one, in experiments with giant unilamellar vesicle (GUV) membranes using a pure DOPC (18:1) or a [3:1] molar mixture of DOPC (18:1) and DietherPC (18:1). The membranes composed of pure DOPC (18:1) or a mixture of DOPC (18:1) and DietherPC (18:1) did not exhibit distinguishable phases; instead, they formed a homogeneous L_d phase membrane. The overall effect of HDL particle interaction and the associated cargo transfer can be determined indirectly using an indicator lipid (i.e., DPPE-ASR). DPPE-ASR is used as an L_d-phase [9] marker, though the lipid can likewise be used to indirectly determine overall changes in the membrane due to an altered diffusion behavior [50]. Noteworthily, no difference regarding the lateral diffusion constant of the indicator lipid was observed before HDL interaction (Figure 2A).

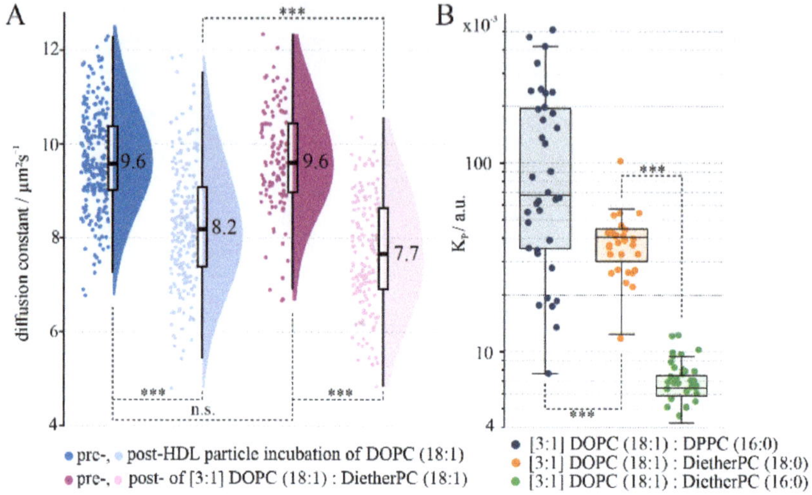

Figure 2. An HDL particle interaction with different lipid membranes. (**A**) shows the scatter, box and distribution plots of the lateral diffusion constant of DPPE-ASR within a GUV membrane of DOPC (18:1) and DOPC (18:1): DietherPC (18:1) [3:1] pre and post incubation with an HDL particle solution. Each individual point depicts the diffusion constant obtained from one measurement on the top GUV membrane: pre HDL incubation (dark blue) (DOPC (18:1) preHDL = 9.6 $\mu m^2/s \pm 0.1\ \mu m^2/s$, N = 237) and (dark magenta) (DOPC (18:1): DietherPC (18:1) preHDL = 9.6 $\mu m^2/s \pm 0.1\ \mu m^2/s$, N = 114), post HDL incubation (light blue) (DOPC (18:1) postHDL = 8.2 $\mu m^2/s \pm 0.1\ \mu m^2/s$, N = 208) and (light magenta) (DOPC (18:1): DietherPC (18:1) postHDL = 7.7 $\mu m^2/s \pm 0.1\ \mu m^2/s$, N = 153) (n.s. not significant, *** $p < 0.001$@ $\alpha = 5\%$). (**B**) shows the boxplot chart of the calculated partitioning coefficient K_p (phase 1: L_d phase (i.e., DOPC); phase 2: gel phase (i.e., DPPC, DietherPC)) regarding HDLap localization in respect of three different (i.e., lipid composition) PSLBs. (Median ± SE) DOPC (18:1): DPPC (16:0) [3:1] (blue): $K_p = 68 \cdot 10^{-3} \pm 22 \cdot 10^{-3}$ a.u. (N = 36); DOPC (18:1): DietherPC (18:0) [3:1] (orange): $K_p = 40 \cdot 10^{-3} \pm 3 \cdot 10^{-3}$ a.u. (N= 30); DOPC (18:1): DietherPC (16:0) [3:1] (green): $K_p = 6 \cdot 10^{-3} \pm 3 \cdot 10^{-4}$ a.u. (N = 30) (*** $p < 0.001$ @ $\alpha = 5\%$).

However, adding HDL particles yielded statistically significant differences: while in DOPC (18:1) membranes the indicator lipid's lateral diffusion constant decreased by 15%, in DOPC (18:1)/DietherPC (18:1) [3:1] lipid mixtures it decreased by 20%. In general, the decrease in diffusion constant can be attributed to HDLap anchoring in the membrane and to the alteration of the target membrane's composition through cargo transfer (e.g., the increased Chol content of the target membrane) [37]. The more pronounced decline in the DOPC (18:1)/DietherPC (18:1) [3:1] lipid mixture relative to pure DOPC (18:1) may be attributed to an augmented interaction with HDL, encompassing binding and cargo transfer. An alternative explanation could be that the quantitatively similar interaction has a greater effect on the DietherPC (18:1) than on DOPC (18:1) lipids.

3.3. *HDL Particles Strongly Interact with Saturated Ether Lipid Phases*

The impact of the chain length and glycerol linkage between the fatty acid chain and the head group on HDL particle interaction was investigated using planar membranes composed of two-component mixtures. Their two-dimensional nature facilitates the analysis of lipoprotein particle partitioning. When DOPC (18:1) is mixed with DPPC (16:0), a two-phase system is formed consisting of a DOPC (18:1) L_d phase and a DPPC (16:0) gel phase. After the addition of HDL particles, the calculated partition coefficient of the HDLap fluorescence signals was $K_p = 68{\cdot}10^{-3} \pm 22{\cdot}10^{-3}$ a.u. (Figure 2B, Figure S3, first row). In comparison to our measurement of a DOPC/bSM/Chol mixture (Figure 1B), wherein HDLap predominantly interacted with the L_d phase, the partition coefficient is nearly two orders of magnitude lower for the DOPC/DPPC mixture. This shows that the disparity in HDLap signal density (i.e., K_p value) between the two phases is more pronounced for the DOPC/DPPC PSLB than previously observed for the DOPC/bSM/Chol PSLB. This may be due to the decreased area per lipid of DPPC [51,52] and thus the stronger hydrogen bond network within the gel phase. Diether lipids with different chain lengths were compared to evaluate any influences of the fatty acid chain length. Similarly to DPPC (16:0) [53], saturated DietherPC lipids form gel phases. In general, a higher disparity in HDLap signal density was observed for both DietherPC lipids compared to DPPC (16:0) (Figure S3, middle and bottom illustration, K_p (DietherPC (18:0)) = $40{\cdot}10^{-3} \pm 3{\cdot}10^{-3}$ a.u., K_p (DietherPC (16:0)) = $6{\cdot}10^{-3} \pm 3{\cdot}10^{-4}$ a.u. Figure 2B). Additionally, the observation from previous measurement (Figure 2A) is supported, as the linkage of the glycerol region influences HDL interaction. However, an equally decisive factor is the length of the fatty acid chains. A comparison of the partition coefficient of DietherPC (18:0) with DietherPC (16:0) reveals a significant difference (Figure 2B). In summary, enhanced interactions are observed in two-phase lipid mixtures when ether linkages (DietherPC (18/16:0)) are present compared to ester linkages (DOPC (18:1)). Additionally, these interactions are further amplified by shorter fatty acid chain lengths (DietherPC (18:0) vs. DietherPC (16:0)).

3.4. A Multi-Correlative Analysis of Individual Lipid Properties Reveals Differences in HDL Particle Interactions and Cargo Transfer

To attain a more precise understanding of how the glycerol region mobility and thus its hydrogen network influences the interaction and subsequent cargo (i.e., Chol-BodipyFL) transfer of HDL particles, lipid vesicles consisting of DOPC (18:1), DOPC (18:1) with varying Chol content (10% and 20% Chol), POPC (16:0–18:1), DPPC (16:0), and DietherPC (16:0) were investigated. The cholesterol concentrations in DOPC membranes were not increased further, as this resulted in relatively unstable and non-reproducible target membranes, most likely caused by the incipient asymmetry of the lipid bilayers [54]. The integration of HDL-associated components (Chol-BodipyFL and HDLap-Atto 647) and the localization of the HDL particles within GUVs, produced from different lipids or lipid mixtures, (Figure S4A,B) was analyzed with confocal microscopy and correlated with Laurdan's GP value and TDFS experiments [55], both yielding information on glycerol region mobility [56]. The GP value of Laurdan is an empirical steady-state ratiometric parameter that reports on both the hydrogen bond network and the mobility of carbonyl groups in the membrane. Nonetheless, the GP value can be taken as an indicator of glycerol region mobility but not as an indicator of the extent of the hydrogen bond network, "or water penetration", of a fully hydrated, fluid L_d lipid bilayer, as often found in the literature [55]. Laurdan, which is located near the DOPC (18:1) sn-1 carbonyl group [57], makes it possible to sense the mechano-elastic properties of the linkage region of the lipid bilayer. With increasing GP value and thus decreasing glycerol region mobility, we found for L_o and L_d phase lipids a significant decrease in the average HDLap-Atto 647 and Chol-BodipyFL signals.

To quantify the initial membrane environment's influence on HDL particle interaction, we performed TDFS experiments [55], which measure the spectral shift during the relaxation of a fluorophore caused by its local environment (Figure 3A). Analysis of the frequency shift $\Delta\nu$ and the relaxation time τ_r yields independent information on the probe's environment polarity and mobility, i.e., the level and strength of the hydrogen bond network in its vicinity, respectively. Although the changes in $\Delta\nu$ are close to the error, the measured data predict an increase in the hydrogen bond network for the lipid bilayers in the L_d phase. Interestingly, the increase in $\Delta\nu$ by the addition of Chol indicates increased probe polarity although the variations are very small and close to the error. This also confirms that interpreting GP values as changes in the level of the hydrogen bond network is incorrect [55]. With increasing GP value and thus decreasing glycerol region mobility, we found for L_o- and L_d-phase lipids a marked decrease in the average HDLap-Atto 647 and Chol-BodipyFL signals (Figure 3B). In contrast to this trend, HDL particle interaction was highest for the DietherPC (16:0) gel phase, whereas the overall lowest value was detectable with DPPC (16:0). The estimated relaxation times for the DietherPC (16:0) and DPPC (16:0) lipids evidence their significantly slower relaxation, i.e., restricted glycerol region mobility, experienced in these gel-phase samples in comparison to all others in the L_d phase. GP correlates with the magnitude of τ_r. Therefore, a higher GP and higher magnitudes of τ_r signify a slower relaxation process, i.e., the reduced mobility of the sn-1 carbonyls of the lipids in the L_d phase. This can be related to an increase in the strength of the hydrogen bond network at the level of the sn-1 carbonyls in the L_d phase.

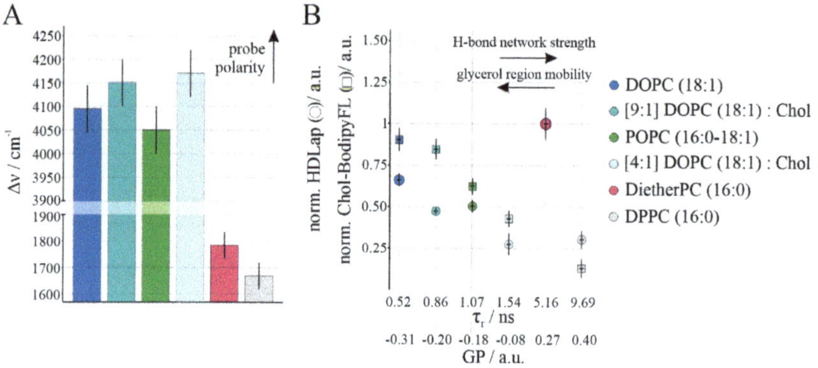

Figure 3. An HDL particle interaction with lipid membranes of different probe polarity/GP value/hydrogen bond network strength. (**A**) The bar chart depicts the total amount of fluorescence shift $\Delta\nu$ of the Laurdan emission for different GUV membrane compositions. The intrinsic uncertainty for the $\Delta\nu$ parameter is 50 cm^{-1}. The $\Delta\nu$ parameter is directly related to the hydrogen bond network level of sn-1 carbonyls. A higher magnitude of fluorescence shift reflects a higher probe polarity. (**B**) shows the relaxation time τ_r/GP value for different GUV membranes (Supplemental Table S1) in relation to normalized (i.e., both DietherPC (16:0) values were set to 1) HDL-associated protein (HDLap) Atto 647 (○) and the normalized Chol-BodipyFL (□) signal per pixel (Supplemental Figure S4). A higher magnitude of GP value and relaxation time τ_r reflect a lower glycerol region mobility and stronger hydrogen bond network, respectively. Error bars indicate the standard error of the mean.

However, the value for the estimated maxima for the DietherPC (16:0) is about half of the value determined for the DPPC (16:0) bilayer. Apparently, even being in the gel phase the glycerol region of the ether lipids showed a remarkably high mobility. Ether lipids lack the carbonyl groups that strongly contribute to the hydrogen bond network in the glycerol region of lipids. Thus, the hydrogen bond network in the glycerol region between the alkyl chain and the headgroup is an essential factor for mediating HDL particle interaction with lipid bilayers. However, when comparing protein interaction to cargo transfer, it becomes clear that the ratio reverses for gel-phase lipids, regardless of the glycerol binding site. This indicates that cargo transfer is favored by increased fluidity and reduced mobility in the glycerol region.

4. Discussion/Conclusions

Lipid composition is an important mediator of membrane fusion as it alters the physical properties of the membrane as well as the structure, organization, and dynamics of the fusion proteins [58]. In fact, some lipids bind directly to proteins and control their behavior; for example, by affecting membrane curvature [59]. Additionally, the membrane interface, particularly the dynamics of the hydrogen bond network, plays a decisive role in protein interactions [24,60]. This study focuses on the receptor-independent interaction of HDL particles and plane PC lipid membranes. The impact of specific properties of the fatty acid chains, including their length and degree of saturation, and of the glycerol linkage (ester vs. ether) was evaluated. In conjunction with ternary membrane mixtures, the relationship between HDL interaction and glycerol region mobility, and the hydrogen bond network, and consequently lipid membrane polarity, was subjected to a comprehensive investigation. Three-component mixtures demonstrate that HDL particles predominantly interact and transfer their cargo in the fluid phase, primarily composed of DOPC (18:1). Lipids with identical fatty acid chain lengths and degrees of saturation, but with an ester (i.e., DOPC (18:1)) instead of an ether (i.e., DietherPC (18:1)) glycerol linkage, exhibit a diminished interaction with HDL particles. This indicates that lipids with a stronger hydrogen bond network interact preferentially with HDL particles. In this particular case, it might be that ether lipids within a DOPC membrane facilitate the initial contact and interaction with the HDL particle and subsequently facilitate cargo exchange within the DOPC membrane. In general, saturated gel-phase lipids (i.e., DPPC (16:0)) exhibit even stronger hydrogen bond network characteristics. However, there were significant differences in the interaction between ester- and ether-linked lipids, with the ether bonds leading to a stronger interaction due to decreased glycerol region mobility. Ether lipids lack the carbonyl groups that strongly contribute to the hydrogen bond network in the glycerol region of lipids. Thus, the hydrogen bond network in the glycerol region between the alkyl chain and the headgroup is an essential factor for mediating HDL particle interactions with membranes. Both gel-phase and fluid-phase membranes show interactions with HDL particles. However, there are significant differences in the nature of these interactions. While binding occurs preferentially in gel-phase membranes, fluid-phase membranes facilitate cargo transfer. Cholesterol in the target membrane reduces the interaction with HDL particles by increasing the packing density and stiffness of the membrane, thereby modulating its fluidity. Cholesterol is small enough to fill the space between lipid head groups and thus influences the overall stiffness and fluidity of membrane components [61]. Recently, a study conducted by Yesylevskyy et al. [62] showed by atomistic simulation that the order parameter of lipid tails decreases significantly in curved membranes, whereas the area per lipid increases in the convex monolayer and decreases in the concave monolayer. The concomitant lipid composition based on membrane curvature massively influences the interaction. In addition to the cholesterol redistribution within the membrane, curvature modulates phase separation in lipid bilayer membranes [63]. In addition, lipid properties such as saturation state, headgroup size, and chain length, play an essential role in protein interaction [64]. A less noticed property of lipids is their hydrophilicity/hydrophilicity within the membrane assembly [65]. Moreover, increased Chol content reduces the lipid's hydrogen bond network [66], reduces passive permeability, and increases the GP value [28]. In addition to the hydrophobic and van der Waals forces, the 3-OH group of Chol to PC lipids forms H-bonds with sn-2 carbonyl and phosphate groups in the glycerol linkage region of the bilayer. Membrane fluidity and hydrogen bond network strength change between different membrane phases [67]. Moreover, the length of the fatty acid chain also appears to influence the interaction, likely due to a decrease in membrane rigidity and thus increased mobility [68]. Lipid transfer from lipoprotein particles to membranes depends on the properties of the lipid itself and the contact process with the plasma membrane. Moreover, membrane curvature affects lipid exposure and, consequently, the contact process. In future studies, we aim to focus more closely on the influence of membrane curvature in conjunction with the corresponding lipid composition. Consequently, the plasma membrane itself yields a high binding capacity for HDL particles.

In summary, the influence of chain length and the degree of saturation of the fatty acid chains, as well as the cholesterol content in the membrane, could be discerned. Major factors for increased interaction between lipid membranes and HDL particles are properties such as glycerol region mobility. By increasing the accessible surface area in the glycerol region, interaction occurs preferentially. However, the data indicate that neither the hydrogen bond network nor the glycerol region mobility alone are sufficient to fully explain the interaction with HDL particles.

Supplementary Materials: The following supporting information can be downloaded at: https://www.mdpi.com/article/10.3390/membranes14120261/s1, Supplementary Figure S1: Overview of the structure of the PC lipids used, Figure S2:Tracking analysis of the HDL-associated protein–Atto 647 signal and of Chol-BodipyFL signal, Figure S3: Partitioning of HDL-associated protein (HDLap) on PSLBs with different lipid compositions, Figure S4: HDL particle interaction with lipid membranes. Supplemental Table S1: Data of the different GUV membranes presented in Figure 3.

Author Contributions: The manuscript was written through the contributions of B.P., M.A. (Markus Axmann), H.S., M.H. and E.S. B.P., M.A. (Markus Axmann), H.S., G.J.S. and E.S., obtained funding and performed supervision. B.P., E.S., M.A. (Markus Axmann), M.A. (Mariana Amaro), T.S., A.H. and F.W. performed experiments and data analysis. All authors have read and agreed to the published version of the manuscript.

Funding: This research was funded in part by the Austrian Science Fund Projects (P33481-B, P 29110). This project has received funding from the European Union's Horizon 2020 research and innovation programme under grant agreement No. 101017902. E.S. is supported by a Swedish Research Council Starting Grant (grant no. 2020-02682).

Institutional Review Board Statement: Not applicable.

Data Availability Statement: The data presented in this study are available from the corresponding author upon request. The data are not publicly available due to restrictions of the involved scientists and facilities.

Acknowledgments: We wish to thank Ingrid Hassl for performing the HDL particle preparations.

Conflicts of Interest: The authors declare no conflicts of interest.

References

1. Heberle, F.A.; Feigenson, G.W. Phase Separation in Lipid Membranes. *Cold Spring Harb. Perspect. Biol.* **2011**, *3*, a004630. [CrossRef] [PubMed]
2. Janiak, M.J.; Small, D.M.; Shipley, G.G. Temperature and Compositional Dependence of the Structure of Hydrated Dimyristoyl Lecithin. *J. Biol. Chem.* **1979**, *254*, 6068–6078. [CrossRef] [PubMed]
3. Korlach, J.; Schwille, P.; Webb, W.W.; Feigenson, G.W. Characterization of Lipid Bilayer Phases by Confocal Microscopy and Fluorescence Correlation Spectroscopy. *Proc. Natl. Acad. Sci. USA* **1999**, *96*, 8461–8466. [CrossRef]
4. Ulrich, A.S.; Sami, M.; Watts, A. Hydration of DOPC Bilayers by Differential Scanning Calorimetry. *Biochim. Biophys. Acta (BBA)-Biomembr.* **1994**, *1191*, 225–230. [CrossRef]
5. Hjort Ipsen, J.; Karlström, G.; Mourtisen, O.G.; Wennerström, H.; Zuckermann, M.J. Phase Equilibria in the Phosphatidylcholine-Cholesterol System. *Biochim. Biophys. Acta (BBA)-Biomembr.* **1987**, *905*, 162–172. [CrossRef]
6. Sezgin, E.; Levental, I.; Grzybek, M.; Schwarzmann, G.; Mueller, V.; Honigmann, A.; Belov, V.N.; Eggeling, C.; Coskun, Ü.; Simons, K.; et al. Partitioning, Diffusion, and Ligand Binding of Raft Lipid Analogs in Model and Cellular Plasma Membranes. *Biochim. Biophys. Acta (BBA)-Biomembr.* **2012**, *1818*, 1777–1784. [CrossRef]
7. Kaiser, H.-J.; Lingwood, D.; Levental, I.; Sampaio, J.L.; Kalvodova, L.; Rajendran, L.; Simons, K. Order of Lipid Phases in Model and Plasma Membranes. *Proc. Natl. Acad. Sci. USA* **2009**, *106*, 16645–16650. [CrossRef]
8. Schachter, I.; Paananen, R.O.; Fábián, B.; Jurkiewicz, P.; Javanainen, M. The Two Faces of the Liquid Ordered Phase. *J. Phys. Chem. Lett.* **2022**, *13*, 1307–1313. [CrossRef]
9. Sych, T.; Gurdap, C.O.; Wedemann, L.; Sezgin, E. How Does Liquid-Liquid Phase Separation in Model Membranes Reflect Cell Membrane Heterogeneity? *Membranes* **2021**, *11*, 323. [CrossRef]
10. Shimshick, E.J.; McConnell, H.M. Lateral Phase Separation in Phospholipid Membranes. *Biochemistry* **1973**, *12*, 2351–2360. [CrossRef]
11. Semrau, S.; Idema, T.; Schmidt, T.; Storm, C. Membrane-Mediated Interactions Measured Using Membrane Domains. *Biophys. J.* **2009**, *96*, 4906–4915. [CrossRef] [PubMed]
12. Dean, J.M.; Lodhi, I.J. Structural and Functional Roles of Ether Lipids. *Protein Cell* **2018**, *9*, 196–206. [CrossRef] [PubMed]
13. Braverman, N.E.; Moser, A.B. Functions of Plasmalogen Lipids in Health and Disease. *Biochim. Biophys. Acta (BBA)-Mol. Basis Dis.* **2012**, *1822*, 1442–1452. [CrossRef]
14. Pike, L.J.; Han, X.; Chung, K.-N.; Gross, R.W. Lipid Rafts Are Enriched in Arachidonic Acid and Plasmenylethanolamine and Their Composition Is Independent of Caveolin-1 Expression: A Quantitative Electrospray Ionization/Mass Spectrometric Analysis. *Biochemistry* **2002**, *41*, 2075–2088. [CrossRef]
15. Rodemer, C.; Thai, T.-P.; Brugger, B.; Kaercher, T.; Werner, H.; Nave, K.-A.; Wieland, F.; Gorgas, K.; Just, W.W. Inactivation of Ether Lipid Biosynthesis Causes Male Infertility, Defects in Eye Development and Optic Nerve Hypoplasia in Mice. *Hum. Mol. Genet.* **2003**, *12*, 1881–1895. [CrossRef]

16. Mandel, H.; Sharf, R.; Berant, M.; Wanders, R.J.A.; Vreken, P.; Aviram, M. Plasmalogen Phospholipids Are Involved in HDL-Mediated Cholesterol Efflux: Insights from Investigations with Plasmalogen-Deficient Cells. *Biochem. Biophys. Res. Commun.* **1998**, *250*, 369–373. [CrossRef]
17. Vance, J.E. Lipoproteins Secreted by Cultured Rat Hepatocytes Contain the Antioxidant 1-Alk-1-Enyl-2-Acylglycerophosphoethanolamine. *Biochim. Biophys. Acta (BBA)-Lipids Lipid Metab.* **1990**, *1045*, 128–134. [CrossRef]
18. an Meer, G.; Voelker, D.R.; Feigenson, W.G. Membrane Lipids: Where They Are and How They Behave. *Nat. Rev. Mol. Cell Biol.* **2009**, *9*, 112–124. [CrossRef]
19. Raghava, S.; Giorda, K.M.; Romano, F.B.; Heuck, A.P.; Hebert, D.N. The SV40 Late Protein VP4 Is a Viroporin That Forms Pores to Disrupt Membranes for Viral Release. *PLoS Pathog.* **2011**, *7*, e1002116. [CrossRef]
20. Diomede, L.; Bizzi, A.; Magistrelli, A.; Modest, E.J.; Salmona, M.; Noseda, A. Role of Cell Cholesterol in Modulating Antineoplastic Ether Lipid Uptake, Membrane Effects and Cytotoxicity. *Int. J. Cancer* **1990**, *46*, 341–346. [CrossRef]
21. Corradi, V.; Sejdiu, B.I.; Mesa-Galloso, H.; Abdizadeh, H.; Noskov, S.Y.; Marrink, S.J.; Tieleman, D.P. Emerging Diversity in Lipid–Protein Interactions. *Chem. Rev.* **2019**, *119*, 5775–5848. [CrossRef] [PubMed]
22. Escribá, P.V. Membrane-Lipid Therapy: A New Approach in Molecular Medicine. *Trends Mol. Med.* **2006**, *12*, 34–43. [CrossRef] [PubMed]
23. Escribá, P.V.; González-Ros, J.M.; Goñi, F.M.; Kinnunen, P.K.J.; Vigh, L.; Sánchez-Magraner, L.; Fernández, A.M.; Busquets, X.; Horváth, I.; Barceló-Coblijn, G. Membranes: A Meeting Point for Lipids, Proteins and Therapies. *J. Cell. Mol. Med.* **2008**, *12*, 829–875. [CrossRef] [PubMed]
24. Karathanou, K.; Bondar, A.-N. Dynamic Water Hydrogen-Bond Networks at the Interface of a Lipid Membrane Containing Palmitoyl-Oleoyl Phosphatidylglycerol. *J. Membr. Biol.* **2018**, *251*, 461–473. [CrossRef] [PubMed]
25. Murzyn, K.; Zhao, W.; Karttunen, M.; Kurdziel, M.; Róg, T. Dynamics of Water at Membrane Surfaces: Effect of Headgroup Structure. *Biointerphases* **2006**, *1*, 98–105. [CrossRef]
26. Berkowitz, M.L.; Vácha, R. Aqueous Solutions at the Interface with Phospholipid Bilayers. *Acc. Chem. Res.* **2012**, *45*, 74–82. [CrossRef]
27. Nagao, M.; Kelley, E.G.; Ashkar, R.; Bradbury, R.; Butler, P.D. Probing Elastic and Viscous Properties of Phospholipid Bilayers Using Neutron Spin Echo Spectroscopy. *J. Phys. Chem. Lett.* **2017**, *8*, 4679–4684. [CrossRef]
28. Orlikowska-Rzeznik, H.; Krok, E.; Chattopadhyay, M.; Lester, A.; Piatkowski, L. Laurdan Discerns Lipid Membrane Hydration and Cholesterol Content. *J. Phys. Chem. B* **2023**, *127*, 3382–3391. [CrossRef]
29. Doole, F.T.; Kumarage, T.; Ashkar, R.; Brown, M.F. Cholesterol Stiffening of Lipid Membranes. *J. Membr. Biol.* **2022**, *255*, 385–405. [CrossRef]
30. Harris, F.M.; Best, K.B.; Bell, J.D. Use of Laurdan Fluorescence Intensity and Polarization to Distinguish between Changes in Membrane Fluidity and Phospholipid Order. *Biochim. Biophys. Acta (BBA)-Biomembr.* **2002**, *1565*, 123–128. [CrossRef]
31. Scollo, F.; Evci, H.; Amaro, M.; Jurkiewicz, P.; Sykora, J.; Hof, M. What Does Time-Dependent Fluorescence Shift (TDFS) in Biomembranes (and Proteins) Report on? *Front. Chem.* **2021**, *9*, 738350. [CrossRef] [PubMed]
32. Santos, N.C.; Prieto, M.; Castanho, M.A.R.B. Quantifying Molecular Partition into Model Systems of Biomembranes: An Emphasis on Optical Spectroscopic Methods. *Biochim. Biophys. Acta (BBA)-Biomembr.* **2003**, *1612*, 123–135. [CrossRef]
33. Rusu, L.; Gambhir, A.; McLaughlin, S.; Rädler, J. Fluorescence Correlation Spectroscopy Studies of Peptide and Protein Binding to Phospholipid Vesicles. *Biophys. J.* **2004**, *87*, 1044–1053. [CrossRef]
34. Baumgart, T.; Hunt, G.; Farkas, E.R.; Webb, W.W.; Feigenson, G.W. Fluorescence Probe Partitioning between Lo/Ld Phases in Lipid Membranes. *Biochim. Biophys. Acta (BBA)-Biomembr.* **2007**, *1768*, 2182–2194. [CrossRef]
35. Li, Z.; Mintzer, E.; Bittman, R. First Synthesis of Free Cholesterol−BODIPY Conjugates. *J. Org. Chem.* **2006**, *71*, 1718–1721. [CrossRef]
36. Thomas, F.A.; Visco, I.; Petrášek, Z.; Heinemann, F.; Schwille, P. Introducing a Fluorescence-Based Standard to Quantify Protein Partitioning into Membranes. *Biochim. Biophys. Acta (BBA)-Biomembr.* **2015**, *1848*, 2932–2941. [CrossRef]
37. Plochberger, B.; Röhrl, C.; Preiner, J.; Rankl, C.; Brameshuber, M.; Madl, J.; Bittman, R.; Ros, R.; Sezgin, E.; Eggeling, C.; et al. HDL Particles Incorporate into Lipid Bilayers–a Combined AFM and Single Molecule Fluorescence Microscopy Study. *Sci. Rep.* **2017**, *7*, 15886. [CrossRef]
38. Kim, H.M.; Choo, H.; Jung, S.; Ko, Y. A Two-Photon Fluorescent Probe for Lipid Raft Imaging: C-Laurdan. *ChemBioChem* **2007**, *8*, 553–559. [CrossRef]
39. Schumaker, V.N.; Puppione, D.L. Sequential Flotation Ultracentrifugation. *Methods Enzymol.* **1986**, *128*, 155–170.
40. Boban, Z.; Mardešić, I.; Subczynski, W.K.; Raguz, M. Giant Unilamellar Vesicle Electroformation: What to Use, What to Avoid, and How to Quantify the Results. *Membranes* **2021**, *11*, 860. [CrossRef]
41. Parini, P.; Johansson, L.; Bröijersén, A.; Angelin, B.; Rudling, M. Lipoprotein Profiles in Plasma and Interstitial Fluid Analyzed with an Automated Gel-filtration System. *Eur. J. Clin. Investig.* **2006**, *36*, 98–104. [CrossRef] [PubMed]
42. Waithe, D.; Clausen, M.P.; Sezgin, E.; Eggeling, C. FoCuS-Point: Software for STED Fluorescence Correlation and Time-Gated Single Photon Counting. *Bioinformatics* **2016**, *32*, 958–960. [CrossRef] [PubMed]
43. Parasassi, T.; De Stasio, G.; D'Ubaldo, A.; Gratton, E. Phase Fluctuation in Phospholipid Membranes Revealed by Laurdan Fluorescence. *Biophys. J.* **1990**, *57*, 1179–1186. [CrossRef]

44. Horng, M.L.; Gardecki, J.A.; Papazyan, A.; Maroncelli, M. Subpicosecond Measurements of Polar Solvation Dynamics: Coumarin 153 Revisited. *J. Phys. Chem.* **1995**, *99*, 17311–17337. [CrossRef]
45. Fee, R.S.; Maroncelli, M. Estimating the Time-Zero Spectrum in Time-Resolved Emmsion Measurements of Solvation Dynamics. *Chem. Phys.* **1994**, *183*, 235–247. [CrossRef]
46. Wieser, S.; Schütz, G.J. Tracking Single Molecules in the Live Cell Plasma Membrane-Do's and Don't's. *Methods* **2008**, *46*, 131–140. [CrossRef]
47. Simons, K.; Gerl, M.J. Revitalizing Membrane Rafts: New Tools and Insights. *Nat. Rev. Mol. Cell Biol.* **2010**, *11*, 688–699. [CrossRef]
48. Dietrich, C.; Bagatolli, L.A.; Volovyk, Z.N.; Thompson, N.L.; Levi, M.; Jacobson, K.; Gratton, E. Lipid Rafts Reconstituted in Model Membranes. *Biophys. J.* **2001**, *80*, 1417–1428. [CrossRef]
49. Yang, S.-T.; Kiessling, V.; Tamm, L.K. Line Tension at Lipid Phase Boundaries as Driving Force for HIV Fusion Peptide-Mediated Fusion. *Nat. Commun.* **2016**, *7*, 11401. [CrossRef]
50. Sych, T.; Schlegel, J.; Barriga, H.M.G.; Ojansivu, M.; Hanke, L.; Weber, F.; Beklem Bostancioglu, R.; Ezzat, K.; Stangl, H.; Plochberger, B.; et al. High-Throughput Measurement of the Content and Properties of Nano-Sized Bioparticles with Single-Particle Profiler. *Nat. Biotechnol.* **2023**, *42*, 587–590. [CrossRef]
51. Kelley, E.G.; Butler, P.D.; Ashkar, R.; Bradbury, R.; Nagao, M. Scaling Relationships for the Elastic Moduli and Viscosity of Mixed Lipid Membranes. *Proc. Natl. Acad. Sci. USA* **2020**, *117*, 23365–23373. [CrossRef] [PubMed]
52. Lindblom, G.; Orädd, G. Lipid Lateral Diffusion and Membrane Heterogeneity. *Biochim. Biophys. Acta (BBA)-Biomembr.* **2009**, *1788*, 234–244. [CrossRef] [PubMed]
53. Nagle, J.F.; Cognet, P.; Dupuy, F.G.; Tristram-Nagle, S. Structure of Gel Phase DPPC Determined by X-Ray Diffraction. *Chem. Phys. Lipids* **2019**, *218*, 168–177. [CrossRef] [PubMed]
54. Adhyapak, P.R.; Panchal, S.V.; Murthy, A.V.R. Cholesterol Induced Asymmetry in DOPC Bilayers Probed by AFM Force Spectroscopy. *Biochim. Biophys. Acta (BBA)-Biomembr.* **2018**, *1860*, 953–959. [CrossRef]
55. Amaro, M.; Sachl, R.; Jurkiewicz, P.; Coutinho, A.; Prieto, M.; Hof, M. Time-Resolved Fluorescence in Lipid Bilayers: Selected Applications and Advantages over Steady State. *Biophys. J.* **2014**, *107*, 2751–2760. [CrossRef]
56. Pruchnik, H.; Kral, T.; Hof, M. Interaction of Newly Platinum(II) with Tris(2-Carboxyethyl)Phosphine Complex with DNA and Model Lipid Membrane. *J. Membr. Biol.* **2017**, *250*, 461–470. [CrossRef]
57. Jurkiewicz, P.; Olz, A.; Langner, M.; Hof, M.; Heyro, J. V Headgroup Hydration and Mobility of DOTAP/DOPC Bilayers: A Fluorescence Solvent Relaxation Study. *Langmuir* **2006**, *22*, 8741–8749. [CrossRef]
58. Meher, G.; Chakraborty, H. Membrane Composition Modulates Fusion by Altering Membrane Properties and Fusion Peptide Structure. *J. Membr. Biol.* **2019**, *252*, 261–272. [CrossRef]
59. Wang, W.; Yang, L.; Huang, H.W. Evidence of Cholesterol Accumulated in High Curvature Regions: Implication to the Curvature Elastic Energy for Lipid Mixtures. *Biophys. J.* **2007**, *92*, 2819–2830. [CrossRef]
60. Bondar, A.-N.; White, S.H. Hydrogen Bond Dynamics in Membrane Protein Function. *Biochim. Biophys. Acta (BBA)-Biomembr.* **2012**, *1818*, 942–950. [CrossRef]
61. Needham, D.; Nunn, R.S. Elastic Deformation and Failure of Lipid Bilayer Membranes Containing Cholesterol. *Biophys. J.* **1990**, *58*, 997–1009. [CrossRef] [PubMed]
62. Yesylevskyy, S.O.; Rivel, T.; Ramseyer, C. The Influence of Curvature on the Properties of the Plasma Membrane. Insights from Atomistic Molecular Dynamics Simulations. *Sci. Rep.* **2017**, *7*, 16078. [CrossRef] [PubMed]
63. Parthasarathy, R.; Yu, C.; Groves, J.T. Curvature-Modulated Phase Separation in Lipid Bilayer Membranes. *Langmuir* **2006**, *22*, 5095–5099. [CrossRef] [PubMed]
64. Larsen, J.B.; Kennard, C.; Pedersen, S.L.; Jensen, K.J.; Uline, M.J.; Hatzakis, N.S.; Stamou, D. Membrane Curvature and Lipid Composition Synergize to Regulate N-Ras Anchor Recruitment. *Biophys. J.* **2017**, *113*, 1269–1279. [CrossRef]
65. Campelo, F.; McMahon, H.T.; Kozlov, M.M. The Hydrophobic Insertion Mechanism of Membrane Curvature Generation by Proteins. *Biophys. J.* **2008**, *95*, 2325–2339. [CrossRef]
66. Pandit, S.A.; Bostick, D.; Berkowitz, M.L. Complexation of Phosphatidylcholine Lipids with Cholesterol. *Biophys. J.* **2004**, *86*, 1345–1356. [CrossRef]
67. M'Baye, G.; Mély, Y.; Duportail, G.; Klymchenko, A.S. Liquid Ordered and Gel Phases of Lipid Bilayers: Fluorescent Probes Reveal Close Fluidity but Different Hydration. *Biophys. J.* **2008**, *95*, 1217–1225. [CrossRef]
68. Bianchetti, G.; Azoulay-Ginsburg, S.; Keshet-Levy, N.Y.; Malka, A.; Zilber, S.; Korshin, E.E.; Sasson, S.; De Spirito, M.; Gruzman, A.; Maulucci, G. Investigation of the Membrane Fluidity Regulation of Fatty Acid Intracellular Distribution by Fluorescence Lifetime Imaging of Novel Polarity Sensitive Fluorescent Derivatives. *Int. J. Mol. Sci.* **2021**, *22*, 3106. [CrossRef]

membranes

MDPI

Review

Modulation of Biological Membranes Using Small-Molecule Compounds to Counter Toxicity Caused by Amyloidogenic Proteins

Raina Marie Seychell [1], Adam El Saghir [1] and Neville Vassallo [1,2,*]

[1] Department of Physiology and Biochemistry, Faculty of Medicine and Surgery, University of Malta, MSD 2080 Msida, Malta
[2] Centre for Molecular Medicine and Biobanking, University of Malta, MSD 2080 Msida, Malta
* Correspondence: neville.vassallo@um.edu.mt

Abstract: The transition of peptides or proteins along a misfolding continuum from soluble functional states to pathological aggregates, to ultimately deposit as amyloid fibrils, is a process that underlies an expanding group of human diseases—collectively known as protein-misfolding disorders (PMDs). These include common and debilitating conditions, such as Alzheimer's disease, Parkinson's disease, and type-2 diabetes. Compelling evidence has emerged that the complex interplay between the misfolded proteins and biological membranes is a key determinant of the pathogenic mechanisms by which harmful amyloid entities are formed and exert their cytotoxicity. Most efforts thus far to develop disease-modifying treatments for PMDs have largely focused on anti-aggregation strategies: to neutralise, or prevent the formation of, toxic amyloid species. Herein, we review the critical role of the phospholipid membrane in mediating and enabling amyloid pathogenicity. We consequently propose that the development of small molecules, which have the potential to uniquely modify the physicochemical properties of the membrane and make it more resilient against damage by misfolded proteins, could provide a novel therapeutic approach in PMDs. By way of an example, natural compounds shown to intercalate into lipid bilayers and inhibit amyloid–lipid interactions, such as the aminosterols, squalamine and trodusquamine, cholesterol, ubiquinone, and select polyphenols, are discussed. Such a strategy would provide a novel approach to counter a wide range of toxic biomolecules implicit in numerous human amyloid pathologies.

Keywords: protein-misfolding diseases; amyloid; protein aggregation; lipid membranes; small molecules; therapeutics

1. Introduction

Protein-misfolding disorders (PMDs) are a growing challenge in modern medicine, marked by their complex pathogenesis, lack of effective treatments and increasing prevalence. These clinical syndromes, broadly described as proteinopathies, are characterised by aberrant protein folding, self-assembly and, ultimately, deposition of large insoluble amyloid assemblies. In healthy living systems, proteins adopt various conformational states, which are governed by complex kinetic and thermodynamic equilibria, together with intricate quality control mechanisms, regulating and maintaining proteome integrity and cellular health. However, in the event of erroneous folding, some proteins can acquire toxic, non-native conformations, leading to cellular damage and disease progression [1,2].

A large group of human disorders have been linked to this toxic sequence of protein misfolding and amyloid deposition. They include degenerative, and highly debilitating diseases that burden ageing populations, most notably Alzheimer's disease (AD), Parkinson's disease (PD), type-2 diabetes mellitus (T2DM), as well as a number of systemic conditions [1,3]. A summarised list of amyloidogenic proteins associated with human diseases is provided in Table 1. Despite the absence of similarities amongst the amyloid-associated proteins in the primary sequence or function, they all share a common pathogenic pathway involving the restructuring of proteins into pathological aggregates, with eventual formation of amyloid fibrils [4,5].

Table 1. Selected amyloidogenic proteins and peptides with their associated human diseases (reviewed in [1,3,6]).

Peptide/Protein	Associated Disease/s
Amyloid-β peptide (Aβ)	Alzheimer's disease
Atrial natriuretic factor	Atrial amyloidosis
Calcitonin	Medullary carcinoma of the thyroid
Fragments of immunoglobulin light chains	Light-chain amyloidosis
Huntingtin protein (htt)	Huntington's disease (HD)
Human islet amyloid polypeptide (hIAPP)/amylin	Type-2 diabetes mellitus
Lysozyme	Lysozyme amyloidosis
N-terminal fragments of prolactin	Pituitary prolactinoma
Prion protein (PrP)	Creutzfeldt–Jakob disease Spongiform encephalopathies
Microtubule-associated protein (tau)	Alzheimer's disease Frontotemporal dementia
Transthyretin	Senile systemic amyloidosis
α-synuclein (α-syn)	Parkinson's disease Multiple system atrophy

Amyloidogenesis commonly involves what are referred to as intrinsically disordered peptides (IDPs). The amyloid-β (Aβ), amylin (also known as the human islet amyloid polypeptide, hIAPP), α-synuclein (α-syn), tau and prion proteins all belong to the amyloidogenic IDP family; these species are largely unstructured in their native form and can adopt a continuum of transient conformations in the aqueous phase [7]. The course of fibril assembly by amyloid proteins follows a nucleation-dependent polymerisation pathway. The kinetics of amyloid formation generally exhibit a sigmoidal pattern, comprising three distinct phases of growth; the initial lag phase is followed by rapid growth and towards completion, the curve reaches a plateau [8]. The initial step in this process entails the formation of transient oligomeric species, which are dynamic and heterogeneous in structure [9]. The lag phase is predominantly populated by monomeric precursors, which aggregate into critical nuclei via distinct phases of growth; the initial lag phase is followed by rapid growth and towards completion, the curve reaches a plateau. The growth-competent nuclei act as a template for the accumulation of additional protein molecules [10,11]. During the elongation phase, monomers and small oligomeric assemblies continue to interact, while the prefibrillar aggregates rapidly extend and develop into ordered protofibrils. Finally, the saturation phase is reached, wherein the slope of the kinetic curve reaches a plateau due to monomer consumption and protofibrils assemble into end-stage fibrils [5,11]. The resulting amyloid fibrils exhibit a common structural motif in the fibrillar state, described as the cross-β-spine, comprised of a highly ordered structure rich in intermolecular β-strands arranged perpendicular to the main fibril axis [1,12].

In this complicated cascade of events, soluble oligomeric assemblies are generated, which are generally described as the primary cytotoxic species along the aggregation pathway. The mechanisms of oligomer-mediated cellular toxicity include the generation of metal and ionic dyshomeostasis, induction of elevated calcium levels, perturbation of mitochondrial function and generation of reactive oxygen species (ROS), amongst others [13]. Compared to amyloid fibrils, these low-molecular-weight conformers possess distinct molecular features that enhance their potential to induce toxicity, such as their small size, exposed hydrophobic patches and unsaturated edge-strands [14,15]. These physicochemical features allow promiscuity in interactions with a range of cellular components. Particularly, oligomers possess a high affinity for biological membranes and have the capacity to insert into lipid bilayers due to their amphipathic nature [7,13]. Furthermore, accumulating evidence indicates that amyloid formation and cytotoxicity are membraneassisted processes, with lipid bilayers presenting a platform for initiating the conversion of the IDPs into β-sheet-rich aggregates that are conductive to fibril formation [16,17].

2. Toxic Interplay Between Amyloid Proteins and Membranes

Lipid membranes serve as a highly complex hydrophobic boundary between the homeostatic intracellular space and the highly fluctuating extracellular environment, as well as providing a functional barrier between subcellular compartments. Phospholipids are the primary constituent of biological membranes, together with sphingolipids and sterols [18,19]. Due to the inherent complexity of membranes, many researchers utilise structurally and compositionally simplified membrane models, including monolayers, micelles, liposomes, and supported lipid bilayers (SBLMs), to probe the interactions between amyloidogenic assemblies and membranes [20].

The interaction of amyloid proteins with phospholipid bilayers is essentially a two-way interaction [21,22]. On the one hand, the detrimental effects of amyloid aggregates are strongly associated with their interactions with biomembranes [23–26]. On the other hand, lipid bilayers can also fulfil a catalytic role in the generation of fibrillar amyloids, with a growing body of research suggesting that the toxic cascade of amyloid assembly commences at membrane surfaces [27–31]. Hence, the protein–lipid environment plays a fundamental role in the kinetics of both fibrillogenesis and cytotoxicity, with the interaction between amyloid-forming peptides and lipid membranes leading to a mutually destructive perturbation of both parties.

The bilayer composition exerts a profound influence on the routes of adsorption, incorporation, orientation, structure, and assembly pathways of amyloid peptides [32,33]. Most notably, negatively charged headgroups of phospholipids, like phosphatidic acid (PA), phosphatidylserine (PS), and phosphatidylinositol (PI), strongly favour aggregation on membrane surfaces. The influence of anionic phospholipids on membrane binding, aggregation and disruption has been well documented for the hIAPP [31,34,35], htt [36], α-syn [37,38], Aβ [39,40], and tau [41,42] proteins. Other membrane components are also important for amyloid–membrane interactions [40]. For example, variations in the relative size of the phospholipid headgroup and acyl chains can give rise to local membrane defects and intrinsic curvature, which in turn influence the extent of the peptide alignment and adsorption at the membrane surface [43]. Smaller headgroups accelerate the kinetics of Aβ40 fibril assembly, shortening the lag time for primary nucleation [40]. Recent evidence indicates that the lipid bilayer thickness may also influence Aβ membrane-associated aggregation. Interestingly, membrane bilayers with the shorter-chain lipid dilauroyl phosphatidylcholine (DLPC) remodelled preformed Aβ fibrils into toxic misfolded aggregates, suggesting that pathologically thin bilayers may have a role in Aβ aggregation [44]. Experiments using quartz crystal microbalance–dissipation (QCM-D) confirmed that the hydrophobic thickness of supported bilayers impacts the membrane remodelling triggered upon adsorption of Aβ [45].

The presence of a high curvature [46–48] and packing defects [49–51] are reported to facilitate membrane association and permeabilisation events. Aβ has been reported to preferentially associate with small unilamellar vesicles (SUVs, ~15–50 nm) characterised by higher surface curvature, and such architecture accelerates the nucleation step [47,52]. Due to the presence of more water-accessible hydrophobic regions, SUVs most likely trigger the aggregation process by concentrating Aβ monomers sufficiently to initiate nucleation [47,48]. The binding of Aβ to membranes is also profoundly influenced by lipid packing defects and electrostatic interactions with the lipid headgroups. Packing defects are induced by the incorporation of lipids, which do not possess a cylindrical morphology. Specifically, no significant binding was observed between Aβ40 and membranes composed of 1,2-dioleoyl-sn-glycero-3-phospho-(1′-rac-glycerol) (DOPC; a cylindrical and zwitterionic lipid), in contrast to significant early binding events with 1,2-dioleoyl-sn-glycero-3-phospho-(1′-rac-glycerol) (DOPG; a cylindrical and anionic lipid) and PI (an inverted conical and anionic lipid) [51]. αS has also been described to possess an affinity for lipid-packing defects present on highly curved lipid surfaces as well as in planar membranes with cone-shaped lipids [50,53,54].

hIAPP is described as a curvature-sensing and curvature-inducing peptide possessing an affinity for high curvature regions of bilayers and, in turn, causing further negative curvature strain within membranes. Supporting this, hIAPP is demonstrated to preferentially localise to mitochondrial cristae, which are highly curved structures, rich in PE and CL [46].

Membrane destabilisation is described as a key patho-mechanism for PMDs. Compelling evidence provides support for the hypothesis that cytotoxicity is predominantly triggered by the interaction of prefibrillar amyloids with cellular lipid bilayers. This interaction leads to bilayer perturbation, inducing membrane leakage, disrupting ionic homeostasis, and triggering cellular dysfunction, oxidative stress and apoptosis [55–59]. Although it is generally accepted that the disruptive effects of amyloid-forming proteins are strongly correlated with their ability to interact with and disturb biological membranes, the exact molecular mechanisms governing these deleterious interactions remain elusive [60]. The three primary mechanistic models that have been proposed to describe the mechanisms by which amyloid assemblies cause membrane damage are (i) the creation of stable transmembrane protein pores, (ii) the destabilisation of the bilayer through a "carpet model", and (iii) the removal of lipids via detergent-like membrane dissolution [60,61]. Furthermore, the insertion of amyloid peptides into lipid bilayers can disrupt the structural integrity of the membrane by inducing local bending deformations and curvature stress [32,62,63]. In addition to the interactions of early aggregate species with membranes, one should also consider the effects of amyloid fibrils on bilayer integrity. A fibrillar amyloid induces membrane damage via its elongation on the membrane surface through secondary nucleation, which has been especially well documented for hIAPP [64,65]. It is highly probable that the mechanisms in operation are not mutually exclusive; instead, membrane disruption is instigated by the combined effects of multiple models [61,66].

3. Small-Molecule Amyloid Inhibitors: The Case for Modulating Lipid Membranes

The complex pathophysiology of amyloid diseases, involving a highly heterogenous population of protein aggregate species that can exert their cytotoxic effects through a wide variety mechanisms, renders the development of effective treatments for these disorders exceptionally challenging. As a result, different therapeutic modalities have been explored to address these disorders, with most efforts to date being focused on the discovery of potent anti-aggregation agents. In this regard, several natural and synthetic small molecules have been identified that can target different aspects of the aggregation cascade, including stabilisation of the native conformation; inhibition of monomer aggregation into toxic oligomers; redirecting the aggregation pathway into a non-toxic oligomer population; structural modification of the oligomeric conformers; and accelerating aggregation into high-molecular-weight fibrillar species, thereby reducing the toxic oligomer population [67–71]. Yet other approaches focus on enhancing the clearance of misassembled and misfolded proteins [13,72]. Despite promising findings in laboratory settings, however, clinical outcomes involving anti-amyloid strategies have been largely disappointing. In this respect, the Food and Drug Administration (FDA) approvals of the antibodies lecanemab and aducanumab for the treatment of AD have arguably bolstered support for anti-amyloid-focused therapeutic strategies [73].

As discussed above, the toxicity of aggregates derived from amyloidogenic proteins is largely mediated by their interactions with the cellular *milieu*. In particular, direct association with, and perturbation of, biomembranes are the mechanisms that appear to be most commonly identified when amyloid toxicity is addressed in vitro. Further, amyloid protein–lipid interactions may instigate misfolding and assembly of the intrinsically disordered peptides, thereby potentially exacerbating aggregation. It follows from the above considerations that modulating the physical state of lipid membranes might act as an effective countermeasure against the membrane-active properties of the misfolded proteins. Many small-molecule compounds, both natural and synthetic, are known to interact with and insert into biological membranes, thereby inducing alterations to the physicochemical properties of the lipidic bilayer; for example, changes in the membrane surface charge, membrane fluidity, bilayer thickness, and membrane bending stiffness [74,75]. Thus, in addition to "classical" inhibitory mechanisms of protein aggregation, anti-amyloid agents might also include those compounds that uniquely intercalate into lipid bilayers to safeguard membranes from toxic biomolecules implicit in numerous human proteinopathies (Figure 1). The following section shall explore a number of naturally derived molecules that have been demonstrated to exhibit this mechanism. Such molecules include aminosterols, cholesterol and polyphenolic compounds (Figure 2).

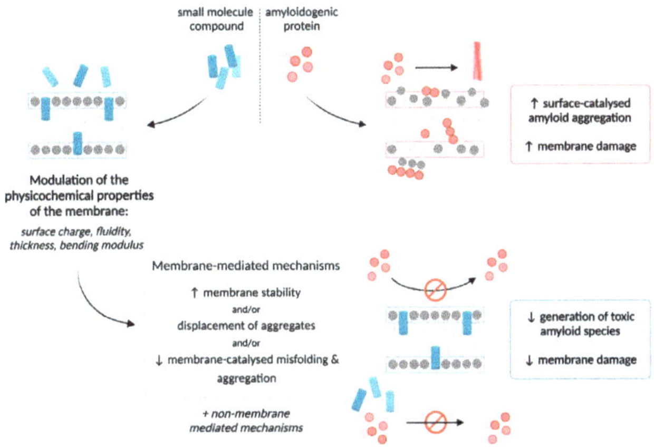

Figure 1. Key protective mechanisms of membrane-mediated protection against amyloid toxicity by small molecules. The schematic illustrates multiple mechanisms through which therapeutic compounds can be exploited to modulate phospholipid membranes and confer protection against bilayer disruption by toxic amyloid assemblies. Specifically, by altering the physicochemical properties of phospholipid bilayers, these small molecules can impede the misfolding, accumulation and insertion events of amyloidogenic peptides, thereby mitigating amyloid toxicity.

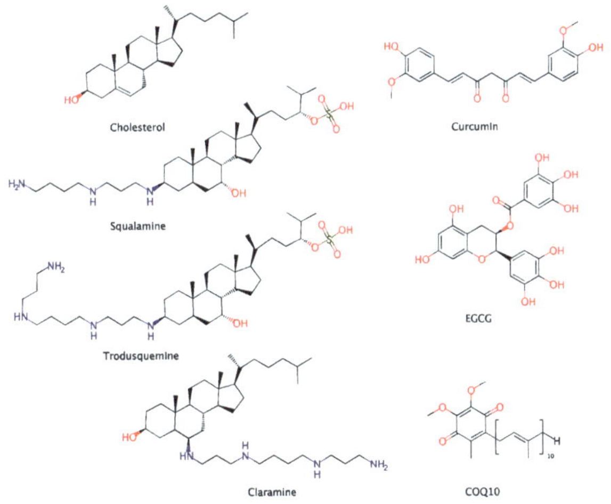

Figure 2. Structures of the natural compounds reviewed for their membrane-mediated protective activities against amyloid toxicity.

3.1. Aminosterols

3.1.1. Squalamine and Trodusquemine

Squalamine (SQ) and Trodusquemine (TRO) represent an intriguing pair of natural aminosterols, originally isolated from the dogfish shark, *Squalus acanthias*. As will be discussed in more detail below, they have recently been extensively investigated for their potential in combating amyloid disorders through their ability to bind biological membranes [76]. Both SQ and TRO have successfully completed phase-1 clinical trials, demonstrating favourable safety and tolerability profiles (NCT00606112 and NCT00139282, respectively).

SQ and TRO are structurally very similar, and they are both cationic amphipathic aminosterols. Thus, they share a hydrophobic sterol domain with a sulphated side chain whilst exhibiting distinct hydrophilic polyamine moieties: spermine (possessing four amine groups) in TRO and spermidine (possessing three amine groups) in SQ. This results in TRO carrying a greater net positive charge (+3) compared to SQ (+2) at a physiological pH [77,78]. The cationic polyamine side chains confer the ability to neutralise the negative charges on the inner leaflet of eukaryotic membranes, consequently displacing proteins electrostatically bound to the bilayer [79]. In agreement with this mechanism, the difference of a single positive charge between the two aminosterols appears to exert a significant impact on their respective efficacies, with TRO demonstrating a stronger protective effect relative to equivalent concentrations of SQ. Moreover, the architecture of these cationic steroids, featuring a negatively charged sulphate group on one side of the sterol core and a positively charged polyamine on the other, is a configuration known to significantly stabilise lipid membranes [78].

Nuclear magnetic resonance (NMR) experiments suggest that SQ effectively outcompetes α-syn for binding sites on lipid vesicles, thereby mitigating α-syn aggregation and reducing oligomer-mediated toxicity [80,81]. A decline in protein–vesicle interactions was observed without any direct binding between SQ and α-syn—thus favouring the notion that the protective effects of the compound are primarily mediated via its membrane interactions. Competitive displacement of α-syn was further evidenced by a dose-dependent reduction in the α-helical content of α-syn upon incubation with lipid vesicles and SQ. In a *Caenorhabditis elegans* (*C. elegans*) model of PD, SQ significantly reduced the formation of intracellular inclusions and improved motility without affecting α-syn expression [80]. Moreover, in PD mouse models and aged wild-type mice, SQ has been shown to restore enteric neuron excitability and restore colonic motility [82,83]. Beyond these cellular and animal models, SQ has demonstrated notable efficacy in human clinical trials, including a phase-2b clinical trial (KARMET, NCT03781791), showing improvements in PD-related constipation, hallucinations, and dementia [84].

The other aminosterol, TRO, is notable for its efficacy at lower doses and its ability to cross the blood–brain barrier, enhancing its therapeutic potential in neurodegenerative diseases [76,85,86]. Trodusquemine stably integrates with high affinity into the interface between the hydrophilic and hydrophobic portions of the bilayer, exposing both charged sulphate and spermine groups to the aqueous environment. This insertion induces three key physicochemical alterations to the bilayer: a reduction in the total negative charge, an increased mechanical resistance to indentation, and a reorganisation of the lipid constituents [77]. Overall, these modifications "stiffen" biological membranes, making them more resistant to perturbation by misfolded protein oligomers [78]. Perni and coworkers [85] described a dose-dependent inhibition of lipid-induced α-syn assembly by TRO in the presence of anionic 1,2-dimyristoyl-sn-glycero-3-phospho-L-serine (DMPS) vesicles. While in the absence of TRO all the α-syn was bound to the vesicles in an α-helical conformation, circular dichroism (CD) experiments revealed that TRO shifted the signal of α-syn to that characteristic of a random coil, indicating the displacement of the protein, akin to SQ. Data suggest that the inhibitory mechanism exhibited by TRO appears to involve not only membrane displacement of monomeric α-syn, as seen with SQ, but also interaction with intermediate aggregation species. While TRO was reported to not influence fibril elongation events, this aminosterol significantly decreased the rate of secondary nucleation of α-syn, implying a selective mechanism targeting the early stages of amyloidogenesis [85,87]. Furthermore, confocal microscopy revealed that TRO inhibited α-syn oligomer toxicity in human neuroblastoma (SH-SY5Y) cells; this is attributed to the ability of TRO to displace and directly interact with oligomers, as suggested by the synthetic vesicle investigations.

In the same investigation [85], Perni et al. also conducted in vivo studies using *C. elegans* models of PD and reported reduced intracellular inclusions and improved motility and longevity upon TRO administration.

The membrane-safeguarding effects of TRO extend beyond α-syn, as the aminosterol has also been shown to suppress the toxicity of oligomers of Aβ peptides ($A\beta_{42}$ and $A\beta_{40}$) and of the amyloidogenic N-terminal domain of the *Escherichia coli* (*E. coli*) HypF protein (HypF-N). At protective concentrations, TRO completely prevented oligomers from binding to the bilayer [87,88]. Interestingly, while TRO slowed down the lipid-induced aggregation of α-syn, it accelerated the amyloid fibril formation of $A\beta_{42}$ by predominantly enhancing secondary nucleation events. This acceleration shifted the assembly process towards the generation of mature fibrils, hence depleting cytotoxic oligomeric intermediates. These observations were also replicated in vivo, wherein TRO administration in *C. elegans* models of AD was associated with increased aggregate formation but reduced $A\beta_{42}$-induced toxicity [89].

In conclusion, although α-syn, amyloid-β, and HypF-N follow distinct aggregation kinetics, both aminosterols demonstrate efficacy in suppressing amyloid toxicity through a shared mechanism of outcompeting and displacing oligomeric aggregates from cell membranes.

3.1.2. Claramine

Claramine, a blood–brain barrier-permeable steroid polyamine, is structurally analogous to TRO, having a spermine side chain attached to a sterol ring. Once again, this configuration enables the interaction with cellular membranes and modulation of their physicochemical properties. In a recent investigation by Kreiser et al. [90], claramine was found to neutralise the cytotoxicity of two pore-forming biological toxins, namely melittin from honeybee venom and α-hemolysin from *Staphylococcus aureus*. Claramine effectively prevented these porating agents from binding to cell membranes, without any detectable alterations to their overall structures. Importantly, this illustrates that the protective effects of claramine are mediated primarily through its modulation of membrane properties, rather than by direct interaction with the pore-forming proteins themselves, akin to the other aminosterols discussed [90]. With regards to amyloid proteins, it was reported that claramine effectively inhibited the aggregation of α-syn. In a well-established *C. elegans* model of PD, claramine administration resulted in a marked reduction in α-syn inclusions, accompanied by improved motility. These effects were attributed to the ability of this sterol to alter the phase behaviour of α-syn, particularly by slowing the rate of primary nucleation, a critical initial step in the aggregation process [91].

Based on these collective observations, it can be inferred that the effects of aminosterols in promoting membrane resilience against permeabilising biological toxins and misfolded oligomers can be ascribed mainly to their ability to regulate the physicochemical properties of cell membranes [76]. It is therefore tempting to speculate regarding a broad protective mechanism, which might be applicable to various human pathologies where membrane integrity is compromised. Additionally, the proven safety profiles and favourable pharmacokinetic properties of aminosterols, as supported by various clinical trials, further support their potential as therapeutic agents.

3.2. *Cholesterol*

Cholesterol is a physiologically very important component of lipid membranes that plays a key role in modulating membrane fluidity, permeability, and hydrophobicity [92,93]. This amphipathic molecule is composed of three well-defined regions: a small polar hydroxyl group, a rigid steroid ring, and a flexible iso-octyl side chain. Within membrane bilayers, cholesterol intercalates such that the polar hydroxyl group is positioned between the phospholipid polar head groups while the plate-like steroid ring extends into the hydrophobic core of the bilayer, wherein it imparts a condensing effect on the membrane [92,94]. Cholesterol fulfils a particularly important role in the brain, which accounts for 20–25% of the total body cholesterol [95,96]. Dysregulation of cholesterol homeostasis, particularly in the neuronal plasma membrane, has been implicated in synaptic loss, impaired synaptic plasticity, and the pathological changes observed in various neurodegenerative diseases, including AD, PD, and Huntington's disease (HD) [96–98].

The interplay between cholesterol, protein aggregation and aggregate toxicity is complex and conflicting results have been reported [61]. Several studies have attested that cholesterol enhances the aggregation propensity of amyloidogenic peptides, including Aβ [99–101], α-syn [102,103], hIAPP [104], htt [105] and transthyretin [106]. Cholesterol-containing membranes have been shown to significantly accelerate amyloid-β aggregation and enhance its membrane-binding affinity [99–101]. Specifically, under time-lapse atomic force microscopy (AFM), the presence of cholesterol in 1-palmitoyl-2-oleoyl-*sn*-glycero-3-phospho-choline/1-palmitoyl-2-oleoyl-*sn*-glycero-3-phospho-L-serine (POPC/POPS) lipid bilayers led to the faster appearance of $A\beta_{42}$ aggregates compared to non-cholesterol bilayers, implying an enhanced aggregation propensity [99]. Similarly, in 1,2-dimyristoyl-*sn*-glycero-3-phosphocholine (DMPC) vesicles, cholesterol increased the nucleation rate of $A\beta_{42}$ peptides by catalysing the primary nucleation step [101]. Computational modelling also demonstrated that lipid membranes harbouring cholesterol manifested a stronger affinity to $A\beta_{42}$, both in zwitterionic POPC [100] and anionic POPC/POPS [99] bilayers. Yu and Zheng [100] attributed the enhanced membrane-binding of Aβ peptides to cholesterol-induced alterations in the structure and dynamics of the lipid bilayer. Further, it was reported that cholesterol, together with the ganglioside content, promoted the conformational transition from an α-helix- to a β-sheet-rich structure [107].

In contrast, other substantial evidence supports a *protective* role for cholesterol against amyloid-β toxicity, describing cholesterol-rich bilayers to be less prone to destabilisation and perforation. Fernández-Pérez et al. [108] examined the impact of the cholesterol content on neuronal membranes, reporting that increasing membrane cholesterol levels increased the number of membrane-associated Aβ clusters but prevented membrane perforation and neurotoxicity. Additionally, high cholesterol content in neuroblastoma cell membranes prevented Aβ peptide binding, with the cholesterol content being inversely correlated with the membrane-perturbing effects of the Aβ oligomers [109]. Molecular dynamics (MD) simulations also support this notion, with cholesterol-rich raft model membranes being found to be more resistant to $A\beta_{42}$ tetramer insertion than pure PC membranes [110]. Enhanced membrane resilience to amyloid-β is attributed to cholesterol-induced physicochemical changes in the bilayer, specifically a more packed and rigid environment that likely impedes membrane permeabilisation [108,109].

Similar to amyloid-β, the interplay between cholesterol, hIAPP and membranes is also complex [111]. Cholesterol has been reported to slow hIAPP aggregation and reduce dye leakage from PC and PS lipid vesicles, although this protective effect was overcome by high concentrations of anionic lipids [34]. MD simulations of POPC/POPS membranes also indicate that the addition of cholesterol caused a weaker insertion of hIAPP into the bilayer and made the membrane less vulnerable to hIAPP-induced poration [112]. However, whilst cholesterol suppressed membrane damage in raft-free anionic POPC/POPS vesicles, in raft-containing vesicles, cholesterol conversely exacerbated hIAPP fibrillisation and membrane leakage [104]. These findings suggest that the influence of cholesterol on hIAPP-mediated membrane leakage is to a significant extent attributed to its influence on the physicochemical properties of the lipid membrane, rather than via direct interactions with the peptide itself. By investigating different sterols, Zhang et al. [113] established that sterols which led to a more tightly packed membrane environment resulted in lower vesicle binding of hIAPP, decreased membrane leakage, and slower peptide assembly. This hypothesis is supported by other investigations showing that increased membrane rigidity caused by cholesterol impedes pore formation by melittin [114].

α-syn has a high affinity for cholesterol, mediated by residues 67–78 in the aggregation-prone non-amyloid-β component (NAC) domain of the protein [115]. In fact, cholesterol increased the propensity of α-syn to induce bilayer defects and was required for the formation of cytotoxic amyloid transmembrane pores [116]. However, cholesterol may also influence α-syn–bilayer interactions via alteration of the membrane environment, with contrasting effects. For instance, increasing the amount of cholesterol in DPPC vesicles weakened the binding of α-syn with the vesicles as a result of cholesterol blocking membrane defects [117]. As indicated by recent MD simulations, such an effect can be ascribed to a decrease in bilayer fluidity induced by cholesterol [102]. In agreement, other investigations have also reported that cholesterol inhibits the binding of α-syn to membranes comprising either zwitterionic or anionic lipids [103,118]. The impact of cholesterol on α-syn binding and fibrillation thus varies in different lipid environments.

Overall, while the relationship between the cholesterol membrane content and amyloid–membrane interactions remains an area of active investigation, the protective effects of cholesterol are often linked to its membrane-stabilising properties, which reduce the propensity for amyloidogenic peptides to penetrate the bilayer [119].

3.3. *Polyphenols*

Polyphenols are a class of phytochemicals with considerable structural diversity, essentially characterised by the presence of multiple aromatic rings with two or more phenol units. They are primarily found in herbal teas, fruits, and vegetables and are categorised according to the number of phenol rings and the elements connecting them [120]. Polyphenolic compounds show considerable promise as effective anti-amyloid agents and possess various biologically relevant properties, including antioxidant, anti-inflammatory, and antimicrobial effects. Several classes of polyphenols, including flavonoids, curcuminoids, and phenolic acids, have attracted intensive interest for their therapeutic potential for amyloid-related diseases due to their ability to inhibit the protein aggregation cascade [121–126]. Polyphenols that effectively antagonise amyloid formation share structural similarities. Such a capability is thought to be attributed to their phenolic rings, which disturb π-stacking interactions between monomeric units in aggregate species and form π–π interactions with aromatic residues present in amyloid-forming proteins [127,128]. However, the anti-aggregative effects of polyphenols on amyloid assembly are usually probed in vitro in aqueous solutions, which does not resemble the physiological environment in which such aggregates exist. This is especially significant in light of the substantial influence lipid membranes have on the aggregation cascade [129].

Polyphenols are able to interact with all the macromolecules, including lipids. They are generally localised close to the polar headgroups of the bilayer, just beneath the lipid–water interface. However, these interactions are greatly influenced by the structures of both the polyphenol and the lipid environment [130]. Numerous reports propose that the biological effects of polyphenols are attributable to their impact on membrane characteristics, such as fluidity [131,132].

3.3.1. Curcumin

Curcumin is a widely researched lipophilic polyphenol abundantly found in the turmeric plant *Curcuma longa* [133]. Curcumin has attracted substantial interest for its potential therapeutic applications, including in proteinopathies [134]. Thus, curcumin potently interferes with the aggregation of the major amyloidogenic proteins, including amyloid-β [135,136], tau [137], α-syn [138], and hIAPP [139]. Notably, curcumin is reported to redirect the assembly of Aβ and α-syn towards the generation of non-toxic oligomers [136,140] and even disaggregate the preformed fibrillar aggregates of these proteins [135,137,140], thereby reducing the levels of neurotoxic species.

The breadth of curcumin's interactions has led to speculation that in addition to its antioxidant, anti-inflammatory and anti-aggregative properties, the beneficial effects of this small molecule may also be membrane-mediated [141]. Due to its lipophilic nature, curcumin can distribute into cellular membranes, where it is reported to promote membrane cohesion and stability [142]. The orientation and distribution of curcumin within phospholipid bilayers is concentration-dependent. At lower concentrations, curcumin orients itself flat on the membrane surface, binding to the lipid–water interface through

H- bonding with the phosphate headgroup, and acts as a protective steric barrier or "carpet", reducing membrane fluidity and water permeability. Upon increasing the curcumin concentration, the molecule penetrates into the acyl chain region, where it fluidises the membrane and counters peptide insertion [141,143,144]. Curcumin thus reduced the rate and extent of Aβ insertion into anionic lipid monolayers, leading to decreased Aβ-induced bilayer disruption in a dose-dependent manner [145]. Furthermore, curcumin embedded into anionic bilayers reduced the accumulation of fibrillar Aβ oligomers on the membrane by almost 70% [146]. Along the same lines, research by Khondker and co-workers demonstrated that curcumin, by altering the structural parameters of the membrane, significantly reduced the formation of $A\beta_{25-35}$ peptide plaques by an unsaturated anionic membrane model [147]. These and other investigations have provided further support for curcumin as a small-molecule polyphenol that inhibits amyloid self-assembly and membrane toxicity by a combination of membrane and protein interactions.

3.3.2. Epigallocatechin Gallate (EGCG)

The green tea flavonoid EGCG is another well-studied polyphenol against amyloid protein self-assembly. Several studies have reported that EGCG effectively inhibits the aggregation of a number of disease-related peptides, including Aβ [148,149], tau [150], α-syn [122,151], hIAPP [152], and huntingtin [36], amongst others. Since this compound has exhibited promising therapeutic potential for multiple PMDs, EGCG has entered clinical trials for the treatment of neurodegenerative diseases. However, despite the positive outcomes observed in in vitro and animal models, few to no positive results have been reported to date in clinical trials [153].

EGCG is a gallic acid ester featuring seven hydroxyl groups distributed among three aromatic rings. Integration into lipid bilayers occurs mostly through its gallic acid ester group [148,154]. The presence of galloyl moieties in polyphenols enhances their lipophilicity, enabling galloylated catechins like EGCG to penetrate lipid bilayers and disturb the organisation of the tail region, in turn imparting morphological and fluidic alterations to membranes [131,155].

EGCG has been studied for its potential to inhibit amyloid aggregation and mitigate toxicity at lipid interfaces. Preincubation of $A\beta_{42}$ with EGCG lessened the bilayer disruption of negatively charged vesicles, suggesting that EGCG attenuated the peptides' membrane interaction [156]. This is consistent with recent MD simulations, which reported that in the presence of anionic membranes composed of POPC and POPG (1-palmitoyl-2-oleoyl-*sn*-glycero-3-phosphoglycerol) lipids, EGCG altered the binding dynamics between protofibrillar $A\beta_{42}$ and the bilayer. Further, EGCG slowed the adsorption of $A\beta_{42}$ and hindered its assembly, albeit less effectively than in a membrane-free environment. The numbers of contacts formed between EGCG and lipid bilayers were even greater than those between EGCG and $A\beta_{42}$ protofibrils, implying that EGCG interacts preferentially and strongly with membranes [148]. In an analogous manner, EGCG associated primarily with the bilayer to weaken the H-bonding and hydrophobic interactions between protofibrillar α-syn and the membrane, hence limiting α-syn membrane-binding and enhancing membrane integrity [151]. Interestingly, EGCG retained its inhibitory effect on α-syn aggregation, as well as its capacity to reduce the concentration of oligomeric α-syn, in the presence of both zwitterionic and anionic liposomes [157]. Similarly, EGCG also prevented aggregation of the htt protein into amyloid fibrils even in the presence of vesicles made from the synthetic lipid POPC or from total brain lipid extract [158].

However, the efficacy of EGCG in membrane environments has not always been consistent. EGCG less effectively inhibited $A\beta_{40}$ fibrillisation upon incubation with large unilamellar vesicles (LUVs), with PCPG anionic vesicles reducing the inhibitory efficacy of EGCG more markedly in comparison to zwitterionic PC vesicles [157]. Mass spectrometry experiments point to a likely partitioning of EGCG into the lipid bilayer, hence reducing the availability of this inhibitor to bind Aβ. In concordance, another study reported a significant reduction in the ability of EGCG to inhibit hIAPP fibrillogenesis in the presence of negatively charged dimyristoyl-phosphatidylglycerol (DPPG) lipid bilayers [159]. This was attributed to diminished π–π stacking interactions between EGCG and the hydrophobic binding sites of the peptide, which are sequestered within the lipid phase and thus inaccessible to EGCG. Taken together, these mechanistic insights highlight the possibility of reduced inhibitor efficacy in a lipid environment due to competing interactions between the inhibitor compound, the peptide and the membrane—hence further underscoring the influence of lipid membranes during therapeutic design.

3.4. *Ubiquinones*

Ubiquinones are a group of lipid-soluble antioxidant molecules located primarily in the membranes of Gram-negative bacteria and in the inner mitochondrial membranes of eukaryotic cells. Ubiquinones are composed of a quinone headgroup connected to a lipophilic isoprenoid chain of species-specific length, confining these molecules to lipid-rich environments [160]. In humans, the chain comprises ten isoprenoid units and is thus referred to as ubiquinone-10, or coenzyme Q10 (CoQ10). CoQ10 can be considered a mobile component of the mitochondrial electron transport chain, shuttling electrons in a ubiquinone (oxidised) and ubiquinol (reduced) redox cycle [161]. It therefore plays a significant role in cellular metabolism, participating in various aspects of mitochondrial energy production. Further, CoQ10 protects mitochondrial membrane lipids and proteins from free radical-induced oxidative injury [162,163]. While the exact position of Q10 in the bilayers remains to be fully elucidated, it is generally accepted that the isoprenoid chain is embedded within the mid-plane of the hydrophobic core, while the quinone ring alternates between the polar headgroup region and the mid-plane [164,165].

The influence of CoQ10 on the mechanical properties and permeability of lipid membranes may be compared to that of cholesterol, with studies suggesting that the inclusion of CoQ10 in the phospholipid leaflet imparts a concentration-dependent increase in the phospholipid order and a pronounced membrane-rigidifying effect [166,167]. In agreement with this, the levels of endogenous ubiquinones in bacteria increased dramatically under prolonged osmotic stress, implying a potential stability-enhancing role of ubiquinones within membranes [168]. This "stiffening" effect on lipid bilayers has been corroborated by other investigations employing biomimetic membranes [166,169].

Studies have reported that CoQ10 confers cyto-protection against Aβ through multiple mechanisms, including the inhibition of oxidative stress and increased expression of proteins related to neuronal cell survival [170], as well as membrane-mediated mechanisms delaying Aβ incorporation into membranes [171,172]. The protective effect of CoQ10 does not appear to be mediated by a simple chemical interaction with the Aβ peptide, since the simultaneous addition of CoQ10 and Aβ did not confer protection against Aβ toxicity in endothelial cells. However, pre-treatment of cultured cells with CoQ10 prevented Aβ-induced damage. CoQ10 was reported to prevent the incorporation of Aβ into the plasmalemma and mitochondrial membranes, reducing its accumulation within the cells. Additionally, CoQ10 decreased the production of ROS and prevented cell death [171,172]. Notably, the protective role of CoQ10 may be partially attributed to its ability to shield the membrane. It is hypothesised that CoQ10 acts as a "cellular armour", impeding the insertion of Aβ into bilayers via steric hinderance, in turn reducing the formation of Aβ channels in the membrane. Furthermore, considering the oxidative stress [173] and mitochondrial dysfunction [174] hypotheses of AD, the antioxidant, neuroprotective, and anti-inflammatory properties of CoQ10 appear promising for the amelioration of the oxidative stress and neuroinflammation associated with AD [175]. This is further supported by evidence suggesting that serum CoQ10 levels may be inversely associated with the risk of developing dementia [176], and other reports demonstrating the protective effects of CoQ10 against Aβ neurotoxicity [170]. These findings underscore the need for further investigation into the mechanisms of action of CoQ10, particularly its ability to stabilise membranes and prevent the damaging insertion of other amyloidogenic peptides, whilst highlighting the critical role of bilayer stabilisation in mitigating amyloid toxicity.

4. Conclusions

In conclusion, the reviewed literature provides strong support for the notion that regulating the physical state of lipid membranes may be a promising therapeutic strategy to suppress the toxicity of amyloidogenic assemblies in PMDs. Considering the central role of membranes in amyloid pathology, the identification of small molecules that have the potential to modulate peptide–membrane interactions to (i) mitigate membrane-catalysed misfolding and assembly, and (ii) preserve membrane integrity, is an important step in the development of effective treatments for these debilitating disorders (Table 2). However, evidence also suggests that compound interaction with lipid membranes does not necessarily confer protection, as evidenced by EGCG, a well-established anti-amyloidogenic compound. The conflicting data and gaps in our understanding underscore the necessity for a broader investigation into the fundamental mechanisms by which these compounds interact with both peptides and membranes. Elucidating these complex interactions in future studies will be crucial for informing the design of more targeted and potent small-molecule therapeutics, which could address the membrane-mediated aspects of amyloid toxicity.

Table 2. List of compounds and associated peptide/membrane models.

Compound	Protein/Peptide	Membrane Model	References
Squalamine	α-syn	DMPS vesicles DOPE/DOPS/DOPC (5:3:2) vesicles	[80,81]
Trodusquemine	α-syn	DMPS vesicles DOPE/DOPS/DOPC (5:3:2) vesicles Neuronal membranes (SH-SY5Y cells)	[85,87]
	Aβ, HypF-N	DOPE/DOPS/DOPC (5:3:2) vesicles DOPC/SM (2:1) vesicles with 1% CHOL and 1% GM1 Neuronal membranes (SH-SY5Y cells)	[87–89]
Cholesterol	α-syn	POPC, POPC/POPG, and POPC/POPE vesicles DPPC, DOPC and DOPC/SM vesicles DOPE/DOPS/DOPC (5:3:2) vesicles DOPC, DOPC/DOPE/DOPG (4:3:1) vesicles POPC and POPC/POPS bilayers	[116]
	Aβ	DMPC vesicles Raft model membranes Plasma membranes (HEK-293 cells, SH-SY5Y cells, and hippocampal neurons)	[99–101,108–110]
	hIAPP	DOPC and DOPC/DOPS bilayers DOPC and POPC lipid vesicles with DOPG, POPS or SM POPC/POPS (7:3) vesicles DOPC/DPPC (1:2) vesicles	[112]
Curcumin	Aβ	DMPC, DOPC, DMPG and POPC/DMPS (97:3) vesicles DMPG and DMPG/DMPC monolayers Neuronal membranes (SH-SY5Y cells)	[141,143–145]
EGCG	α-syn	POPC/POPG bilayers POPC and POPC/POPG vesicles	[151,157]
	Aβ	POPC/POPG (7:3) bilayers POPC, POPC/POPG and DMPC/DMPG vesicles	[156,157]
	hIAPP	DPPG bilayers	[159]
	htt	POPC vesicles Total brain lipid extract	[158]
CoQ10	Aβ	Endothelial cell membranes	[171,172]

Abbreviations: *CHOL*, cholesterol; *DMPC*, 1,2-dimyristoyl-*sn*-glycero-3-phosphocholine; *DMPG*, 1,2-dimyristoyl-*sn*-glycero-3-phospho-(1′-rac-glycerol); *DMPS*, 1,2-dimyristoyl-*sn*-glycero-3-phosphoserine; *DOPC*, 1,2-dioleoyl-*sn*-glycero-3-phosphocholine; *DOPE*, 1,2-dioleoyl-*sn*-glycero-3-phosphoethanolamine; *DOPG*, 1,2-dioleoyl-*sn*-glycero-3-phospho-(1′-rac-glycerol); *DOPS*, 1,2-dioleoyl-*sn*-glycero-3-phosphoserine; *DPPC*, 1,2-dipalmitoyl-*sn*-glycero-3-phosphocholine; *GM1*, monosialotetrahexosylganglioside; *POPC*, 1-palmitoyl-2-oleoyl-*sn*-glycero-3-phosphocholine; *POPE*, 1-palmitoyl-2-oleoyl-*sn*-glycero-3-phosphoethanolamine; *POPG*, 1-palmitoyl-2-oleoyl-*sn*-glycero-3-phospho-(1′-rac-glycerol); *POPS*, 1-palmitoyl-2-oleoyl-*sn*-glycero-3-phosphoserine; *SM*, sphingomyelin.

Author Contributions: Conceptualisation, N.V.; writing—original draft preparation, R.M.S. and A.E.S.; writing—review and editing, N.V. All authors have read and agreed to the published version of the manuscript.

Funding: RMS is supported by the Malta Council of Science and Technology (MCST) Research Excellence Programme REP-2023-028 (MITOLIPID). The Article Processing Charges were partly funded by the Internationalisation Partnership and Awards Scheme Plus (IPAS+) of the MCST, awarded to NV (IPAS-2023-060; MPIMS-DM).

Acknowledgments: We are grateful to Christian Griesinger (Max Planck Institute for Multidisciplinary Sciences, Göttingen, Germany) for his collaboration on the small-molecule therapeutics in the MITOLIPID and MPIMS-DM grants.

Conflicts of Interest: The authors declare no conflicts of interest.

References

1. Chiti, F.; Dobson, C.M. Protein Misfolding, Amyloid Formation, and Human Disease: A Summary of Progress Over the Last Decade. *Annu. Rev. Biochem.* **2017**, *86*, 27–68. [CrossRef] [PubMed]
2. Yadav, K.; Yadav, A.; Vashistha, P.; Pandey, V.P.; Dwivedi, U.N. Protein Misfolding Diseases and Therapeutic Approaches. *Curr. Protein Pept. Sci.* **2019**, *20*, 1226–1245. [CrossRef] [PubMed]
3. Khan, A.N.; Khan, R.H. Protein misfolding and related human diseases: A comprehensive review of toxicity, proteins involved, and current therapeutic strategies. *Int. J. Biol. Macromol.* **2022**, *223*, 143–160. [CrossRef] [PubMed]
4. Landreh, M.; Sawaya, M.R.; Hipp, M.S.; Eisenberg, D.S.; Wuthrich, K.; Hartl, F.U. The formation, function and regulation of amyloids: Insights from structural biology. *J. Intern. Med.* **2016**, *280*, 164–176. [CrossRef]
5. Almeida, Z.L.; Brito, R.M.M. Structure and Aggregation Mechanisms in Amyloids. *Molecules* **2020**, *25*, 1195. [CrossRef]
6. Basha, S.; Mukunda, D.C.; Rodrigues, J.; Gail D'Souza, M.; Gangadharan, G.; Pai, A.R.; Mahato, K.K. A comprehensive review of protein misfolding disorders, underlying mechanism, clinical diagnosis, and therapeutic strategies. *Ageing Res. Rev.* **2023**, *90*, 102017. [CrossRef]
7. Scollo, F.; La Rosa, C. Amyloidogenic Intrinsically Disordered Proteins: New Insights into Their Self-Assembly and Their Interaction with Membranes. *Life* **2020**, *10*, 144. [CrossRef]
8. Linse, B.; Linse, S. Monte Carlo simulations of protein amyloid formation reveal origin of sigmoidal aggregation kinetics. *Mol. Biosyst.* **2011**, *7*, 2296–2303. [CrossRef]
9. Dear, A.J.; Meisl, G.; Saric, A.; Michaels, T.C.T.; Kjaergaard, M.; Linse, S.; Knowles, T.P.J. Identification of on- and off-pathway oligomers in amyloid fibril formation. *Chem. Sci.* **2020**, *11*, 6236–6247. [CrossRef]
10. Chatani, E.; Yamamoto, N. Recent progress on understanding the mechanisms of amyloid nucleation. *Biophys. Rev.* **2018**, *10*, 527–534. [CrossRef]
11. Michaels, T.C.T.; Saric, A.; Habchi, J.; Chia, S.; Meisl, G.; Vendruscolo, M.; Dobson, C.M.; Knowles, T.P.J. Chemical Kinetics for Bridging Molecular Mechanisms and Macroscopic Measurements of Amyloid Fibril Formation. *Annu. Rev. Phys. Chem.* **2018**, *69*, 273–298. [CrossRef] [PubMed]
12. Nelson, R.; Sawaya, M.R.; Balbirnie, M.; Madsen, A.O.; Riekel, C.; Grothe, R.; Eisenberg, D. Structure of the cross-beta spine of amyloid-like fibrils. *Nature* **2005**, *435*, 773–778. [CrossRef] [PubMed]
13. Rinauro, D.J.; Chiti, F.; Vendruscolo, M.; Limbocker, R. Misfolded protein oligomers: Mechanisms of formation, cytotoxic effects, and pharmacological approaches against protein misfolding diseases. *Mol. Neurodegener.* **2024**, *19*, 20. [CrossRef] [PubMed]
14. Vivoli Vega, M.; Cascella, R.; Chen, S.W.; Fusco, G.; De Simone, A.; Dobson, C.M.; Cecchi, C.; Chiti, F. The Toxicity of Misfolded Protein Oligomers Is Independent of Their Secondary Structure. *ACS Chem. Biol.* **2019**, *14*, 1593–1600. [CrossRef] [PubMed]
15. Mannini, B.; Mulvihill, E.; Sgromo, C.; Cascella, R.; Khodarahmi, R.; Ramazzotti, M.; Dobson, C.M.; Cecchi, C.; Chiti, F. Toxicity of protein oligomers is rationalized by a function combining size and surface hydrophobicity. *ACS Chem. Biol.* **2014**, *9*, 2309–2317. [CrossRef]
16. Banerjee, S.; Hashemi, M.; Zagorski, K.; Lyubchenko, Y.L. Interaction of Abeta42 with Membranes Triggers the Self-Assembly into Oligomers. *Int. J. Mol. Sci.* **2020**, *21*, 1129. [CrossRef]
17. Rawat, A.; Langen, R.; Varkey, J. Membranes as modulators of amyloid protein misfolding and target of toxicity. *Biochim. Biophys. Acta Biomembr.* **2018**, *1860*, 1863–1875. [CrossRef]
18. Casares, D.; Escriba, P.V.; Rossello, C.A. Membrane Lipid Composition: Effect on Membrane and Organelle Structure, Function and Compartmentalization and Therapeutic Avenues. *Int. J. Mol. Sci.* **2019**, *20*, 2167. [CrossRef]
19. Harayama, T.; Riezman, H. Understanding the diversity of membrane lipid composition. *Nat. Rev. Mol. Cell Biol.* **2018**, *19*, 281–296. [CrossRef]
20. Butterfield, S.M.; Lashuel, H.A. Amyloidogenic protein-membrane interactions: Mechanistic insight from model systems. *Angew. Chem. Int. Ed. Engl.* **2010**, *49*, 5628–5654. [CrossRef]
21. Relini, A.; Cavalleri, O.; Rolandi, R.; Gliozzi, A. The two-fold aspect of the interplay of amyloidogenic proteins with lipid membranes. *Chem. Phys. Lipids* **2009**, *158*, 1–9. [CrossRef] [PubMed]
22. Canale, C.; Oropesa-Nunez, R.; Diaspro, A.; Dante, S. Amyloid and membrane complexity: The toxic interplay revealed by AFM. *Semin. Cell Dev. Biol.* **2018**, *73*, 82–94. [CrossRef] [PubMed]

23. Ciudad, S.; Puig, E.; Botzanowski, T.; Meigooni, M.; Arango, A.S.; Do, J.; Mayzel, M.; Bayoumi, M.; Chaignepain, S.; Maglia, G.; et al. Abeta(1-42) tetramer and octamer structures reveal edge conductivity pores as a mechanism for membrane damage. *Nat. Commun.* **2020**, *11*, 3014. [CrossRef]
24. Cascella, R.; Perni, M.; Chen, S.W.; Fusco, G.; Cecchi, C.; Vendruscolo, M.; Chiti, F.; Dobson, C.M.; De Simone, A. Probing the Origin of the Toxicity of Oligomeric Aggregates of alpha-Synuclein with Antibodies. *ACS Chem. Biol.* **2019**, *14*, 1352–1362. [CrossRef]
25. Yasumoto, T.; Takamura, Y.; Tsuji, M.; Watanabe-Nakayama, T.; Imamura, K.; Inoue, H.; Nakamura, S.; Inoue, T.; Kimura, A.; Yano, S.; et al. High molecular weight amyloid beta(1–42) oligomers induce neurotoxicity via plasma membrane damage. *FASEB J.* **2019**, *33*, 9220–9234. [CrossRef]
26. Quist, A.; Doudevski, I.; Lin, H.; Azimova, R.; Ng, D.; Frangione, B.; Kagan, B.; Ghiso, J.; Lal, R. Amyloid ion channels: A common structural link for protein-misfolding disease. *Proc. Natl. Acad. Sci. USA* **2005**, *102*, 10427–10432. [CrossRef] [PubMed]
27. Lyubchenko, Y.L. Amyloid B-Protein Aggregation at Physiologically Relevant Concentrations. A Critical Role of Membranes. *Alzheimers Res. Ther. Open Access* **2020**, *3*, 114. [CrossRef]
28. Gu, L.; Guo, Z. Lipid membranes induce structural conversion from amyloid oligomers to fibrils. *Biochem. Biophys. Res. Commun.* **2021**, *557*, 122–126. [CrossRef]
29. Majewski, J.; Jones, E.M.; Vander Zanden, C.M.; Biernat, J.; Mandelkow, E.; Chi, E.Y. Lipid membrane templated misfolding and self-assembly of intrinsically disordered tau protein. *Sci. Rep.* **2020**, *10*, 13324. [CrossRef]
30. Galvagnion, C.; Buell, A.K.; Meisl, G.; Michaels, T.C.; Vendruscolo, M.; Knowles, T.P.; Dobson, C.M. Lipid vesicles trigger alpha-synuclein aggregation by stimulating primary nucleation. *Nat. Chem. Biol.* **2015**, *11*, 229–234. [CrossRef]
31. Elenbaas, B.O.W.; Khemtemourian, L.; Killian, J.A.; Sinnige, T. Membrane-Catalyzed Aggregation of Islet Amyloid Polypeptide Is Dominated by Secondary Nucleation. *Biochemistry* **2022**, *61*, 1465–1472. [CrossRef] [PubMed]
32. Liu, Y.; Ren, B.; Zhang, Y.; Sun, Y.; Chang, Y.; Liang, G.; Xu, L.; Zheng, J. Molecular simulation aspects of amyloid peptides at membrane interface. *Biochim. Biophys. Acta Biomembr.* **2018**, *1860*, 1906–1916. [CrossRef] [PubMed]
33. Mirdha, L. Aggregation Behavior of Amyloid Beta Peptide Depends Upon the Membrane Lipid Composition. *J. Membr. Biol.* **2024**, *257*, 151–164. [CrossRef]
34. Zhang, X.; St Clair, J.R.; London, E.; Raleigh, D.P. Islet Amyloid Polypeptide Membrane Interactions: Effects of Membrane Composition. *Biochemistry* **2017**, *56*, 376–390. [CrossRef]
35. Nguyen, N.; Lewis, A.; Pham, T.; Sikazwe, D.; Cheng, K.H. Exploring the Role of Anionic Lipid Nanodomains in the Membrane Disruption and Protein Folding of Human Islet Amyloid Polypeptide Oligomers on Lipid Membrane Surfaces Using Multiscale Molecular Dynamics Simulations. *Molecules* **2023**, *28*, 4191. [CrossRef]
36. Beasley, M.; Groover, S.; Valentine, S.J.; Legleiter, J. Lipid headgroups alter huntingtin aggregation on membranes. *Biochim. Biophys. Acta Biomembr.* **2021**, *1863*, 183497. [CrossRef]
37. Stockl, M.; Fischer, P.; Wanker, E.; Herrmann, A. Alpha-synuclein selectively binds to anionic phospholipids embedded in liquid-disordered domains. *J. Mol. Biol.* **2008**, *375*, 1394–1404. [CrossRef]
38. Kiskis, J.; Horvath, I.; Wittung-Stafshede, P.; Rocha, S. Unraveling amyloid formation paths of Parkinson's disease protein alpha-synuclein triggered by anionic vesicles. *Q. Rev. Biophys.* **2017**, *50*, e3. [CrossRef]
39. Kandel, N.; Zheng, T.; Huo, Q.; Tatulian, S.A. Membrane Binding and Pore Formation by a Cytotoxic Fragment of Amyloid beta Peptide. *J. Phys. Chem. B* **2017**, *121*, 10293–10305. [CrossRef]
40. Khursheed, A.; Viles, J.H. Impact of Membrane Phospholipids and Exosomes on the Kinetics of Amyloid-beta Fibril Assembly. *J. Mol. Biol.* **2024**, *436*, 168464. [CrossRef]
41. Fanni, A.M.; Vander Zanden, C.M.; Majewska, P.V.; Majewski, J.; Chi, E.Y. Membrane-mediated fibrillation and toxicity of the tau hexapeptide PHF6. *J. Biol. Chem.* **2019**, *294*, 15304–15317. [CrossRef] [PubMed]
42. Elbaum-Garfinkle, S.; Ramlall, T.; Rhoades, E. The role of the lipid bilayer in tau aggregation. *Biophys. J.* **2010**, *98*, 2722–2730. [CrossRef] [PubMed]
43. McMahon, H.T.; Boucrot, E. Membrane curvature at a glance. *J. Cell Sci.* **2015**, *128*, 1065–1070. [CrossRef]
44. Korshavn, K.J.; Satriano, C.; Lin, Y.; Zhang, R.; Dulchavsky, M.; Bhunia, A.; Ivanova, M.I.; Lee, Y.H.; La Rosa, C.; Lim, M.H.; et al. Reduced Lipid Bilayer Thickness Regulates the Aggregation and Cytotoxicity of Amyloid-beta. *J. Biol. Chem.* **2017**, *292*, 4638–4650. [CrossRef]
45. Meker, S.; Chin, H.; Sut, T.N.; Cho, N.J. Amyloid-beta Peptide Triggers Membrane Remodeling in Supported Lipid Bilayers Depending on Their Hydrophobic Thickness. *Langmuir* **2018**, *34*, 9548–9560. [CrossRef] [PubMed]
46. Kegulian, N.C.; Sankhagowit, S.; Apostolidou, M.; Jayasinghe, S.A.; Malmstadt, N.; Butler, P.C.; Langen, R. Membrane Curvature-sensing and Curvature-inducing Activity of Islet Amyloid Polypeptide and Its Implications for Membrane Disruption. *J. Biol. Chem.* **2015**, *290*, 25782–25793. [CrossRef]
47. Terakawa, M.S.; Yagi, H.; Adachi, M.; Lee, Y.H.; Goto, Y. Small liposomes accelerate the fibrillation of amyloid beta (1–40). *J. Biol. Chem.* **2015**, *290*, 815–826. [CrossRef]
48. Terakawa, M.S.; Lin, Y.; Kinoshita, M.; Kanemura, S.; Itoh, D.; Sugiki, T.; Okumura, M.; Ramamoorthy, A.; Lee, Y.H. Impact of membrane curvature on amyloid aggregation. *Biochim. Biophys. Acta Biomembr.* **2018**, *1860*, 1741–1764. [CrossRef]
49. Ermilova, I.; Lyubartsev, A.P. Modelling of interactions between Abeta(25–35) peptide and phospholipid bilayers: Effects of cholesterol and lipid saturation. *RSC Adv.* **2020**, *10*, 3902–3915. [CrossRef]

50. Ouberai, M.M.; Wang, J.; Swann, M.J.; Galvagnion, C.; Guilliams, T.; Dobson, C.M.; Welland, M.E. alpha-Synuclein senses lipid packing defects and induces lateral expansion of lipids leading to membrane remodeling. *J. Biol. Chem.* **2013**, *288*, 20883–20895. [CrossRef]
51. Tiwari, A.; Pradhan, S.; Sannigrahi, A.; Mahakud, A.K.; Jha, S.; Chattopadhyay, K.; Biswas, M.; Saleem, M. Interplay of lipid head group and packing defects in driving amyloid-beta-mediated myelin-like model membrane deformation. *J. Biol. Chem.* **2023**, *299*, 104653. [CrossRef] [PubMed]
52. Sugiura, Y.; Ikeda, K.; Nakano, M. High Membrane Curvature Enhances Binding, Conformational Changes, and Fibrillation of Amyloid-beta on Lipid Bilayer Surfaces. *Langmuir* **2015**, *31*, 11549–11557. [CrossRef] [PubMed]
53. Nuscher, B.; Kamp, F.; Mehnert, T.; Odoy, S.; Haass, C.; Kahle, P.J.; Beyer, K. Alpha-synuclein has a high affinity for packing defects in a bilayer membrane: A thermodynamics study. *J. Biol. Chem.* **2004**, *279*, 21966–21975. [CrossRef]
54. Middleton, E.R.; Rhoades, E. Effects of curvature and composition on alpha-synuclein binding to lipid vesicles. *Biophys. J.* **2010**, *99*, 2279–2288. [CrossRef]
55. Angelova, P.R.; Ludtmann, M.H.; Horrocks, M.H.; Negoda, A.; Cremades, N.; Klenerman, D.; Dobson, C.M.; Wood, N.W.; Pavlov, E.V.; Gandhi, S.; et al. Ca^{2+} is a key factor in alpha-synuclein-induced neurotoxicity. *J. Cell Sci.* **2016**, *129*, 1792–1801. [CrossRef] [PubMed]
56. Li, M.H.; Zhang, X.; London, E.; Raleigh, D.P. Impact of Ca(2+) on membrane catalyzed IAPP amyloid formation and IAPP induced vesicle leakage. *Biochim. Biophys. Acta Biomembr.* **2023**, *1865*, 184161. [CrossRef]
57. Ghio, S.; Camilleri, A.; Caruana, M.; Ruf, V.C.; Schmidt, F.; Leonov, A.; Ryazanov, S.; Griesinger, C.; Cauchi, R.J.; Kamp, F.; et al. Cardiolipin Promotes Pore-Forming Activity of Alpha-Synuclein Oligomers in Mitochondrial Membranes. *ACS Chem. Neurosci.* **2019**, *10*, 3815–3829. [CrossRef]
58. Cheignon, C.; Tomas, M.; Bonnefont-Rousselot, D.; Faller, P.; Hureau, C.; Collin, F. Oxidative stress and the amyloid beta peptide in Alzheimer's disease. *Redox Biol.* **2018**, *14*, 450–464. [CrossRef]
59. Evangelisti, E.; Wright, D.; Zampagni, M.; Cascella, R.; Fiorillo, C.; Bagnoli, S.; Relini, A.; Nichino, D.; Scartabelli, T.; Nacmias, B.; et al. Lipid rafts mediate amyloid-induced calcium dyshomeostasis and oxidative stress in Alzheimer's disease. *Curr. Alzheimer Res.* **2013**, *10*, 143–153. [CrossRef]
60. Tempra, C.; Scollo, F.; Pannuzzo, M.; Lolicato, F.; La Rosa, C. A unifying framework for amyloid-mediated membrane damage: The lipid-chaperone hypothesis. *Biochim. Biophys. Acta Proteins Proteom.* **2022**, *1870*, 140767. [CrossRef]
61. Sciacca, M.F.M.; Tempra, C.; Scollo, F.; Milardi, D.; La Rosa, C. Amyloid growth and membrane damage: Current themes and emerging perspectives from theory and experiments on Abeta and hIAPP. *Biochim. Biophys. Acta Biomembr.* **2018**, *1860*, 1625–1638. [CrossRef] [PubMed]
62. Pannuzzo, M.; Milardi, D.; Raudino, A.; Karttunen, M.; La Rosa, C. Analytical model and multiscale simulations of Abeta peptide aggregation in lipid membranes: Towards a unifying description of conformational transitions, oligomerization and membrane damage. *Phys. Chem. Chem. Phys.* **2013**, *15*, 8940–8951. [CrossRef]
63. Pannuzzo, M.; Raudino, A.; Milardi, D.; La Rosa, C.; Karttunen, M. alpha-helical structures drive early stages of self-assembly of amyloidogenic amyloid polypeptide aggregate formation in membranes. *Sci. Rep.* **2013**, *3*, 2781. [CrossRef] [PubMed]
64. Engel, M.F.; Khemtemourian, L.; Kleijer, C.C.; Meeldijk, H.J.; Jacobs, J.; Verkleij, A.J.; de Kruijff, B.; Killian, J.A.; Hoppener, J.W. Membrane damage by human islet amyloid polypeptide through fibril growth at the membrane. *Proc. Natl. Acad. Sci. USA* **2008**, *105*, 6033–6038. [CrossRef]
65. Elenbaas, B.O.W.; Kremsreiter, S.M.; Khemtemourian, L.; Killian, J.A.; Sinnige, T. Fibril elongation by human islet amyloid polypeptide is the main event linking aggregation to membrane damage. *BBA Adv.* **2023**, *3*, 100083. [CrossRef]
66. Andreasen, M.; Lorenzen, N.; Otzen, D. Interactions between misfolded protein oligomers and membranes: A central topic in neurodegenerative diseases? *Biochim. Biophys. Acta* **2015**, *1848*, 1897–1907. [CrossRef]
67. Scannevin, R.H. Therapeutic strategies for targeting neurodegenerative protein misfolding disorders. *Curr. Opin. Chem. Biol.* **2018**, *44*, 66–74. [CrossRef]
68. Habchi, J.; Chia, S.; Limbocker, R.; Mannini, B.; Ahn, M.; Perni, M.; Hansson, O.; Arosio, P.; Kumita, J.R.; Challa, P.K.; et al. Systematic development of small molecules to inhibit specific microscopic steps of Abeta42 aggregation in Alzheimer's disease. *Proc. Natl. Acad. Sci. USA* **2017**, *114*, E200–E208. [CrossRef]
69. Tortora, P.; Aprile, F.A. Natural Inhibitors of Amyloid Aggregation. *Int. J. Mol. Sci.* **2023**, *24*, 13310. [CrossRef]
70. Wagner, J.; Ryazanov, S.; Leonov, A.; Levin, J.; Shi, S.; Schmidt, F.; Prix, C.; Pan-Montojo, F.; Bertsch, U.; Mitteregger-Kretzschmar, G.; et al. Anle138b: A novel oligomer modulator for disease-modifying therapy of neurodegenerative diseases such as prion and Parkinson's disease. *Acta Neuropathol.* **2013**, *125*, 795–813. [CrossRef]
71. Kreiser, R.P.; Wright, A.K.; Block, N.R.; Hollows, J.E.; Nguyen, L.T.; LeForte, K.; Mannini, B.; Vendruscolo, M.; Limbocker, R. Therapeutic Strategies to Reduce the Toxicity of Misfolded Protein Oligomers. *Int. J. Mol. Sci.* **2020**, *21*, 8651. [CrossRef] [PubMed]
72. Giorgetti, S.; Greco, C.; Tortora, P.; Aprile, F.A. Targeting Amyloid Aggregation: An Overview of Strategies and Mechanisms. *Int. J. Mol. Sci.* **2018**, *19*, 2677. [CrossRef] [PubMed]
73. van Dyck, C.H.; Swanson, C.J.; Aisen, P.; Bateman, R.J.; Chen, C.; Gee, M.; Kanekiyo, M.; Li, D.; Reyderman, L.; Cohen, S.; et al. Lecanemab in Early Alzheimer's Disease. *N. Engl. J. Med.* **2023**, *388*, 9–21. [CrossRef] [PubMed]
74. Pineda, A.; Burre, J. Modulating membrane binding of alpha-synuclein as a therapeutic strategy. *Proc. Natl. Acad. Sci. USA* **2017**, *114*, 1223–1225. [CrossRef]

75. Andrade, S.; Ramalho, M.J.; Loureiro, J.A.; Pereira, M.C. Interaction of natural compounds with biomembrane models: A biophysical approach for the Alzheimer's disease therapy. *Colloids Surf. B Biointerfaces* **2019**, *180*, 83–92. [CrossRef]
76. Errico, S.; Lucchesi, G.; Odino, D.; Osman, E.Y.; Cascella, R.; Neri, L.; Capitini, C.; Calamai, M.; Bemporad, F.; Cecchi, C.; et al. Quantitative Attribution of the Protective Effects of Aminosterols against Protein Aggregates to Their Chemical Structures and Ability to Modulate Biological Membranes. *J. Med. Chem.* **2023**, *66*, 9519–9536. [CrossRef]
77. Errico, S.; Lucchesi, G.; Odino, D.; Muscat, S.; Capitini, C.; Bugelli, C.; Canale, C.; Ferrando, R.; Grasso, G.; Barbut, D.; et al. Making biological membrane resistant to the toxicity of misfolded protein oligomers: A lesson from trodusquemine. *Nanoscale* **2020**, *12*, 22596–22614. [CrossRef]
78. Barletti, B.; Lucchesi, G.; Muscat, S.; Errico, S.; Barbut, D.; Danani, A.; Zasloff, M.; Grasso, G.; Chiti, F.; Caminati, G. Reorganization of the outer layer of a model of the plasma membrane induced by a neuroprotective aminosterol. *Colloids Surf. B Biointerfaces* **2023**, *222*, 113115. [CrossRef]
79. Kazakova, O.; Giniyatullina, G.; Babkov, D.; Wimmer, Z. From Marine Metabolites to the Drugs of the Future: Squalamine, Trodusquemine, Their Steroid and Triterpene Analogues. *Int. J. Mol. Sci.* **2022**, *23*, 1075. [CrossRef]
80. Perni, M.; Galvagnion, C.; Maltsev, A.; Meisl, G.; Muller, M.B.; Challa, P.K.; Kirkegaard, J.B.; Flagmeier, P.; Cohen, S.I.; Cascella, R.; et al. A natural product inhibits the initiation of alpha-synuclein aggregation and suppresses its toxicity. *Proc. Natl. Acad. Sci. USA* **2017**, *114*, E1009–E1017. [CrossRef]
81. Limbocker, R.; Staats, R.; Chia, S.; Ruggeri, F.S.; Mannini, B.; Xu, C.K.; Perni, M.; Cascella, R.; Bigi, A.; Sasser, L.R.; et al. Squalamine and Its Derivatives Modulate the Aggregation of Amyloid-beta and alpha-Synuclein and Suppress the Toxicity of Their Oligomers. *Front. Neurosci.* **2021**, *15*, 680026. [CrossRef] [PubMed]
82. West, C.L.; Mao, Y.K.; Delungahawatta, T.; Amin, J.Y.; Farhin, S.; McQuade, R.M.; Diwakarla, S.; Pustovit, R.; Stanisz, A.M.; Bienenstock, J.; et al. Squalamine Restores the Function of the Enteric Nervous System in Mouse Models of Parkinson's Disease. *J. Parkinsons Dis.* **2020**, *10*, 1477–1491. [CrossRef]
83. West, C.L.; McVey Neufeld, K.A.; Mao, Y.K.; Stanisz, A.M.; Forsythe, P.; Bienenstock, J.; Barbut, D.; Zasloff, M.; Kunze, W.A. Identification of SSRI-evoked antidepressant sensory signals by decoding vagus nerve activity. *Sci. Rep.* **2021**, *11*, 21130. [CrossRef]
84. Camilleri, M.; Subramanian, T.; Pagan, F.; Isaacson, S.; Gil, R.; Hauser, R.A.; Feldman, M.; Goldstein, M.; Kumar, R.; Truong, D.; et al. Oral ENT-01 Targets Enteric Neurons to Treat Constipation in Parkinson Disease: A Randomized Controlled Trial. *Ann. Intern. Med.* **2022**, *175*, 1666–1674. [CrossRef] [PubMed]
85. Perni, M.; Flagmeier, P.; Limbocker, R.; Cascella, R.; Aprile, F.A.; Galvagnion, C.; Heller, G.T.; Meisl, G.; Chen, S.W.; Kumita, J.R.; et al. Multistep Inhibition of alpha-Synuclein Aggregation and Toxicity in Vitro and in Vivo by Trodusquemine. *ACS Chem. Biol.* **2018**, *13*, 2308–2319. [CrossRef]
86. Lantz, K.A.; Hart, S.G.; Planey, S.L.; Roitman, M.F.; Ruiz-White, I.A.; Wolfe, H.R.; McLane, M.P. Inhibition of PTP1B by trodusquemine (MSI-1436) causes fat-specific weight loss in diet-induced obese mice. *Obesity* **2010**, *18*, 1516–1523. [CrossRef] [PubMed]
87. Limbocker, R.; Mannini, B.; Ruggeri, F.S.; Cascella, R.; Xu, C.K.; Perni, M.; Chia, S.; Chen, S.W.; Habchi, J.; Bigi, A.; et al. Trodusquemine displaces protein misfolded oligomers from cell membranes and abrogates their cytotoxicity through a generic mechanism. *Commun. Biol.* **2020**, *3*, 435. [CrossRef] [PubMed]
88. Errico, S.; Ramshini, H.; Capitini, C.; Canale, C.; Spaziano, M.; Barbut, D.; Calamai, M.; Zasloff, M.; Oropesa-Nunez, R.; Vendruscolo, M.; et al. Quantitative Measurement of the Affinity of Toxic and Nontoxic Misfolded Protein Oligomers for Lipid Bilayers and of its Modulation by Lipid Composition and Trodusquemine. *ACS Chem. Neurosci.* **2021**, *12*, 3189–3202. [CrossRef]
89. Limbocker, R.; Chia, S.; Ruggeri, F.S.; Perni, M.; Cascella, R.; Heller, G.T.; Meisl, G.; Mannini, B.; Habchi, J.; Michaels, T.C.T.; et al. Trodusquemine enhances Abeta(42) aggregation but suppresses its toxicity by displacing oligomers from cell membranes. *Nat. Commun.* **2019**, *10*, 225. [CrossRef]
90. Kreiser, R.P.; Wright, A.K.; Sasser, L.R.; Rinauro, D.J.; Gabriel, J.M.; Hsu, C.M.; Hurtado, J.A.; McKenzie, T.L.; Errico, S.; Albright, J.A.; et al. A Brain-Permeable Aminosterol Regulates Cell Membranes to Mitigate the Toxicity of Diverse Pore-Forming Agents. *ACS Chem. Neurosci.* **2022**, *13*, 1219–1231. [CrossRef]
91. Dada, S.T.; Toprakcioglu, Z.; Cali, M.P.; Rontgen, A.; Hardenberg, M.C.; Morris, O.M.; Mrugalla, L.K.; Knowles, T.P.J.; Vendruscolo, M. Pharmacological inhibition of alpha-synuclein aggregation within liquid condensates. *Nat. Commun.* **2024**, *15*, 3835. [CrossRef] [PubMed]
92. Subczynski, W.K.; Pasenkiewicz-Gierula, M.; Widomska, J.; Mainali, L.; Raguz, M. High Cholesterol/Low Cholesterol: Effects in Biological Membranes: A Review. *Cell Biochem. Biophys.* **2017**, *75*, 369–385. [CrossRef] [PubMed]
93. Maxfield, F.R.; van Meer, G. Cholesterol, the central lipid of mammalian cells. *Curr. Opin. Cell Biol.* **2010**, *22*, 422–429. [CrossRef] [PubMed]
94. Doole, F.T.; Kumarage, T.; Ashkar, R.; Brown, M.F. Cholesterol Stiffening of Lipid Membranes. *J. Membr. Biol.* **2022**, *255*, 385–405. [CrossRef]
95. Petrov, A.M.; Kasimov, M.R.; Zefirov, A.L. Brain Cholesterol Metabolism and Its Defects: Linkage to Neurodegenerative Diseases and Synaptic Dysfunction. *Acta Naturae* **2016**, *8*, 58–73. [CrossRef]

96. Shin, K.C.; Ali Moussa, H.Y.; Park, Y. Cholesterol imbalance and neurotransmission defects in neurodegeneration. *Exp. Mol. Med.* **2024**, *56*, 1685–1690. [CrossRef]
97. Dai, L.; Zou, L.; Meng, L.; Qiang, G.; Yan, M.; Zhang, Z. Cholesterol Metabolism in Neurodegenerative Diseases: Molecular Mechanisms and Therapeutic Targets. *Mol. Neurobiol.* **2021**, *58*, 2183–2201. [CrossRef]
98. Vance, J.E. Dysregulation of cholesterol balance in the brain: Contribution to neurodegenerative diseases. *Dis. Model. Mech.* **2012**, *5*, 746–755. [CrossRef]
99. Banerjee, S.; Hashemi, M.; Zagorski, K.; Lyubchenko, Y.L. Cholesterol in Membranes Facilitates Aggregation of Amyloid beta Protein at Physiologically Relevant Concentrations. *ACS Chem. Neurosci.* **2021**, *12*, 506–516. [CrossRef]
100. Yu, X.; Zheng, J. Cholesterol promotes the interaction of Alzheimer beta-amyloid monomer with lipid bilayer. *J. Mol. Biol.* **2012**, *421*, 561–571. [CrossRef]
101. Habchi, J.; Chia, S.; Galvagnion, C.; Michaels, T.C.T.; Bellaiche, M.M.J.; Ruggeri, F.S.; Sanguanini, M.; Idini, I.; Kumita, J.R.; Sparr, E.; et al. Cholesterol catalyses Abeta42 aggregation through a heterogeneous nucleation pathway in the presence of lipid membranes. *Nat. Chem.* **2018**, *10*, 673–683. [CrossRef] [PubMed]
102. Qi, Z.; Wan, M.; Zhang, J.; Li, Z. Influence of Cholesterol on the Membrane Binding and Conformation of alpha-Synuclein. *J. Phys. Chem. B* **2023**, *127*, 1956–1964. [CrossRef]
103. Jakubec, M.; Barias, E.; Furse, S.; Govasli, M.L.; George, V.; Turcu, D.; Iashchishyn, I.A.; Morozova-Roche, L.A.; Halskau, O. Cholesterol-containing lipid nanodiscs promote an alpha-synuclein binding mode that accelerates oligomerization. *FEBS J.* **2021**, *288*, 1887–1905. [CrossRef] [PubMed]
104. Sciacca, M.F.; Lolicato, F.; Di Mauro, G.; Milardi, D.; D'Urso, L.; Satriano, C.; Ramamoorthy, A.; La Rosa, C. The Role of Cholesterol in Driving IAPP-Membrane Interactions. *Biophys. J.* **2016**, *111*, 140–151. [CrossRef]
105. Stonebraker, A.R.; Beasley, M.; Massinople, S.; Wunder, M.; Li, P.; Valentine, S.J.; Legleiter, J. Cholesterol impacts the formation of huntingtin/lipid complexes and subsequent aggregation. *Protein Sci.* **2023**, *32*, e4642. [CrossRef]
106. Ali, A.; Zhaliazka, K.; Dou, T.; Holman, A.P.; Kurouski, D. Cholesterol and Sphingomyelin Uniquely Alter the Rate of Transthyretin Aggregation and Decrease the Toxicity of Amyloid Fibrils. *J. Phys. Chem. Lett.* **2023**, *14*, 10886–10893. [CrossRef]
107. Kakio, A.; Nishimoto, S.I.; Yanagisawa, K.; Kozutsumi, Y.; Matsuzaki, K. Cholesterol-dependent formation of GM1 ganglioside-bound amyloid beta-protein, an endogenous seed for Alzheimer amyloid. *J. Biol. Chem.* **2001**, *276*, 24985–24990. [CrossRef]
108. Fernandez-Perez, E.J.; Sepulveda, F.J.; Peters, C.; Bascunan, D.; Riffo-Lepe, N.O.; Gonzalez-Sanmiguel, J.; Sanchez, S.A.; Peoples, R.W.; Vicente, B.; Aguayo, L.G. Effect of Cholesterol on Membrane Fluidity and Association of Abeta Oligomers and Subsequent Neuronal Damage: A Double-Edged Sword. *Front. Aging Neurosci.* **2018**, *10*, 226. [CrossRef]
109. Cecchi, C.; Nichino, D.; Zampagni, M.; Bernacchioni, C.; Evangelisti, E.; Pensalfini, A.; Liguri, G.; Gliozzi, A.; Stefani, M.; Relini, A. A protective role for lipid raft cholesterol against amyloid-induced membrane damage in human neuroblastoma cells. *Biochim. Biophys. Acta* **2009**, *1788*, 2204–2216. [CrossRef]
110. Brown, A.M.; Bevan, D.R. Molecular Dynamics Simulations of Amyloid beta-Peptide (1–42): Tetramer Formation and Membrane Interactions. *Biophys. J.* **2016**, *111*, 937–949. [CrossRef]
111. Tan, Q.; Liu, H.; Duan, M.; Huo, S. Interplay between human islet amyloid polypeptide aggregates and micro-heterogeneous membranes. *Biochim. Biophys. Acta Biomembr.* **2021**, *1863*, 183691. [CrossRef] [PubMed]
112. Christensen, M.; Berglund, N.A.; Schiott, B. The Effect of Cholesterol on Membrane-Bound Islet Amyloid Polypeptide. *Front. Mol. Biosci.* **2021**, *8*, 657946. [CrossRef]
113. Zhang, X.; London, E.; Raleigh, D.P. Sterol Structure Strongly Modulates Membrane-Islet Amyloid Polypeptide Interactions. *Biochemistry* **2018**, *57*, 1868–1879. [CrossRef]
114. Allende, D.; Simon, S.A.; McIntosh, T.J. Melittin-induced bilayer leakage depends on lipid material properties: Evidence for toroidal pores. *Biophys. J.* **2005**, *88*, 1828–1837. [CrossRef] [PubMed]
115. Fantini, J.; Carlus, D.; Yahi, N. The fusogenic tilted peptide (67–78) of alpha-synuclein is a cholesterol binding domain. *Biochim. Biophys. Acta* **2011**, *1808*, 2343–2351. [CrossRef]
116. Pan, J.; Dalzini, A.; Khadka, N.K.; Aryal, C.M.; Song, L. Lipid Extraction by alpha-Synuclein Generates Semi-Transmembrane Defects and Lipoprotein Nanoparticles. *ACS Omega* **2018**, *3*, 9586–9597. [CrossRef]
117. O'Leary, E.I.; Jiang, Z.; Strub, M.P.; Lee, J.C. Effects of phosphatidylcholine membrane fluidity on the conformation and aggregation of N-terminally acetylated alpha-synuclein. *J. Biol. Chem.* **2018**, *293*, 11195–11205. [CrossRef]
118. Man, W.K.; De Simone, A.; Barritt, J.D.; Vendruscolo, M.; Dobson, C.M.; Fusco, G. A Role of Cholesterol in Modulating the Binding of alpha-Synuclein to Synaptic-Like Vesicles. *Front. Neurosci.* **2020**, *14*, 18. [CrossRef]
119. Rudajev, V.; Novotny, J. Cholesterol as a key player in amyloid beta-mediated toxicity in Alzheimer's disease. *Front. Mol. Neurosci.* **2022**, *15*, 937056. [CrossRef]
120. Singla, R.K.; Dubey, A.K.; Garg, A.; Sharma, R.K.; Fiorino, M.; Ameen, S.M.; Haddad, M.A.; Al-Hiary, M. Natural Polyphenols: Chemical Classification, Definition of Classes, Subcategories, and Structures. *J. AOAC Int.* **2019**, *102*, 1397–1400. [CrossRef]
121. Caruana, M.; Cauchi, R.; Vassallo, N. Putative Role of Red Wine Polyphenols against Brain Pathology in Alzheimer's and Parkinson's Disease. *Front. Nutr.* **2016**, *3*, 31. [CrossRef] [PubMed]

122. Caruana, M.; Hogen, T.; Levin, J.; Hillmer, A.; Giese, A.; Vassallo, N. Inhibition and disaggregation of alpha-synuclein oligomers by natural polyphenolic compounds. *FEBS Lett.* **2011**, *585*, 1113–1120. [CrossRef] [PubMed]
123. Gauci, A.J.; Caruana, M.; Giese, A.; Scerri, C.; Vassallo, N. Identification of polyphenolic compounds and black tea extract as potent inhibitors of lipid membrane destabilization by Abeta(4)(2) aggregates. *J. Alzheimers Dis.* **2011**, *27*, 767–779. [CrossRef]
124. Mahboob, A.; Senevirathne, D.K.L.; Paul, P.; Nabi, F.; Khan, R.H.; Chaari, A. An investigation into the potential action of polyphenols against human Islet Amyloid Polypeptide aggregation in type 2 diabetes. *Int. J. Biol. Macromol.* **2023**, *225*, 318–350. [CrossRef]
125. Kabir, E.R.; Chowdhury, N.M.; Yasmin, H.; Kabir, M.T.; Akter, R.; Perveen, A.; Ashraf, G.M.; Akter, S.; Rahman, M.H.; Sweilam, S.H. Unveiling the Potential of Polyphenols as Anti-Amyloid Molecules in Alzheimer's Disease. *Curr. Neuropharmacol.* **2023**, *21*, 787–807. [CrossRef]
126. Grabska-Kobylecka, I.; Szpakowski, P.; Krol, A.; Ksiazek-Winiarek, D.; Kobylecki, A.; Glabinski, A.; Nowak, D. Polyphenols and Their Impact on the Prevention of Neurodegenerative Diseases and Development. *Nutrients* **2023**, *15*, 3454. [CrossRef] [PubMed]
127. Caruana, M.; Camilleri, A.; Farrugia, M.Y.; Ghio, S.; Jakubickova, M.; Cauchi, R.J.; Vassallo, N. Extract from the Marine Seaweed Padina pavonica Protects Mitochondrial Biomembranes from Damage by Amyloidogenic Peptides. *Molecules* **2021**, *26*, 1444. [CrossRef]
128. Ngoungoure, V.L.; Schluesener, J.; Moundipa, P.F.; Schluesener, H. Natural polyphenols binding to amyloid: A broad class of compounds to treat different human amyloid diseases. *Mol. Nutr. Food Res.* **2015**, *59*, 8–20. [CrossRef]
129. Nie, T.; Cooper, G.J.S. Mechanisms Underlying the Antidiabetic Activities of Polyphenolic Compounds: A Review. *Front. Pharmacol.* **2021**, *12*, 798329. [CrossRef]
130. Karonen, M. Insights into Polyphenol-Lipid Interactions: Chemical Methods, Molecular Aspects and Their Effects on Membrane Structures. *Plants* **2022**, *11*, 1809. [CrossRef]
131. Wang, R.; Peng, J.; Shi, X.; Cao, S.; Xu, Y.; Xiao, G.; Li, C. Change in membrane fluidity induced by polyphenols is highly dependent on the position and number of galloyl groups. *Biochim. Biophys. Acta Biomembr.* **2022**, *1864*, 184015. [CrossRef] [PubMed]
132. Margina, D.; Ilie, M.; Manda, G.; Neagoe, I.; Mocanu, M.; Ionescu, D.; Gradinaru, D.; Ganea, C. Quercetin and epigallocatechin gallate effects on the cell membranes biophysical properties correlate with their antioxidant potential. *Gen. Physiol. Biophys.* **2012**, *31*, 47–55. [CrossRef]
133. Hewlings, S.J.; Kalman, D.S. Curcumin: A Review of Its Effects on Human Health. *Foods* **2017**, *6*, 92. [CrossRef] [PubMed]
134. Radbakhsh, S.; Barreto, G.E.; Bland, A.R.; Sahebkar, A. Curcumin: A small molecule with big functionality against amyloid aggregation in neurodegenerative diseases and type 2 diabetes. *Biofactors* **2021**, *47*, 570–586. [CrossRef]
135. Liang, F.; Wan, Y.; Schaak, D.; Ward, J.; Shen, X.; Tanzi, R.E.; Zhang, C.; Quan, Q. Nanoplasmonic fiber tip probe detects significant reduction of intracellular Alzheimer's disease-related oligomers by curcumin. *Sci. Rep.* **2017**, *7*, 5722. [CrossRef] [PubMed]
136. Thapa, A.; Jett, S.D.; Chi, E.Y. Curcumin Attenuates Amyloid-beta Aggregate Toxicity and Modulates Amyloid-beta Aggregation Pathway. *ACS Chem. Neurosci.* **2016**, *7*, 56–68. [CrossRef]
137. Rane, J.S.; Bhaumik, P.; Panda, D. Curcumin Inhibits Tau Aggregation and Disintegrates Preformed Tau Filaments in vitro. *J. Alzheimers Dis.* **2017**, *60*, 999–1014. [CrossRef]
138. Li, J.F.; Jiang, Z.Q.; Cao, S.; Zhang, M.X.; Wang, L.H.; Liu, J.; Lu, Y.H.; Wang, H.Y.; Hong, X.J.; Wang, Z.G.; et al. Curcumin Inhibits alpha-Synuclein Aggregation by Acting on Liquid-Liquid Phase Transition. *Foods* **2024**, *13*, 1287. [CrossRef]
139. Pithadia, A.S.; Bhunia, A.; Sribalan, R.; Padmini, V.; Fierke, C.A.; Ramamoorthy, A. Influence of a curcumin derivative on hIAPP aggregation in the absence and presence of lipid membranes. *Chem. Commun.* **2016**, *52*, 942–945. [CrossRef]
140. Ahsan, N.; Mishra, S.; Jain, M.K.; Surolia, A.; Gupta, S. Curcumin Pyrazole and its derivative (N-(3-Nitrophenylpyrazole) Curcumin inhibit aggregation, disrupt fibrils and modulate toxicity of Wild type and Mutant alpha-Synuclein. *Sci. Rep.* **2015**, *5*, 9862. [CrossRef]
141. Alsop, R.J.; Dhaliwal, A.; Rheinstadter, M.C. Curcumin Protects Membranes through a Carpet or Insertion Model Depending on Hydration. *Langmuir* **2017**, *33*, 8516–8524. [CrossRef] [PubMed]
142. Pedrosa, M.; Maldonado-Valderrama, J.; Galvez-Ruiz, M.J. Interactions between curcumin and cell membrane models by Langmuir monolayers. *Colloids Surf. B Biointerfaces* **2022**, *217*, 112636. [CrossRef]
143. Gudyka, J.; Ceja-Vega, J.; Ivanchenko, K.; Morocho, Z.; Panella, M.; Gamez Hernandez, A.; Clarke, C.; Perez, E.; Silverberg, S.; Lee, S. Concentration-Dependent Effects of Curcumin on Membrane Permeability and Structure. *ACS Pharmacol. Transl. Sci.* **2024**, *7*, 1546–1556. [CrossRef]
144. Sahu, A.K.; Mishra, A.K. Curcumin-Induced Membrane Property Changes in DMPC Multilamellar Vesicles and the Effects of Membrane-Destabilizing Molecules on Curcumin-Loaded Multilamellar Vesicles. *Langmuir* **2021**, *37*, 12753–12766. [CrossRef] [PubMed]
145. Thapa, A.; Vernon, B.C.; De la Pena, K.; Soliz, G.; Moreno, H.A.; Lopez, G.P.; Chi, E.Y. Membrane-mediated neuroprotection by curcumin from amyloid-beta-peptide-induced toxicity. *Langmuir* **2013**, *29*, 11713–11723. [CrossRef]
146. Sallaberry, C.A.; Voss, B.J.; Stone, W.B.; Estrada, F.; Bhatia, A.; Soto, J.D.; Griffin, C.W.; Vander Zanden, C.M. Curcumin Reduces Amyloid Beta Oligomer Interactions with Anionic Membranes. *ACS Chem. Neurosci.* **2023**, *14*, 4026–4038. [CrossRef]
147. Khondker, A.; Alsop, R.J.; Himbert, S.; Tang, J.; Shi, A.C.; Hitchcock, A.P.; Rheinstadter, M.C. Membrane-Modulating Drugs can Affect the Size of Amyloid-beta(25–35) Aggregates in Anionic Membranes. *Sci. Rep.* **2018**, *8*, 12367. [CrossRef] [PubMed]

148. Dong, X.; Tang, Y.; Zhan, C.; Wei, G. Green tea extract EGCG plays a dual role in Abeta(42) protofibril disruption and membrane protection: A molecular dynamic study. *Chem. Phys. Lipids* **2021**, *234*, 105024. [CrossRef]
149. Rho, T.; Choi, M.S.; Jung, M.; Kil, H.W.; Hong, Y.D.; Yoon, K.D. Identification of fermented tea (*Camellia sinensis*) polyphenols and their inhibitory activities against amyloid-beta aggregation. *Phytochemistry* **2019**, *160*, 11–18. [CrossRef]
150. Sonawane, S.K.; Chidambaram, H.; Boral, D.; Gorantla, N.V.; Balmik, A.A.; Dangi, A.; Ramasamy, S.; Marelli, U.K.; Chinnathambi, S. EGCG impedes human Tau aggregation and interacts with Tau. *Sci. Rep.* **2020**, *10*, 12579. [CrossRef]
151. Yang, Z.; Yao, Y.; Zhou, Y.; Li, X.; Tang, Y.; Wei, G. EGCG attenuates alpha-synuclein protofibril-membrane interactions and disrupts the protofibril. *Int. J. Biol. Macromol.* **2023**, *230*, 123194. [CrossRef]
152. Meng, F.; Abedini, A.; Plesner, A.; Verchere, C.B.; Raleigh, D.P. The flavanol (-)-epigallocatechin 3-gallate inhibits amyloid formation by islet amyloid polypeptide, disaggregates amyloid fibrils, and protects cultured cells against IAPP-induced toxicity. *Biochemistry* **2010**, *49*, 8127–8133. [CrossRef]
153. Goncalves, P.B.; Sodero, A.C.R.; Cordeiro, Y. Green Tea Epigallocatechin-3-gallate (EGCG) Targeting Protein Misfolding in Drug Discovery for Neurodegenerative Diseases. *Biomolecules* **2021**, *11*, 767. [CrossRef]
154. Pires, F.; Geraldo, V.P.N.; Rodrigues, B.; Granada-Flor, A.; de Almeida, R.F.M.; Oliveira, O.N., Jr.; Victor, B.L.; Machuqueiro, M.; Raposo, M. Evaluation of EGCG Loading Capacity in DMPC Membranes. *Langmuir* **2019**, *35*, 6771–6781. [CrossRef] [PubMed]
155. Uekusa, Y.; Kamihira-Ishijima, M.; Sugimoto, O.; Ishii, T.; Kumazawa, S.; Nakamura, K.; Tanji, K.; Naito, A.; Nakayama, T. Interaction of epicatechin gallate with phospholipid membranes as revealed by solid-state NMR spectroscopy. *Biochim. Biophys. Acta* **2011**, *1808*, 1654–1660. [CrossRef]
156. Malishev, R.; Nandi, S.; Kolusheva, S.; Levi-Kalisman, Y.; Klarner, F.G.; Schrader, T.; Bitan, G.; Jelinek, R. Toxicity inhibitors protect lipid membranes from disruption by Abeta42. *ACS Chem. Neurosci.* **2015**, *6*, 1860–1869. [CrossRef] [PubMed]
157. Sanders, H.M.; Jovcevski, B.; Marty, M.T.; Pukala, T.L. Structural and mechanistic insights into amyloid-beta and alpha-synuclein fibril formation and polyphenol inhibitor efficacy in phospholipid bilayers. *FEBS J.* **2022**, *289*, 215–230. [CrossRef]
158. Beasley, M.; Stonebraker, A.R.; Hasan, I.; Kapp, K.L.; Liang, B.J.; Agarwal, G.; Groover, S.; Sedighi, F.; Legleiter, J. Lipid Membranes Influence the Ability of Small Molecules To Inhibit Huntingtin Fibrillization. *Biochemistry* **2019**, *58*, 4361–4373. [CrossRef]
159. Engel, M.F.; vandenAkker, C.C.; Schleeger, M.; Velikov, K.P.; Koenderink, G.H.; Bonn, M. The polyphenol EGCG inhibits amyloid formation less efficiently at phospholipid interfaces than in bulk solution. *J. Am. Chem. Soc.* **2012**, *134*, 14781–14788. [CrossRef]
160. Wang, Y.; Hekimi, S. Understanding Ubiquinone. *Trends Cell Biol.* **2016**, *26*, 367–378. [CrossRef]
161. Mantle, D.; Dewsbury, M.; Hargreaves, I.P. The Ubiquinone-Ubiquinol Redox Cycle and Its Clinical Consequences: An Overview. *Int. J. Mol. Sci.* **2024**, *25*, 6765. [CrossRef] [PubMed]
162. Baschiera, E.; Sorrentino, U.; Calderan, C.; Desbats, M.A.; Salviati, L. The multiple roles of coenzyme Q in cellular homeostasis and their relevance for the pathogenesis of coenzyme Q deficiency. *Free Radic. Biol. Med.* **2021**, *166*, 277–286. [CrossRef] [PubMed]
163. Alcazar-Fabra, M.; Navas, P.; Brea-Calvo, G. Coenzyme Q biosynthesis and its role in the respiratory chain structure. *Biochim. Biophys. Acta* **2016**, *1857*, 1073–1078. [CrossRef] [PubMed]
164. Guerra, R.M.; Pagliarini, D.J. Coenzyme Q biochemistry and biosynthesis. *Trends Biochem. Sci.* **2023**, *48*, 463–476. [CrossRef]
165. Braasch-Turi, M.M.; Koehn, J.T.; Crans, D.C. Chemistry of Lipoquinones: Properties, Synthesis, and Membrane Location of Ubiquinones, Plastoquinones, and Menaquinones. *Int. J. Mol. Sci.* **2022**, *23*, 12856. [CrossRef]
166. Agmo Hernandez, V.; Eriksson, E.K.; Edwards, K. Ubiquinone-10 alters mechanical properties and increases stability of phospholipid membranes. *Biochim. Biophys. Acta* **2015**, *1848*, 2233–2243. [CrossRef]
167. Jemiola-Rzeminska, M.; Kruk, J.; Skowronek, M.; Strzalka, K. Location of ubiquinone homologues in liposome membranes studied by fluorescence anisotropy of diphenyl-hexatriene and trimethylammonium-diphenyl-hexatriene. *Chem. Phys. Lipids* **1996**, *79*, 55–63. [CrossRef] [PubMed]
168. Sevin, D.C.; Sauer, U. Ubiquinone accumulation improves osmotic-stress tolerance in Escherichia coli. *Nat. Chem. Biol.* **2014**, *10*, 266–272. [CrossRef]
169. Eriksson, E.K.; Agmo Hernandez, V.; Edwards, K. Effect of ubiquinone-10 on the stability of biomimetic membranes of relevance for the inner mitochondrial membrane. *Biochim. Biophys. Acta Biomembr.* **2018**, *1860*, 1205–1215. [CrossRef]
170. Choi, H.; Park, H.H.; Koh, S.H.; Choi, N.Y.; Yu, H.J.; Park, J.; Lee, Y.J.; Lee, K.Y. Coenzyme Q10 protects against amyloid beta-induced neuronal cell death by inhibiting oxidative stress and activating the P13K pathway. *Neurotoxicology* **2012**, *33*, 85–90. [CrossRef]
171. Duran-Prado, M.; Frontinan, J.; Santiago-Mora, R.; Peinado, J.R.; Parrado-Fernandez, C.; Gomez-Almagro, M.V.; Moreno, M.; Lopez-Dominguez, J.A.; Villalba, J.M.; Alcain, F.J. Coenzyme Q10 protects human endothelial cells from beta-amyloid uptake and oxidative stress-induced injury. *PLoS ONE* **2014**, *9*, e109223. [CrossRef] [PubMed]
172. Frontinan-Rubio, J.; Rabanal-Ruiz, Y.; Duran-Prado, M.; Alcain, F.J. The Protective Effect of Ubiquinone against the Amyloid Peptide in Endothelial Cells Is Isoprenoid Chain Length-Dependent. *Antioxidants* **2021**, *10*, 1806. [CrossRef] [PubMed]
173. Padurariu, M.; Ciobica, A.; Lefter, R.; Serban, I.L.; Stefanescu, C.; Chirita, R. The oxidative stress hypothesis in Alzheimer's disease. *Psychiatr. Danub.* **2013**, *25*, 401–409. [PubMed]
174. Swerdlow, R.H. The Alzheimer's Disease Mitochondrial Cascade Hypothesis: A Current Overview. *J. Alzheimers Dis.* **2023**, *92*, 751–768. [CrossRef]

175. Fisar, Z.; Hroudova, J. CoQ(10) and Mitochondrial Dysfunction in Alzheimer's Disease. *Antioxidants* **2024**, *13*, 191. [CrossRef]
176. Yamagishi, K.; Ikeda, A.; Moriyama, Y.; Chei, C.L.; Noda, H.; Umesawa, M.; Cui, R.; Nagao, M.; Kitamura, A.; Yamamoto, Y.; et al. Serum coenzyme Q10 and risk of disabling dementia: The Circulatory Risk in Communities Study (CIRCS). *Atherosclerosis* **2014**, *237*, 400–403. [CrossRef]

membranes

MDPI

Review

The Role of ABCB1, ABCG2, and SLC Transporters in Pharmacokinetic Parameters of Selected Drugs and Their Involvement in Drug–Drug Interactions

Kajetan Kiełbowski, Małgorzata Król, Estera Bakinowska and Andrzej Pawlik *

Department of Physiology, Pomeranian Medical University, 70-111 Szczecin, Poland; kajetan.kielbowski@onet.pl (K.K.); malgorzatakrol246@gmail.com (M.K.); esterabakinowska@gmail.com (E.B.)
* Correspondence: pawand@poczta.onet.pl

Abstract: Membrane transporters are expressed in a wide range of tissues in the human organism. These proteins regulate the penetration of various substances such as simple ions, xenobiotics, and an extensive number of therapeutics. ABC and SLC drug transporters play a crucial role in drug absorption, distribution, and elimination. Recent decades have shown their contribution to the systemic exposure and tissue penetration of numerous drugs, thereby having an impact on pharmacokinetic and pharmacodynamic parameters. Importantly, the activity and expression of these transporters depend on numerous conditions, including intestinal microbiome profiles or health conditions. Moreover, the combined intake of other drugs or natural agents further affects the functionality of these proteins. In this review, we will discuss the involvement of ABC and SLC transporters in drug disposition. Moreover, we will present current evidence of the potential role of drug transporters as therapeutic targets.

Keywords: drug transporters; ATP binding cassette transporters; solute carrier transporters; p-glycoprotein; drug–drug interactions

1. Introduction

Pharmacology is a broad field of study relating to drugs that can be described using pharmacokinetics (PK) and pharmacodynamics (PD). PK provides information on what happens to the drug after its administration. The acronym LADME refers to the most important elements, comprising liberation, absorption, distribution, metabolism, and elimination. In contrast, PD parameters refer to the influence of the drug on an organism (Figure 1). Understanding the PK and PD parameters of drugs is crucial for them to be safely and effectively administered to humans. Age is an important factor to consider for drug administration, as PK parameters undergo developmental maturation, which has been nicely summarized by van den Anker et al. [1]. Furthermore, the processes of absorption, distribution, and elimination depend greatly upon the activity of cellular transporters. These proteins are responsible for the intake of drugs, but they are also involved in the process of efflux. Single nucleotide polymorphisms (SNPs) are a type of genetic variant that has been investigated for decades. SNPs in drug transporter genes are associated with changes in activity, which translates into modified drug distribution [2]. In this review, we will discuss the involvement of cellular membrane transporters in the distribution and bioavailability of drugs, focusing mainly on ATP-binding cassette (ABC) and solute carrier (SLC) families.

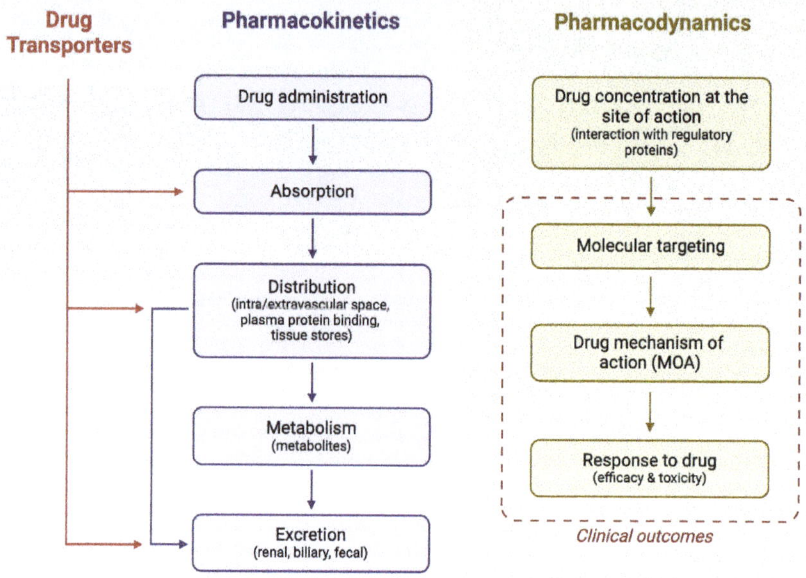

Figure 1. The activity of drug transporters affects pharmacokinetic parameters of drugs, which have a direct influence on their pharmacodynamic properties. Created in BioRender. Kiełbowski, K. (2024) BioRender.com/t19o310.

2. Family of Transporters—Structure and Physiological Role

2.1. ATP-Binding Cassette

The family of ABC proteins involves membrane transporters that are composed of nucleotide binding domains and transmembrane domains. The binding of ATP induces conformational changes that regulate the opening of the transporter. Approximately 50 ABC transporters have been identified in humans [3]. These proteins play numerous roles in human physiology; they are important transporters of ions, organic molecules, peptides, lipids, and drugs. Furthermore, they have been associated with several diseases and pathological conditions, such as cystic fibrosis (ABCC7/CFTR) or drug resistance in the case of P-glycoprotein (P-gp) [4]. As previously mentioned, these transporters perform a variety of tasks, but their involvement in drug distribution is primarily associated with drug efflux. P-gp represents a prominent member of the ABC family and the ABCB subgroup. It is involved in the distribution and elimination of therapeutic agents as it is expressed in the intestine [5], kidneys [6], and the blood–brain barrier (BBB) [7], among others sites.

ABCG2, which is also known as breast cancer resistance protein (BCRP), represents another member of the ABC transporter family. It is involved in the transport of metabolites and uric acid, together with numerous drugs, including conjugates. The structure of BCRP also involves intracellular, transmembrane, and extracellular elements [8]. As the name of the transporter suggests, its presence is associated with chemoresistance in breast cancer tumors [9], including metastatic lesions [10]. The transporter is expressed in various tissues, including the placenta, gastrointestinal tract, and kidneys, among others [11]. The important physiological roles of ABCB1 and ABCG2 have been confirmed in double-knockout rats. In these animals, silencing the genes encoding the above-mentioned transporters induces significant metabolomic and transcriptomic alterations. Analyses of metabolites revealed that amino acids and lipids were significantly altered in cerebrospinal fluid and plasma, respectively. Moreover, the modification changed the transcription of hundreds of genes in the kidneys, liver, and brain. The observed effects could be a result of compensatory mechanisms associated with altered metabolome. However, other pathways were affected as well, such as pathways associated with endoplasmic reticulum stress [12]. The structure of the ABC transporters involves intracellular nucleotide-binding domains and transcellular domains (Figure 2).

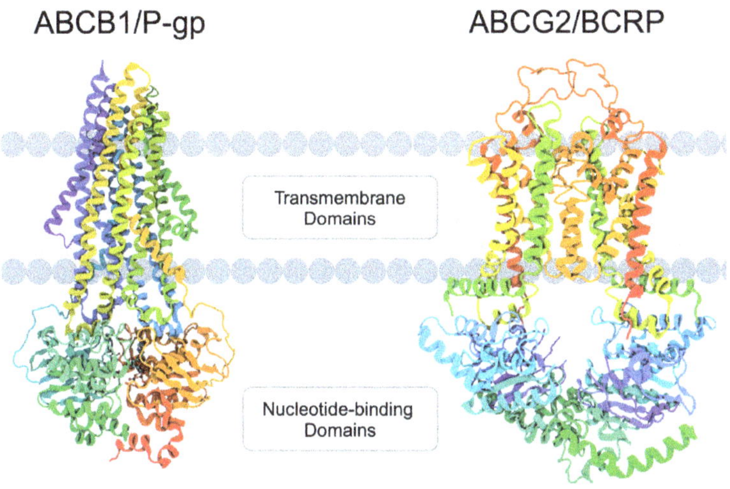

Figure 2. Structure of P-glycoprotein (PDB ID: 6C0V [13–15]) and breast cancer resistance protein (PDB ID: 6VXF [16]). Created in BioRender. Kiełbowski, K. (2024) BioRender.com/s65g631.

2.2. Soluble Carrier Transporters

Solute carrier (SLC) transporters are a diverse family of membrane-bound proteins that include over 66 gene families, and over 430 secondary active transporters have been identified in humans [17]. Some scientific studies report that there are over than 600 members organized into 166 families [18] that play a crucial part in the absorption, distribution, metabolism, and elimination of various pharmacological agents [19,20]. Recent advancements in structural biology, including X-ray crystallography and nuclear magnetic resonance (NMR) spectroscopy, have significantly improved our understanding of the mechanisms and ligand interactions of SLC transporters. The two most prevalent structural folds in SLC proteins are the leucine transporter protein (LeuT)-like fold and the main facilitator superfamily (MFS) fold. The LeuT fold is made up of two five-transmembrane-helix structures, each of which has a bundle and a scaffold domain. The MFS fold is made up of two pseudo-repeats of six-transmembrane-helix structures joined by a cytoplasmic loop [21]. The structural diversity of the SLC family members is remarkable, as they comprise an array of folds that are unlikely to have evolved from one another [22]. Even in cases in which sequence similarities are modest, the folds of SLC members typically identify their processes and reveal their evolutionary linkage. The human SLC superfamily is a classic example of a distantly homologous protein family with substantial sequence diversity and relatively little structural diversity [23]. SLCs are responsible for moving a remarkably wide range of solutes through biological membranes, including medicines, amino acids, lipids, sugars, inorganic ions, and carbohydrates. The majority of these membrane proteins work as coupled symporters, or co-transporters, that propel substrate against its concentration gradient into cells by means of downhill ion (H^+ or Na^+) gradients. Although some members have channel-like characteristics, other members function as antiporters, which usually have a single substrate-binding site and an alternating-access mechanism of transport [24–27].

The SLC transporter superfamily comprises 66 gene families with over 430 identified secondary active transporters in humans [17]. SLC transporters, which include families such as SLC21/SLCO, SLC22, SLC15, and others, play a crucial role in the distribution and PK parameters of drugs within the human body [28].

The H^+-coupled oligopeptide co-transporter family, also known as the solute carrier 15 (SLC15) family of peptide transporters, is a class of membrane transporters that is well known for its critical function in the cellular uptake of di- and tripeptides. These transporters, particularly PEPT1 and PEPT2, facilitate the uptake of small peptides and peptidomimetic drugs across biological membranes [29]. The extracellular domain among twelve transmembrane (TM) helices that fold into an N-terminal domain (NTD, TM1-6) and a C-terminal domain (CTD, TM7-12) play an important role in PepT1 and PepT2 functionality [30]. PepT1 is primarily responsible for mediating intestinal absorption of luminal di/tripeptides, b-lactam antibiotics (cefadroxil), and antiviral medications (valacyclovir) from overall dietary protein digestion [31,32]. SLC15A2 (PepT2) primarily permits renal tubular reuptake of di/tripeptides from ultrafiltration and brain-to-blood efflux of di/tripeptides in the choroid plexus. SLC15A3 (PHT2) and SLC15A4 (PHT1) interact with both di/tripeptides and histidine in the endolysosome membranes of immune cells, and SLC15A5 has an unknown physiological role [33]. A study by Romano at al. found that of the two SLC15 transporters expressed in thyroid follicular cells, namely PEPT2 (SLC15A2) and PHT1 (SLC15A4), only PEPT2 was involved in peptide transport at the plasma membrane, what was the first evidence that peptide transport occurs in thyroid follicular cells [34].

The SLCO/SLC21 superfamily of organic anion transporting polypeptides (OATPs) mediate the transport of a broad range of endo- and xenobiotics, leading to interindividual differences in efficacy and toxicity [35]. OATPs are classified into families and subfamilies based on amino acid sequence similarities, with 36 members identified in humans, rats, and mice [35,36]. These transporters typically have 12 transmembrane domains and function independently of sodium. Genetic variations in SLCO genes can affect OATP function and contribute to the interindividual variability of drug effects [37]. Recent studies have revealed the importance of OATPs in the pharmacology of cancer and autoimmune disorders, with certain members being overexpressed in various solid tumors and potentially influencing tumor development and disease progression [38,39].

The SLC22 transporter family is highly expressed and is involved in the regulation of homeostasis through the transportation of molecular organic compounds, including antioxidants, signaling molecules, and metabolites [40]. It can be divided into organic anion transporters (OATs), organic cation transporters (OCTs), and organic zwitterion/cation transporters (OCTNs) [41]. Recent phylogenetic analysis suggests that SLC22 transporters evolved over 450 million years ago and can be classified into at least six groups [41]. A more recent approach proposes eight functional subgroups, including four new OAT subgroups [42]. These transporters are involved in various physiological processes and diseases, such as primary systemic carnitine deficiency and renal hypouricemia [43]. Different members of the SLC22 family are expressed on the basolateral and apical surfaces of epithelial cells, where they facilitate the movement of molecules between bodily fluids and critical organs such the kidneys, liver, heart, and brain [44]. A recent study presented potential natural product (NP) interactions with other compounds at the transporter site. The researchers found that a variety of NPs, including flavonoids, vitamins, and indoles, were altered in the serum of mice with OAT1 and OAT3 knockouts. A multivariate examination of chemical characteristics indicated that OAT1- and OAT3-dependent NPs could be largely distinguished from one another. Using both protein-binding tests and in vitro transport studies, direct binding to the transporter was verified [45]. For instance, multivariate analyses confirmed the association of SLC22A3 expression with progression-free survival in pancreatic cancer patients. According to a subgroup analysis of patients receiving regimens based on nucleoside analogues, those who showed apical localization of SLC22A3 in tumor cells or negative brush border staining appeared to have a lower overall survival rate [46]. New studies have provided additional information on previously classified transporters. In the study by Redeker and Brockmöller, SLC35G3 and SLC38A10 significantly accelerated the uptake of cations such as clonidine, 3,4-methylenedioxymethamphetamine, and nicotine [47].

The structural complexity and dynamic nature of SLC transporters have long posed challenges for traditional structural biology techniques, such as X-ray crystallography [48]. The advent of cryo-electron microscopy has been a game-changer in this field. Cryo-electron microscopy (cryo-EM) has emerged as a revolutionary tool in the field of structural biology, contributing significantly to understanding of the intricate details of biomolecular structures and their functions, such as exposing multiple transporters' binding and activation mechanisms. Moreover, this microscope has the unique ability to capture the intricate details of membrane protein structures, including their conformational changes and interactions with ligands and their complexes, without the need for crystallization. It has been applied to members of the SLC12A family, including KCCs, NKCC1, NCC, and SLC22A6, elucidating structural elements crucial for ion binding, domain interactions, and overall function [49,50]. The SLC transporter family's druggability has been significantly impacted by the mechanisms of action that cryo-EM has helped to uncover [51].

3. The Involvement of Transporters in Drug Absorption, Distribution, and Elimination

3.1. *ABC Transporters*

Considering the roles of ABC proteins in drug transport and their expression in the intestine, BBB, and kidneys, the expression and functionality of these transporters affect the bioavailability and elimination of therapeutic targets. Recent studies have investigated how various conditions influence ABC transporters. P-gp, also known as ABCB1, transports a wide range of drugs, including immunosuppressants [52], natural agents [53], peptide drugs [54], and cardiological drugs [55], among others.

Experiments in rabbits have shown that the expression of P-gp is different in various segments of the intestine, thus indicating that efflux capability and drug absorbance differ between different intestinal segments [56]. P-gp molecules are transported into the apical parts of enterocytes through an interaction with Myosin Vb (Myo5B). Interestingly, in a recent study by Doolay et al. [57], the authors examined the presence of P-gp in mice and detected these transporters below the cell membrane, which potentially suggests that there is a pool of P-gp that could be rapidly inserted into the membrane [57].

Researchers have demonstrated that various conditions affect the activity of the transporter. For instance, the expression of ABCB1 is dependent on gut microbiome composition. The extensive network of bacterial species found in the intestine play a significant role in human physiology as they affect immune responses [58]. Accordingly, altered composition of the microbiota, a condition known as gut dysbiosis, has been associated with various pathological conditions and diseases [59]. Regarding drug transporter expression, certain metabolites of the gut microbiome suppress the expression of ABCB1. However, this result was not associated with its efflux functionality [52]. A recent study proved that metabolites secreted by *Eggerthella lenta* can inhibit the drug efflux that is mediated by P-gp, thus increasing the bioavailability of digoxin [55]. A combination of several factors involving the gut microbiome could affect both transcriptional expression and functionality, thus influencing both drug efflux and absorption. For instance, some health conditions can affect the absorption efficiency of particular drugs. In an early study by Buchman and colleagues, researchers found significantly elevated expression of P-gp protein in a patient with Crohn's disease [60]. Furthermore, in a more recent study using rats with non-alcoholic steatohepatitis, Li et al. [61] observed a higher expression of P-gp in the small intestines of these animals compared with rats from the control group. This observation was accompanied by that of decreased bioavailability of simvastatin in the study group. Importantly, these differences were not observed when the statin was administered intravenously [61]. Similarly, a higher expression of P-gp was observed in rats with portal hypertension (PH), which is associated with reduced absorbance of octreotide, a somatistatin mimic used to treat PH and PH-related complications [54]. Another condition that affects the intestinal expression of P-gp is acute kidney injury (AKI). In a rat model in which renal injury was induced by cisplatin, AKI significantly downregulated levels of P-gp located in the ileum. However, the authors made a very interesting finding: they observed that reduced expression of the P-gp transporter did not significantly affect the absorbance of Rho-123, a P-gp substrate. On the contrary, absorbance was significantly affected for the other P-gp substrate—gatifloxacin (GFLX) [62]. Thus, this important study demonstrated important physiological properties of P-gp that increase its functionality even when its intestinal expression is reduced. Interestingly, drug transporters are not the only agents that limit the entry of drugs into systemic circulation. CYP3A4 is an enzyme that is expressed in the intestine and takes part in intestinal first-pass metabolism [63]. In a study by Li and collaborators, the authors demonstrated that inhibiting CYP3A4 and not P-gp can improve the absorption of lovastatin [64] (Figure 3).

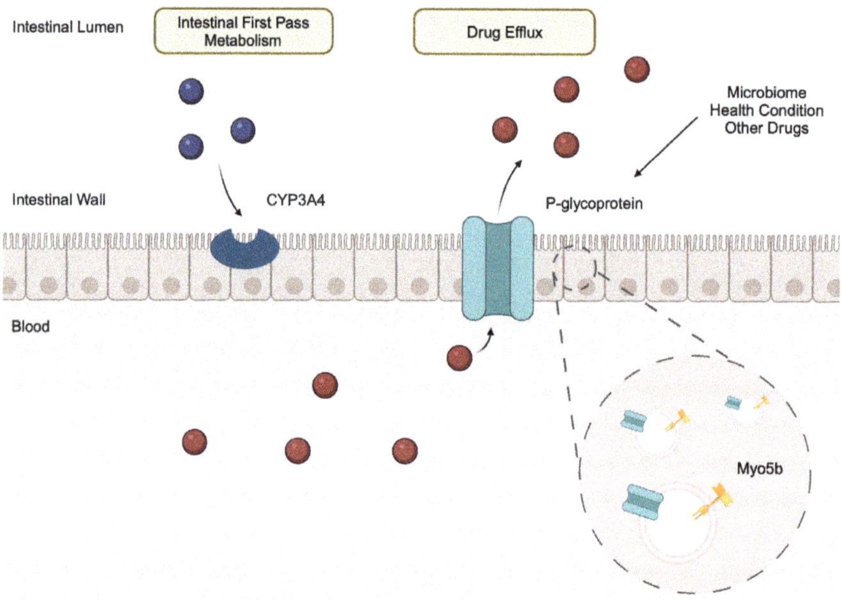

Figure 3. P-glycoprotein is exported to the apical parts of enterocytes, with the help of myosin Vb, where it mediates drug efflux. CYP3A4, which belongs to the family of cytochrome P450 enzymes, is also expressed in the intestinal wall, where it catalyzes first-pass metabolism and acts as a barrier for xenobiotics. Altered health condition and changed microbiome changes the activity of P-gp, which has a direct effect on drug absorption. Furthermore, the presence of other drugs and the occurrence of drug–drug interaction is associated with altered drug absorbance as well. Created in BioRender. Kiełbowski, K. (2024) BioRender.com/a30z627.

P-gp is expressed in other tissues in addition to the intestine, which is also associated with its effects on bioavailability. For instance, considering its role in drug efflux and expression in the BBB, the activity of the P-gp transporter limits the transport of drugs to the central nervous system (CNS); in the field of drug disposition, this is an undesired process as it restricts the ability of antidepressants to reach the CNS [65]. In contrast, the activity of P-gp is thought to be beneficial in Alzheimer's disease (AD), which will be more thoroughly discussed in the following sections. Similar to the intestinal P-gp, particular conditions or substances alter the functionality of the P-gp transporter in the BBB. Copper diacetyl bis (4-methyl-3-thiosemicarbazone) (Cu(ATSM)) was recently demonstrated to enhance P-gp activity in the mouse BBB [66]. Additionally, P-gp regulates intracellular drug concentrations, as it is expressed in lymphocytes [67]. Interestingly, despite being expressed in various tissues, P-gp activity can differentially affect certain therapeutics. For instance, the phosphoinositide 3-kinase delta (PI3Kδ) inhibitor seletalisib has a very good oral bioavailability profile, which indicates low activity of the intestinal efflux mechanisms. However, P-gp significantly restricts access of the drug to the CNS, thus showing high activity towards the drug in the BBB [68]. This finding represents another obstacle for analyzing drug PK parameters, as transporters can demonstrate different activity towards therapeutics depending on the type of tissue in which they are expressed.

Regarding BCRP, the protein is responsible for the transport of numerous agents, including chemotherapeutics and tyrosine kinase inhibitors, sulfasalazine, rosuvastatin, and antileptics, among others [69]. Unlike P-gp, intestinal expression of BCRP does not seem to follow a particular trend; however, an intraindividual variability was observed [70]. Recent studies demonstrate differences between P-gp and BCRP on drug absorption. For instance, BCRP impacts the absorption of topotecan, while P-gp is thought to be linked with limited transport of the drug towards the brain [71]. By contrast, BCRP is involved in the restriction of brain distribution of zonisamide, an antileptic drug. Interestingly, the effects of drug efflux transporters can be surpassed by using a different route of administration. Specifically, intranasal delivery of antileptics is less restricted by the BCRP transporter in mice [72]. Similarly to P-gp, the functionality of BCRP seems to depend on the health condition of individuals. Chronic HCV infection affects the activity of BCRP [73], which can lead to different exposure profiles of BCRP substrates in affected patients. Another important aspect of BCRP activity is pharmacogenetics. The rs2231142 variant affects the activity of the transporter, which is associated with altered drug distribution. The acknowledgment of this clinically relevant genetic variant was described in recently published consortium guidelines linking genetic variants and statin tolerance [74]. In this document, the authors recommend a <20 mg dose of rosuvastatin as a starting dose in patients with a poor function of the ABCG2 gene (c.421AA genotype) [74].

3.2. *SLC Transporters*

SLC transporters play crucial roles in the intestinal absorption and disposition of drugs; in particular, SGLT1 (solute carrier [SLC]5A1) and SGLT2 (SLC5A2) are involved in active glucose transport in the intestine and kidneys [75,76]. Genetic disorders affecting SLC transporters can lead to conditions such as glucose–galactose malabsorption and familial renal glucosuria [77,78]. Drugs and endogenous substances are mostly taken up by hepatic SLC transporters into hepatocytes, whereas their excretion is mediated by efflux transporters [79,80]. Targeting the BBB SLC transporters holds great promise for improving medication delivery to the brain. According to Meszáros et al. [81] and Morris et al. [82], these transporters exhibit distinct patterns at the BBB and are significantly expressed in brain endothelial cells. Al-Majdoub et al. in their study show differences in transporter profiles between healthy and diseased brain tissue [83]. In vitro and in vivo models have demonstrated the enhanced absorption of cargo molecules across the BBB—and increased BBB permeability—delivered by nanoparticles targeting SLC transporters, particularly those with dual ligands [81]. The interplay between various SLC transporters in the kidney is crucial for the excretion and retention of drugs, thus impacting their overall pharmacokinetics and pharmacodynamics. The expression and function of OATs in the kidney can be altered by diseases and drug interactions, potentially leading to unexpected clinical outcomes. For instance, the dromedary camel kidney develops changes in proximal tubule brush borders after long-term dehydration. Studies suggest that ultrastructural changes in the kidney cortex and minor changes in the medulla are connected with SLCs and their mRNA expression levels [84,85]. Placental transporters, particularly those from SLC families play a very significant role in fetal drug exposure and nutrient transfer during pregnancy [86]. Understanding placental drug transport is essential for optimizing pharmacotherapy during pregnancy and predicting fetal drug exposure [87]. Recent studies show that opioid use disorder (OUD) during pregnancy can profoundly affect both the mother and fetus though changes in methylated SLCs that are associated with pathways involved in neonatal opioid withdrawal syndrome (NOWS) [88]. A vital component of the seminiferous epithelium, the blood–testis barrier (BTB) shields developing germ cells from the environment [89]. Drugs and peptides enter the seminiferous epithelium by SLC transporters like Oatp3 and Slc15a1 [89,90]. According to Touré et al. [91], the Slc26a8 (Tat1) transporter is essential for male fertility in mice and sperm terminal differentiation. A range of uptake and efflux transporters are expressed by human Sertoli cells, which regulate the flow of endogenous and foreign chemicals across the BTB [92].

Recent studies highlight the importance of SLC transporters and the connections between one another as therapeutic targets. For instance, SLC13A3 expression is elevated in human liver cancer samples. In mouse cells, β-catenin-activated liver cancer cells undergo autophagic ferroptosis and GSH depletion when SLC13A3 is silenced [93], which indicates that SLC13A3 may be a useful therapeutic target for the treatment of human liver tumors containing GOF CTNNB1 mutations. SLC38A3, SLC38A1, SLC7A6, and SLC1A5 were also linked to hepatocellular carcinoma [94].

4. Drug–Drug Interactions

4.1. ABC Transporters

When describing the effects of drug transporters on drug distribution, it is important to mention drug–drug interactions (DDIs). DDIs occur when the functionality of a transporter is modified to stimulate or suppress the absorbance of a substrate or a victim drug. In recent decades, an extensive number of studies have investigated clinical DDIs involving P-gp, which were elegantly summarized by Gessner et al. [95]. In their state-of-the-art review, these authors presented examples of DDIs associated with P-gp that occur in the intestine and kidneys [96–105] (Table 1). Researchers have recently demonstrated DDIs involving P-gp and other therapeutics. For instance, oral administration of a TRH analogue named rovatirelin together with itraconazole was associated with an increased concentration and AUC of the victim drug. Importantly, the combination of these drugs was also linked to higher concentrations of TSH and thyroid hormones [106]. Therefore, DDIs should be considered in patients receiving several drugs, which is a common situation. Studies investigating DDIs not only describe significant interactions that should bring caution to the planning of combinational therapies, but also demonstrate the safety of such treatments. In a clinical analysis, researchers examined whether remibrutinib, a Bruton's tyrosine kinase inhibitor (BTKi), affects the bioavailability of oral contraceptives (OCs). The authors found that remibrutinib is a weak CYP3A4 and P-gp inhibitor, but its use is not associated with major DDIs, thus confirming that this BTKi can be administered with OCs [107].

These clinical data are preceded by a number of in vitro and preclinical investigations. DDIs significantly affect the bioavailability of drugs, which can enhance or weaken their desired effects. This is especially important in the case of drugs with a narrow therapeutic index or in those with a high potential for causing serious adverse events, such as anticancer therapeutics. Recently, Chu et al. found a significant P-gp-related DDI in the use of encequidar with paclitaxel [108]. Furthermore, another interaction was observed between proton pump inhibitors (PPIs) and cyclin-dependent kinase inhibitors (CDKi), which are used in patients with breast cancer. Desai and colleagues demonstrated that PPIs can block P-gp, which affects the bioavailability of anticancer therapeutics [109]. In addition to known interactions between synthetic drugs, natural agents such as flavonoids are also involved in DDIs. In rats, the combination of repaglinide with genistein was associated with significantly greater absorbance of the diabetic drug, thus demonstrating an inhibitory effect of genistein on P-gp [110].

Table 1. Drug–drug interactions involving intestinal P-glycoprotein and their effects on pharmacokinetic parameters. Selected elements of the table reused from [95] (John Wiley and Sons; Table 1) with permission.

Victim Drug	Perpetrator Drug	PK Change of Victim Drug	Possible Change in Victim Drug Effect or Toxicity	References
Apixaban	Ketoconazole (400 mg/day for 6 days)	AUC ↑ (twofold)	Higher rate of adverse effects	[96,97]
Dabigatran etexilate	Dronedarone (1 × 400 mg)	AUC[a] ↑ (2.1-fold)	Higher rate of adverse effects	[98]
Dabigatran etexilate	Ketoconazole (1 × 400 mg)	AUC[a] ↑ (2.4-fold)	Higher rate of adverse effects	[98]
Edoxaban	Erythromycin (4 × 500 mg/day for 8 days)	AUC ↑ (+85%)	Higher rate of adverse effects	[99]
Edoxaban	Cyclosporin (1 × 500 mg)	AUC ↑ (+73%)	Higher rate of adverse effects	[99]
Digoxin	Quinidine	AUC ↑ (~twofold)	Higher rate of adverse effects	[100–102]
Rivaroxaban	Ritonavir (2 × 600 mg/day for 5 days)	AUC ↑ (2.5-fold)	Higher rate of adverse effects	[102,103]
Apixaban	Rifampin (1 × 600 mg/day for 11 days)	AUC ↓ (−54%)	Reduced therapeutic effects	[96,102]
Dabigatran etexilate	Rifampin (1 × 600 mg/day for 7 days)	AUC[a] ↓ (−67%)	Reduced therapeutic effects	[98,104]
Digoxin	Rifampin (1 × 600 mg/day for 10 days)	AUC ↓ (−30%)	Reduced therapeutic effects	[105]
Rovatirelin	Itraconazole (4 × 50 mg/day for 9 days)	AUC_{last} ↑ (Ratio 2.93)	Higher risk of adverse events	[106]

Note: ↑—Increased; ↓—Decreased; AUC[a]—total dabigatran AUC.

An important issue to consider involves interactions that affect P-gp functionality in the BBB (Figure 4). In a recent publication by Martins et al. [111], the authors demonstrated that P-gp is involved in niraparib excretion from the CNS. Niraparib is a PARP1/2 anticancer therapeutic agent being investigated as a potential treatment agent for glioblastoma [112]. In an in vivo experiment, the use of elacridar, an ABCB1/ABCG2 inhibitor, was associated with a significantly enhanced brain-to-plasma ratio that was similar to one observed in animal models with transporter knockout [111]. Nevertheless, it should be remembered that P-gp is not only a drug transporter—it also regulates the penetration of other substances into the CNS. In an early work by Hanko et al., the authors studied rats to demonstrate that the use of P-gp inhibitors increased the penetration of bilirubin [113].

In addition, DDIs involve the activity of BCRP. BCRP-knockout mice demonstrate significantly greater plasma presence of BCRP substrates, including rosuvastatin and sulfasalazine [114], thus showing the importance of BCRP and the risk of altering substrate drug exposure after co-administration of BCRP inhibitors. Several studies examined DDIs involving statins and BCRP. Firstly, fostamatinib, a tyrosine kinase inhibitor, increases rosuvastatin concentration when they are administered together [115]. Importantly, such studies look for solutions that could be tested in clinical practice. Dong et al. [116] demonstrated that roxadustat, a drug used to treat anemia in patients with chronic kidney disease, could increase concentrations of various statins by interacting with drug transporters, including BCRP. However, the authors simulated the administration of statins 12 h after roxadustat, which inhibited the effects on statin levels when these drugs were introduced simultaneously. Another clinically relevant interaction involves rosuvastatin and finarenone, a novel mineralocorticosteroid inhibitor. By blocking three transporters—BCRP, OATP1B1, and OATP1B3—it contributes to the increased exposure of rosuvastatin. Similarly to roxadustat, separated administration of these two drugs diminished the impact of DDIs [117]. Recently, a clinical trial examined the effect of ticagrelol on rosuvastatin exposure, thus examining a popular combination used in clinical practice. Ticagrelol was found to suppress the activity of several drug transporters, including BCRP. The combination elevated the Cmax and $AUC_{0-\infty}$ of rosuvastatin 2.6-fold [118]. Another condition that could eventually lead to DDIs involving rosuvastatin is gout. The disorder is associated with a history of dyslipidemia [119]. Febuxostat, a xanthine synthase inhibitor also used in the treatment of gout, was also found to induce DDIs involving rosuvastatin and BCRP. Specifically, it increased the Cmax and $AUC_{0-\infty}$ of rosuvastatin approximately 2-fold. By contrast, there were no such findings regarding allopurinol [120]. Therefore, current evidence demonstrates an extensive interaction network between several agents and rosuvastatin, thus increasing the risk of statin-induced myopathy. With rosuvastatin being one of the most commonly used drugs for dyslipidemia, its combination with other agents may require the modification of drug administration, such as the implementation of intervals between drug intakes. These preventive measures could be introduced in patients with increased risk of developing myopathy. For instance, increased age was associated with higher rates of statin-induced myopathy [121]. Currently, researchers are analyzing potential DDIs with BCRP in a number of new drugs, which will increase knowledge about the PK parameters of BCRP substrates [122,123].

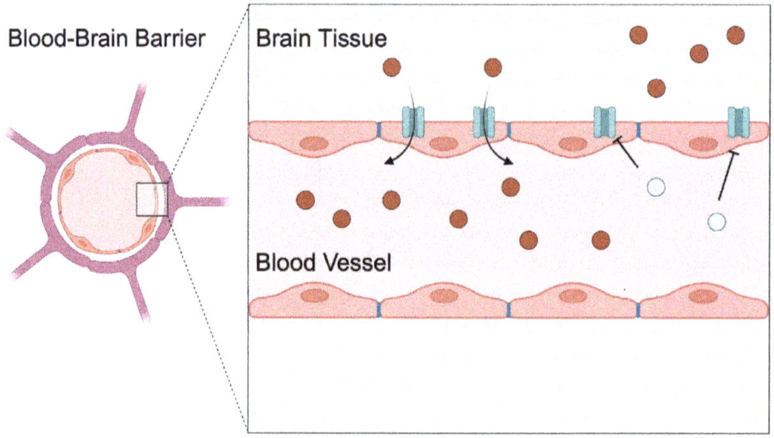

Figure 4. Drugs can inhibit the activity of P-glycoprotein in the blood–brain barrier to suppress the efflux mechanism and increase the concentration of desired therapeutic in the central nervous system. Hypothetically, a purposedly used DDI mechanism or P-gp inhibitor could be implemented to suppress cancer resistance. Created in BioRender. Kiełbowski, K. (2024) BioRender.com/r30a034.

Interestingly, studies have tried to examine whether we could predict the activity of drug transporters and thus potential DDIs. Recently, riboflavin, an endogenous biomarker, was found to indicate the activity of BCRP. Hypothetically, blood levels of riboflavin could be monitored to analyze BCRP functionality and plan treatment strategies adequately to avoid the occurrence of DDIs [124]. Another approach to evaluating the activity of drug transporters is by monitoring non-coding RNA levels. MicroRNA (miRNAs) are small molecules that play a crucial role in regulating gene expression by suppressing the translation of their target mRNA. miR-138-5p and miR-146a-5p target P-gp, and their dysregulated levels were observed in patients with intractable epilepsy, a condition associated with P-gp involvement [125].

4.2. *SLC Transporters*

Over the years, studies demonstrated that various agents could inhibit the activity of SLC transporters, especially OATPs and OATs/OCTs. For instance, DDIs could probably affect patients with inflammatory disorders. OATPs/OATs/OCTs are known transporters of such drugs as non-steroid anti-inflammatory drugs (NSAIDs), as well as immunomodulatory/anticancer agents such as methotrexate (MTX) [126,127]. These groups of SLC transporters were recently found to be inhibited by ruxolitinib, a JAK inhibitor [128]. The family of JAK inhibitors is composed of several agents that are registered for the treatment of various inflammatory conditions. Thus, combinations of ruxolitinib with other immunomodulatory or anti-inflammatory drugs could potentially be associated with impacts on the exposure of the latter drugs, thus changing their profile of efficacy and toxicity. Another drug that affects MTX PK parameters is rifampicin. Hwang and colleagues [129] observed that simultaneous administration of these agents increased the following parameters: Cmax, AUC_{last}, and AUC_{inf}. Mechanistically, these findings involve SLC transporters, as rifampicin is an inhibitor of OATP1B1/1B3. Metformin is another important agent transported through SLC proteins. A concomitant use of metformin with isavuconazole was associated with higher plasma metformin concentrations [130]. Additionally, members of the SLC22 family mediate the transport of antiviral drugs. Inhibiting the activity of transporters while administering antivirals creates DDIs that affect the concentrations of the latter drugs. For instance, uricosuric agents, such as probenecid and lesinurad, were found to increase levels of adefovir, an anti-hepatitis B agent [131]. Similar results were observed when rats were administered with acyclovir and apigenin, a flavonoid compound [132]. Importantly, drug transporters regulate drug distribution and transportation together. Thus, a single agent can be transported through different proteins. At the same time, a single agent can be an inhibitor of several transporters. For instance, isavuconazole is considered to be an inhibitor of P-gp, OCT1, OCT2, and MATE1 [130]. As a result, administration of isavuconazole could affect the PK parameters of a variety of other drugs.

5. Transporters as Therapeutic Targets

5.1. P-Glycoprotein

In the previous sections, we have discussed modifications of P-gp functionality together with DDIs involving this transporter. A wide range of commonly used therapeutics have been found to inhibit the activity of P-gp. These findings demonstrate how P-gp activity affects drug disposition and how we can manipulate this to improve the bioavailability of certain therapeutics or introduce them through oral administration. However, the increased expression and activity of P-gp is considered an important element in cancer treatment resistance [133]. Consequently, the transporter could be targeted to reverse treatment resistance and improve therapeutic outcomes. Research in this direction has been conducted for many years. In this section, we would like to briefly describe the most recent findings on this topic. For example, a combination of chemotherapeutics with P-gp inhibitors has been studied to prevent the efflux of anticancer drugs. Recently, Zhu et al. [134] described a method to develop nanocarriers containing paclitaxel with encequidar. This method could significantly reduce tumors in vivo, and the tumor shrinkage with this treatment was greater than that observed with the use of paclitaxel alone [134]. As taxanes are administered intravenously, the ability to use oral administration would improve the quality of treatment and shorten hospital stays for treated patients. Zhou and colleagues developed a compound that is able to selectively suppress intestinal P-gp. The authors reason that intestinal suppression will increase oral bioavailability but, at the same time, sensitive tissues will be protected from paclitaxel-associated toxicities [135]. In another study, Loos and colleagues [136] showed that cabazitaxel (taxan) could be administered orally in combination with ritonavir (a CYP3A4 inhibitor), whereas the addition of elacridar increased the permeability of the brain to the chemotherapeutic agent. Moreover, studies have also begun to examine the effect of novel anticancer therapeutics on drug transporters. Recently, investigations of KRAS inhibitors were published. KRAS is an oncogene whose aberrant activity is thought to drive tumorigenesis in a wide number of malignancies. Adagrasib is a KRAS G12C inhibitor that has shown anticancer activity in malignancies harbouring the G12C mutation. This drug was recently found to suppress P-gp efflux [137]. Similar findings were recently described regarding BI-2865, a pan-KRAS inhibitor [138].

Importantly, clinical studies investigating the safety and efficacy of P-gp inhibitors are also being performed. Recently, a study examined the 72 h infusion of zosuquidar, a third-generation inhibitor of P-gp, in patients with acute myeloblastic leukaemia who were being treated with daunorubicin and cytarabine. Among 106 patients included in this trial, the overall response rate to treatment was 46%. Neurological adverse events (≥3 grade in 16% of patients) were expected based on previous trials, and these events were resolved by decreasing zosuquidar dosage or discontinuing the treatment [139].

5.2. BCRC Transporter

Similarly to P-gp, researchers have also focused on analyzing inhibitors of BCRP transporters. From the clinical perspective, the potential to modulate BCRP functionality is of significant importance, as it would allow clinicians to increase anticancer treatment. For instance, in an in vitro study, Sun and colleagues [140] demonstrated that suppressing BCRP could significantly improve the anticancer efficacy of lenvatinib in hepatocellular carcinoma. We have discussed studies that examined DDIs involving BCRP. A number of pharmacological agents target and suppress the activity of the transporters. In an interesting analysis by Deng et al. [141], the authors studied 232 FDA-approved agents to find BCRP inhibitors. Researchers identified 75 agents that inhibited BCRP activity, which was measured by monitoring the probe substance transport. However, developing selective ABCG2 inhibitor represents several challenges. For instance, a fungal toxin fumitremorgin C (FTC), which selectively targets BCRP, induces neurotoxic effects. By contrast, its analog, Ko143, was found to be unstable in animal plasma. However, by implementing cryo-EM techniques, Ko143 derivatives were recently found to significantly suppress the activity of BCRP [142]. Importantly, different Kol143 analogs could increase the transport through the BBB in vivo, which creates more opportunities for drugs with limited access to the brain [143]. Furthermore, tetrahydroquinoline/4,5-dihydroisoxazole hybrids were also recently proposed as potent ABCG2 inhibitors. These molecules can bind to the inhibitors' binding region, together with a selective region for substrates [144]. Tariquidar, an agent known to inhibit both ABCB1 and ABCG2, was recently modified in a benzamide core, which formed a compound highly selective towards the BCRP transporter. Moreover, modifications involving ester moiety at the agent further improved the efficacy of these molecules and increased plasma stability [145].

5.3. SLC Transporters

Inhibitors of SLC transporters have been developed, ranging from constrained glutamate analogs to small molecules [146]. Numerous SLC transporters have been demonstrated to be inhibited by multi-kinase inhibitors, with OATPs, OAT3, and OCT1 being especially vulnerable [147]. Other studies have shown that plastics components such as bisphenol A (BPA) inhibit OAT3 transporter in a competitive manner [148]. There is growing interest in developing SLC transporter inhibitors for therapeutic purposes [149]. A recent study by Boeszoermenyi at al. [150] identified feeblin, which disrupts the SLC15A4-TASL adapter module. It causes interruption in the TLR7/8-IRF5 signaling pathway and as a result averts proinflammatory reactions associated with multiple autoimmune disorders. TASL is a protein associated with systemic lupus erythematosus (SLE). The lysosomal proton-coupled amino acid transporter SLC15A4 recruits TASL, which in turn activates IRF5, which consequently causes the transcription of type I interferons and cytokines [151]. Another study by Bruyère at al. revealed that the Janus kinase (JAK) inhibitor ruxolitinib inhibits the activity of OAT3 and OCT2 and blocks OAT1, OAT4, OATP1B1, OATP1B3, OATP2B1, and OCT3 [128]. Although most research has examined potential inhibitors, there is a significant interest in investigating SLC transporter activators as potential treatments for diseases associated with transporter dysfunction. Direct activators, transcriptional regulators, and trafficking modulators are some methods for modifying SLC function [152]. For example, estradiol and testosterone are activators of the expression of OATP1B1 [153], primarily found in the liver and certain tumor cells [154].

6. Conclusions and Future Directions

ABC and SLC transporters play a significant role in drug absorption, distribution, and elimination, thus having a crucial impact on drug bioavailability and systemic exposure. Knowledge about the physiology of drug transporters is important in studies examining PK parameters. It should be remembered that drug transporters rarely act alone, as they rather form a complex system that manages drug distribution. Furthermore, it is important to bear in mind that the impact of drug transporters on bioavailability and pharmacokinetics can be species-specific, as demonstrated by Verscheijden et al. [155]. Consequently, knowledge gained in animal models should be appropriately translated to humans. Furthermore, the occurrence of DDIs should be considered, as they can significantly affect the bioavailability of particular drugs, which translates into greater or weaker actions. Apart from altered PD, this is also linked to an altered safety profile, which needs to be properly examined. However, with increased knowledge about the safety profiles of DDIs, we could boost the bioavailability of certain agents with known poor absorbance. Thus, we could enhance the activity of several drugs or introduce an oral administration route for drugs that currently require intravenous delivery. Additionally, there is a growing field of study investigating drug transporter modulators. These agents could improve the treatment outcomes of oncological patients, benefit pharmacoeconomics, and shorten hospital stays.

Author Contributions: Conceptualization, A.P.; writing—original draft preparation, K.K., M.K., E.B. and A.P.; writing—review and editing, K.K., M.K., E.B. and A.P.; supervision, A.P. All authors have read and agreed to the published version of the manuscript.

Funding: This research received no external funding.

Institutional Review Board Statement: Not applicable.

Data Availability Statement: No new data were created or analyzed in this study. Data sharing is not applicable to this article.

Conflicts of Interest: The authors declare no conflicts of interest.

References

1. van den Anker, J.; Reed, M.D.; Allegaert, K.; Kearns, G.L. Developmental Changes in Pharmacokinetics and Pharmacodynamics. *J. Clin. Pharmacol.* **2018**, *58* (Suppl. S10), S10–S25. [CrossRef] [PubMed]
2. Lai, Y.; Varma, M.; Feng, B.; Stephens, J.C.; Kimoto, E.; El-Kattan, A.; Ichikawa, K.; Kikkawa, H.; Ono, C.; Suzuki, A.; et al. Impact of drug transporter pharmacogenomics on pharmacokinetic and pharmacodynamic variability—Considerations for drug development. *Expert Opin. Drug Metab. Toxicol.* **2012**, *8*, 723–743. [CrossRef] [PubMed]
3. Locher, K.P. Mechanistic diversity in ATP-binding cassette (ABC) transporters. *Nat. Struct. Mol. Biol.* **2016**, *23*, 487–493. [CrossRef] [PubMed]
4. Theodoulou, F.L.; Kerr, I.D. ABC transporter research: Going strong 40 years on. *Biochem. Soc. Trans.* **2015**, *43*, 1033–1040. [CrossRef] [PubMed]
5. Fritz, A.; Busch, D.; Lapczuk, J.; Ostrowski, M.; Drozdzik, M.; Oswald, S. Expression of clinically relevant drug-metabolizing enzymes along the human intestine and their correlation to drug transporters and nuclear receptors: An intra-subject analysis. *Basic. Clin. Pharmacol. Toxicol.* **2019**, *124*, 245–255. [CrossRef]
6. Kanado, Y.; Tsurudome, Y.; Omata, Y.; Yasukochi, S.; Kusunose, N.; Akamine, T.; Matsunaga, N.; Koyanagi, S.; Ohdo, S. Estradiol regulation of P-glycoprotein expression in mouse kidney and human tubular epithelial cells, implication for renal clearance of drugs. *Biochem. Biophys. Res. Commun.* **2019**, *519*, 613–619. [CrossRef]
7. Mossel, P.; Arif, W.M.; De Souza, G.S.; Varela, L.G.; van der Weijden, C.W.J.; Boersma, H.H.; Willemsen, A.T.M.; Boellaard, R.; Elsinga, P.H.; Borra, R.J.H.; et al. Quantification of P-glycoprotein function at the human blood-brain barrier using [(18)F]MC225 and PET. *Eur. J. Nucl. Med. Mol. Imaging* **2023**, *50*, 3917–3927. [CrossRef]
8. Sarkadi, B.; Homolya, L.; Hegedus, T. The ABCG2/BCRP transporter and its variants—From structure to pathology. *FEBS Lett.* **2020**, *594*, 4012–4034. [CrossRef]
9. Chen, Y.; Wang, L.; Zhu, Y.; Chen, Z.; Qi, X.; Jin, L.; Jin, J.; Hua, D.; Ma, X. Breast cancer resistance protein (BCRP)-containing circulating microvesicles contribute to chemoresistance in breast cancer. *Oncol. Lett.* **2015**, *10*, 3742–3748. [CrossRef]
10. Uceda-Castro, R.; Margarido, A.S.; Song, J.Y.; de Gooijer, M.C.; Messal, H.A.; Chambers, C.R.; Nobis, M.; Citirikkaya, C.H.; Hahn, K.; Seinstra, D.; et al. BCRP drives intrinsic chemoresistance in chemotherapy-naive breast cancer brain metastasis. *Sci. Adv.* **2023**, *9*, eabp9530. [CrossRef]

11. Robey, R.W.; To, K.K.; Polgar, O.; Dohse, M.; Fetsch, P.; Dean, M.; Bates, S.E. ABCG2: A perspective. *Adv. Drug Deliv. Rev.* **2009**, *61*, 3–13. [CrossRef] [PubMed]
12. Ganguly, S.; Finkelstein, D.; Shaw, T.I.; Michalek, R.D.; Zorn, K.M.; Ekins, S.; Yasuda, K.; Fukuda, Y.; Schuetz, J.D.; Mukherjee, K.; et al. Metabolomic and transcriptomic analysis reveals endogenous substrates and metabolic adaptation in rats lacking Abcg2 and Abcb1a transporters. *PLoS ONE* **2021**, *16*, e0253852. [CrossRef] [PubMed]
13. Kim, Y.; Chen, J. Molecular structure of human P-glycoprotein in the ATP-bound, outward-facing conformation. *Science* **2018**, *359*, 915–919. [CrossRef] [PubMed]
14. Berman, H.M.; Westbrook, J.; Feng, Z.; Gilliland, G.; Bhat, T.N.; Weissig, H.; Shindyalov, I.N.; Bourne, P.E. The Protein Data Bank. *Nucleic Acids Res.* **2000**, *28*, 235–242. [CrossRef] [PubMed]
15. Sehnal, D.; Bittrich, S.; Deshpande, M.; Svobodova, R.; Berka, K.; Bazgier, V.; Velankar, S.; Burley, S.K.; Koca, J.; Rose, A.S. Mol* Viewer: Modern web app for 3D visualization and analysis of large biomolecular structures. *Nucleic Acids Res.* **2021**, *49*, W431–W437. [CrossRef]
16. Orlando, B.J.; Liao, M. ABCG2 transports anticancer drugs via a closed-to-open switch. *Nat. Commun.* **2020**, *11*, 2264. [CrossRef]
17. Perland, E.; Fredriksson, R. Classification Systems of Secondary Active Transporters. *Trends Pharmacol. Sci.* **2017**, *38*, 305–315. [CrossRef]
18. Gyimesi, G.; Hediger, M.A. Systematic in silico discovery of novel solute carrier-like proteins from proteomes. *PLoS ONE* **2022**, *17*, e0271062. [CrossRef]
19. Kunta, J.R.; Sinko, P.J. Intestinal drug transporters: In vivo function and clinical importance. *Curr. Drug Metab.* **2004**, *5*, 109–124. [CrossRef]
20. Scherrmann, J.M. Transporters in absorption, distribution, and elimination. *Chem. Biodivers.* **2009**, *6*, 1933–1942. [CrossRef]
21. Bai, X.; Moraes, T.F.; Reithmeier, R.A.F. Structural biology of solute carrier (SLC) membrane transport proteins. *Mol. Membr. Biol.* **2017**, *34*, 1–32. [CrossRef] [PubMed]
22. Colas, C.; Ung, P.M.; Schlessinger, A. SLC Transporters: Structure, Function, and Drug Discovery. *Medchemcomm* **2016**, *7*, 1069–1081. [CrossRef] [PubMed]
23. Xie, T.; Chi, X.; Huang, B.; Ye, F.; Zhou, Q.; Huang, J. Rational exploration of fold atlas for human solute carrier proteins. *Structure* **2022**, *30*, 1321–1330. [CrossRef] [PubMed]
24. Hediger, M.A.; Romero, M.F.; Peng, J.B.; Rolfs, A.; Takanaga, H.; Bruford, E.A. The ABCs of solute carriers: Physiological, pathological and therapeutic implications of human membrane transport proteinsIntroduction. *Pflugers. Arch.* **2004**, *447*, 465–468. [CrossRef]
25. Fluman, N.; Ryan, C.M.; Whitelegge, J.P.; Bibi, E. Dissection of mechanistic principles of a secondary multidrug efflux protein. *Mol. Cell* **2012**, *47*, 777–787. [CrossRef]
26. Koepsell, H.; Endou, H. The SLC22 drug transporter family. *Pflugers. Arch.* **2004**, *447*, 666–676. [CrossRef]
27. Schlessinger, A.; Geier, E.; Fan, H.; Irwin, J.J.; Shoichet, B.K.; Giacomini, K.M.; Sali, A. Structure-based discovery of prescription drugs that interact with the norepinephrine transporter, NET. *Proc. Natl. Acad. Sci. USA* **2011**, *108*, 15810–15815. [CrossRef]
28. Palmieri, F. The mitochondrial transporter family SLC25: Identification, properties and physiopathology. *Mol. Asp. Med.* **2013**, *34*, 465–484. [CrossRef]
29. Ganapathy, V.; Miyauchi, S. Transport systems for opioid peptides in mammalian tissues. *AAPS J.* **2005**, *7*, E852–E856. [CrossRef]
30. Shen, J.; Hu, M.; Fan, X.; Ren, Z.; Portioli, C.; Yan, X.; Rong, M.; Zhou, M. Extracellular domain of PepT1 interacts with TM1 to facilitate substrate transport. *Structure* **2022**, *30*, 1035–1041. [CrossRef]
31. Wenzel, U.; Thwaites, D.T.; Daniel, H. Stereoselective uptake of beta-lactam antibiotics by the intestinal peptide transporter. *Br. J. Pharmacol.* **1995**, *116*, 3021–3027. [CrossRef] [PubMed]
32. Tamai, I.; Nakanishi, T.; Hayashi, K.; Terao, T.; Sai, Y.; Shiraga, T.; Miyamoto, K.; Takeda, E.; Higashida, H.; Tsuji, A. The predominant contribution of oligopeptide transporter PepT1 to intestinal absorption of beta-lactam antibiotics in the rat small intestine. *J. Pharm. Pharmacol.* **1997**, *49*, 796–801. [CrossRef]
33. Alexander, S.P.H.; Kelly, E.; Mathie, A.; Peters, J.A.; Veale, E.L.; Armstrong, J.F.; Faccenda, E.; Harding, S.D.; Pawson, A.J.; Sharman, J.L.; et al. THE CONCISE GUIDE TO PHARMACOLOGY 2019/20: Transporters. *Br. J. Pharmacol.* **2019**, *176* (Suppl. S1), S397–S493. [CrossRef]
34. Romano, A.; Barca, A.; Kottra, G.; Daniel, H.; Storelli, C.; Verri, T. Functional expression of SLC15 peptide transporters in rat thyroid follicular cells. *Mol. Cell. Endocrinol.* **2010**, *315*, 174–181. [CrossRef] [PubMed]
35. Hagenbuch, B.; Meier, P.J. Organic anion transporting polypeptides of the OATP/ SLC21 family: Phylogenetic classification as OATP/ SLCO superfamily, new nomenclature and molecular/functional properties. *Pflugers. Arch.* **2004**, *447*, 653–665. [CrossRef] [PubMed]
36. Hagenbuch, B.; Stieger, B. The SLCO (former SLC21) superfamily of transporters. *Mol. Asp. Med.* **2013**, *34*, 396–412. [CrossRef] [PubMed]
37. Konig, J. Uptake transporters of the human OATP family: Molecular characteristics, substrates, their role in drug-drug interactions, and functional consequences of polymorphisms. *Handb. Exp. Pharmacol.* **2011**, *19*, 1–28. [CrossRef]
38. Schulte, R.R.; Ho, R.H. Organic Anion Transporting Polypeptides: Emerging Roles in Cancer Pharmacology. *Mol. Pharmacol.* **2019**, *95*, 490–506. [CrossRef]

39. Obaidat, A.; Roth, M.; Hagenbuch, B. The expression and function of organic anion transporting polypeptides in normal tissues and in cancer. *Annu. Rev. Pharmacol. Toxicol.* **2012**, *52*, 135–151. [CrossRef]
40. Engelhart, D.C.; Azad, P.; Ali, S.; Granados, J.C.; Haddad, G.G.; Nigam, S.K. Drosophila SLC22 Orthologs Related to OATs, OCTs, and OCTNs Regulate Development and Responsiveness to Oxidative Stress. *Int. J. Mol. Sci.* **2020**, *21*, 2002. [CrossRef]
41. Zhu, C.; Nigam, K.B.; Date, R.C.; Bush, K.T.; Springer, S.A.; Saier, M.H., Jr.; Wu, W.; Nigam, S.K. Evolutionary Analysis and Classification of OATs, OCTs, OCTNs, and Other SLC22 Transporters: Structure-Function Implications and Analysis of Sequence Motifs. *PLoS ONE* **2015**, *10*, e0140569. [CrossRef]
42. Engelhart, D.C.; Granados, J.C.; Shi, D.; Saier, M.H., Jr.; Baker, M.E.; Abagyan, R.; Nigam, S.K. Systems Biology Analysis Reveals Eight SLC22 Transporter Subgroups, Including OATs, OCTs, and OCTNs. *Int. J. Mol. Sci.* **2020**, *21*, 1791. [CrossRef] [PubMed]
43. Koepsell, H. The SLC22 family with transporters of organic cations, anions and zwitterions. *Mol. Asp. Med.* **2013**, *34*, 413–435. [CrossRef] [PubMed]
44. Lamhonwah, A.M.; Hawkins, C.E.; Tam, C.; Wong, J.; Mai, L.; Tein, I. Expression patterns of the organic cation/carnitine transporter family in adult murine brain. *Brain Dev.* **2008**, *30*, 31–42. [CrossRef] [PubMed]
45. Falah, K.; Zhang, P.; Nigam, A.K.; Maity, K.; Chang, G.; Granados, J.C.; Momper, J.D.; Nigam, S.K. In Vivo Regulation of Small Molecule Natural Products, Antioxidants, and Nutrients by OAT1 and OAT3. *Nutrients* **2024**, *16*, 2242. [CrossRef] [PubMed]
46. Cervenkova, L.; Vycital, O.; Bruha, J.; Rosendorf, J.; Palek, R.; Liska, V.; Daum, O.; Mohelnikova-Duchonova, B.; Soucek, P. Protein expression of ABCC2 and SLC22A3 associates with prognosis of pancreatic adenocarcinoma. *Sci. Rep.* **2019**, *9*, 19782. [CrossRef]
47. Redeker, K.M.; Brockmoller, J. Several orphan solute carriers functionally identified as organic cation transporters: Substrates specificity compared with known cation transporters. *J. Biol. Chem.* **2024**, *300*, 107629. [CrossRef]
48. Nair, P.C.; Miners, J.O. Molecular dynamics simulations: From structure function relationships to drug discovery. *Silico Pharmacol.* **2014**, *2*, 4. [CrossRef]
49. Trejo, F.; Elizalde, S.; Mercado, A.; Gamba, G.; de losHeros, P. SLC12A cryo-EM: Analysis of relevant ion binding sites, structural domains, and amino acids. *Am. J. Physiol. Cell Physiol.* **2023**, *325*, C921–C939. [CrossRef]
50. Dou, T.; Lian, T.; Shu, S.; He, Y.; Jiang, J. The substrate and inhibitor binding mechanism of polyspecific transporter OAT1 revealed by high-resolution cryo-EM. *Nat. Struct Mol. Biol.* **2023**, *30*, 1794–1805. [CrossRef]
51. Lees, J.A.; Dias, J.M.; Han, S. Applications of Cryo-EM in small molecule and biologics drug design. *Biochem. Soc. Trans.* **2021**, *49*, 2627–2638. [CrossRef] [PubMed]
52. Degraeve, A.L.; Haufroid, V.; Loriot, A.; Gatto, L.; Andries, V.; Vereecke, L.; Elens, L.; Bindels, L.B. Gut microbiome modulates tacrolimus pharmacokinetics through the transcriptional regulation of ABCB1. *Microbiome* **2023**, *11*, 138. [CrossRef] [PubMed]
53. Wang, M.Y.; Yang, M.; Hou, P.Y.; Chen, X.B.; Li, H.G.; Yan, J.X.; Zhang, J.; Zhang, Y.W.; Wu, X.H. Intestinal absorption of pallidifloside D are limited by P-glycoprotein in mice. *Xenobiotica* **2018**, *48*, 739–744. [CrossRef] [PubMed]
54. Sun, X.; Tang, S.; Hou, B.; Duan, Z.; Liu, Z.; Li, Y.; He, S.; Wang, Q.; Chang, Q. Overexpression of P-glycoprotein, MRP2, and CYP3A4 impairs intestinal absorption of octreotide in rats with portal hypertension. *BMC Gastroenterol.* **2021**, *21*, 2. [CrossRef]
55. Kyaw, T.S.; Zhang, C.; Sandy, M.; Trepka, K.; Zhang, S.; Ramirez Hernandez, L.A.; Ramirez, L.; Goh, J.J.N.; Yu, K.; Dimassa, V.; et al. Human gut Actinobacteria boost drug absorption by secreting P-glycoprotein ATPase inhibitors. *iScience* **2024**, *27*, 110122. [CrossRef]
56. Ashmawy, S.M.; El-Gizawy, S.A.; El Maghraby, G.M.; Osman, M.A. Regional difference in intestinal drug absorption as a measure for the potential effect of P-glycoprotein efflux transporters. *J. Pharm. Pharmacol.* **2019**, *71*, 362–370. [CrossRef]
57. Dooley, S.A.; Kolobova, E.; Burman, A.; Kaji, I.; Digrazia, J.R.; Stubler, R.; Goldstein, A.; Packirisamy, C.; Coutts, A.W.; Saqui-Salces, M.; et al. Myosin Vb Traffics P-glycoprotein to the Apical Membrane of Intestinal Epithelial Cells. *Gastroenterology* 2024; *ahead of print*. [CrossRef]
58. Takiishi, T.; Fenero, C.I.M.; Camara, N.O.S. Intestinal barrier and gut microbiota: Shaping our immune responses throughout life. *Tissue Barriers* **2017**, *5*, e1373208. [CrossRef]
59. Zhang, T.; Gao, G.; Kwok, L.Y.; Sun, Z. Gut microbiome-targeted therapies for Alzheimer's disease. *Gut Microbes* **2023**, *15*, 2271613. [CrossRef]
60. Buchman, A.L.; Paine, M.F.; Wallin, A.; Ludington, S.S. A higher dose requirement of tacrolimus in active Crohn's disease may be related to a high intestinal P-glycoprotein content. *Dig. Dis. Sci.* **2005**, *50*, 2312–2315. [CrossRef]
61. Li, Z.; Zhang, J.; Zhang, Y.; Zhou, L.; Zhao, J.; Lyu, Y.; Poon, L.H.; Lin, Z.; To, K.K.W.; Yan, X.; et al. Intestinal absorption and hepatic elimination of drugs in high-fat high-cholesterol diet-induced non-alcoholic steatohepatitis rats: Exemplified by simvastatin. *Br. J. Pharmacol.* **2021**, *178*, 582–599. [CrossRef]
62. Takeda, F.; Oda, M.; Terasaki, M.; Ichimura, Y.; Kojima, H.; Saitoh, H. Downregulated expression of intestinal P-glycoprotein in rats with cisplatin-induced acute kidney injury causes amplification of its transport capacity to maintain "gatekeeper" function. *Toxicol. Appl. Pharmacol.* **2021**, *423*, 115570. [CrossRef] [PubMed]
63. Thummel, K.E. Gut instincts: CYP3A4 and intestinal drug metabolism. *J. Clin. Investig.* **2007**, *117*, 3173–3176. [CrossRef] [PubMed]
64. Li, J.; Di, L.; Cheng, X.; Ji, W.; Piao, H.; Cheng, G.; Zou, M. The characteristics and mechanism of co-administration of lovastatin solid dispersion with kaempferol to increase oral bioavailability. *Xenobiotica* **2020**, *50*, 593–601. [CrossRef] [PubMed]
65. Spieler, D.; Namendorf, C.; Namendorf, T.; Uhr, M. abcb1ab p-glycoprotein is involved in the uptake of the novel antidepressant vortioxetine into the brain of mice. *J. Psychiatr. Res.* **2019**, *109*, 48–51. [CrossRef] [PubMed]

66. Pyun, J.; Koay, H.; Runwal, P.; Mawal, C.; Bush, A.I.; Pan, Y.; Donnelly, P.S.; Short, J.L.; Nicolazzo, J.A. Cu(ATSM) Increases P-Glycoprotein Expression and Function at the Blood-Brain Barrier in C57BL6/J Mice. *Pharmaceutics* **2023**, *15*, 2084. [CrossRef]
67. Ito, K.; Naoi, M.; Nishiyama, K.; Kudo, T.; Tsuda, Y.; MacLean, C.; Ishiguro, N. Impact of P-glycoprotein on intracellular drug concentration in peripheral blood mononuclear cells and K562 cells. *Drug. Metab. Pharmacokinet* **2023**, *49*, 100487. [CrossRef]
68. Nicolas, J.M.; Chanteux, H.; Nicolai, J.; Brouta, F.; Viot, D.; Rosseels, M.L.; Gillent, E.; Bonnaillie, P.; Mathy, F.X.; Long, J.; et al. Role of P-glycoprotein in the brain disposition of seletalisib: Evaluation of the potential for drug-drug interactions. *Eur. J. Pharm. Sci.* **2020**, *142*, 105122. [CrossRef]
69. Mao, Q.; Unadkat, J.D. Role of the breast cancer resistance protein (BCRP/ABCG2) in drug transport--an update. *AAPS J.* **2015**, *17*, 65–82. [CrossRef]
70. Bruyere, A.; Decleves, X.; Bouzom, F.; Ball, K.; Marques, C.; Treton, X.; Pocard, M.; Valleur, P.; Bouhnik, Y.; Panis, Y.; et al. Effect of variations in the amounts of P-glycoprotein (ABCB1), BCRP (ABCG2) and CYP3A4 along the human small intestine on PBPK models for predicting intestinal first pass. *Mol. Pharm.* **2010**, *7*, 1596–1607. [CrossRef]
71. Kawahara, I.; Nishikawa, S.; Yamamoto, A.; Kono, Y.; Fujita, T. Assessment of contribution of BCRP to intestinal absorption of various drugs using portal-systemic blood concentration difference model in mice. *Pharmacol. Res. Perspect.* **2020**, *8*, e00544. [CrossRef]
72. Goncalves, J.; Silva, S.; Gouveia, F.; Bicker, J.; Falcao, A.; Alves, G.; Fortuna, A. A combo-strategy to improve brain delivery of antiepileptic drugs: Focus on BCRP and intranasal administration. *Int. J. Pharm.* **2021**, *593*, 120161. [CrossRef] [PubMed]
73. Pippa, L.F.; Vieira, C.P.; Caris, J.A.; Rocha, A.; Marques, M.P.; Garcia, C.P.; Rezende, R.E.F.; Lanchote, V.L. Effect of Chronic Hepatitis C on the Activity of the Membrane Transporters P-gp and OATP1B1/BCRP on Patients With Different Stages of Hepatic Fibrosis. *Clin. Pharmacol. Ther.* **2023**, *114*, 173–181. [CrossRef] [PubMed]
74. Cooper-DeHoff, R.M.; Niemi, M.; Ramsey, L.B.; Luzum, J.A.; Tarkiainen, E.K.; Straka, R.J.; Gong, L.; Tuteja, S.; Wilke, R.A.; Wadelius, M.; et al. The Clinical Pharmacogenetics Implementation Consortium Guideline for SLCO1B1, ABCG2, and CYP2C9 genotypes and Statin-Associated Musculoskeletal Symptoms. *Clin. Pharmacol. Ther.* **2022**, *111*, 1007–1021. [CrossRef] [PubMed]
75. Poulsen, S.B.; Fenton, R.A.; Rieg, T. Sodium-glucose cotransport. *Curr. Opin. Nephrol. Hypertens.* **2015**, *24*, 463–469. [CrossRef] [PubMed]
76. Wright, E.M.; Hirayama, B.A.; Loo, D.F. Active sugar transport in health and disease. *J. Intern. Med.* **2007**, *261*, 32–43. [CrossRef]
77. Elsas, L.J.; Rosenberg, L.E. Familial renal glycosuria: A genetic reappraisal of hexose transport by kidney and intestine. *J. Clin. Investig.* **1969**, *48*, 1845–1854. [CrossRef]
78. Meeuwisse, G.W.; Melin, K. Glucose-galactose malabsorption. A clinical study of 6 cases. *Acta Pædiatrica* **1969**, *58* (Suppl. S188), 3–18. [CrossRef]
79. Kalliokoski, A.; Niemi, M. Impact of OATP transporters on pharmacokinetics. *Br. J. Pharmacol.* **2009**, *158*, 693–705. [CrossRef]
80. Otsuka, M.; Matsumoto, T.; Morimoto, R.; Arioka, S.; Omote, H.; Moriyama, Y. A human transporter protein that mediates the final excretion step for toxic organic cations. *Proc. Natl. Acad. Sci. USA* **2005**, *102*, 17923–17928. [CrossRef]
81. Meszaros, M.; Porkolab, G.; Kiss, L.; Pilbat, A.M.; Kota, Z.; Kupihar, Z.; Keri, A.; Galbacs, G.; Siklos, L.; Toth, A.; et al. Niosomes decorated with dual ligands targeting brain endothelial transporters increase cargo penetration across the blood-brain barrier. *Eur. J. Pharm. Sci.* **2018**, *123*, 228–240. [CrossRef]
82. Morris, M.E.; Rodriguez-Cruz, V.; Felmlee, M.A. SLC and ABC Transporters: Expression, Localization, and Species Differences at the Blood-Brain and the Blood-Cerebrospinal Fluid Barriers. *AAPS J.* **2017**, *19*, 1317–1331. [CrossRef] [PubMed]
83. Al-Majdoub, Z.M.; Al Feteisi, H.; Achour, B.; Warwood, S.; Neuhoff, S.; Rostami-Hodjegan, A.; Barber, J. Proteomic Quantification of Human Blood-Brain Barrier SLC and ABC Transporters in Healthy Individuals and Dementia Patients. *Mol. Pharm.* **2019**, *16*, 1220–1233. [CrossRef] [PubMed]
84. Damir, H.A.; Ali, M.A.; Adem, M.A.; Amir, N.; Ali, O.M.; Tariq, S.; Adeghate, E.; Greenwood, M.P.; Lin, P.; Alvira-Iraizoz, F.; et al. Effects of long-term dehydration and quick rehydration on the camel kidney: Pathological changes and modulation of the expression of solute carrier proteins and aquaporins. *BMC Vet. Res.* **2024**, *20*, 367. [CrossRef] [PubMed]
85. Huo, X.; Liu, K. Renal organic anion transporters in drug-drug interactions and diseases. *Eur. J. Pharm. Sci.* **2018**, *112*, 8–19. [CrossRef] [PubMed]
86. Walker, N.; Filis, P.; Soffientini, U.; Bellingham, M.; O'Shaughnessy, P.J.; Fowler, P.A. Placental transporter localization and expression in the Human: The importance of species, sex, and gestational age differencesdagger. *Biol. Reprod.* **2017**, *96*, 733–742. [CrossRef]
87. Mao, Q.; Chen, X. An update on placental drug transport and its relevance to fetal drug exposure. *Med. Rev.* **2022**, *2*, 501–511. [CrossRef]
88. Radhakrishna, U.; Radhakrishnan, R.; Uppala, L.V.; Muvvala, S.B.; Prajapati, J.; Rawal, R.M.; Bahado-Singh, R.O.; Sadhasivam, S. Prenatal opioid exposure significantly impacts placental protein kinase C (PKC) and drug transporters, leading to drug resistance and neonatal opioid withdrawal syndrome. *Front. Neurosci.* **2024**, *18*, 1442915. [CrossRef]
89. Su, L.; Mruk, D.D.; Lee, W.M.; Cheng, C.Y. Drug transporters and blood--testis barrier function. *J. Endocrinol.* **2011**, *209*, 337–351. [CrossRef]
90. Su, L.; Zhang, Y.; Cheng, Y.C.; Lee, W.M.; Ye, K.; Hu, D. Slc15a1 is involved in the transport of synthetic F5-peptide into the seminiferous epithelium in adult rat testes. *Sci. Rep.* **2015**, *5*, 16271. [CrossRef]

91. Toure, A.; Lhuillier, P.; Gossen, J.A.; Kuil, C.W.; Lhote, D.; Jegou, B.; Escalier, D.; Gacon, G. The testis anion transporter 1 (Slc26a8) is required for sperm terminal differentiation and male fertility in the mouse. *Hum. Mol. Genet.* **2007**, *16*, 1783–1793. [CrossRef]
92. Hau, R.K.; Wright, S.H.; Cherrington, N.J. Drug Transporters at the Human Blood-Testis Barrier. *Drug. Metab. Dispos.* **2023**, *51*, 560–571. [CrossRef]
93. Zhao, W.; Wang, X.; Han, L.; Zhang, C.; Wang, C.; Kong, D.; Zhang, M.; Xu, T.; Li, G.; Hu, G.; et al. SLC13A3 is a major effector downstream of activated beta-catenin in liver cancer pathogenesis. *Nat. Commun.* **2024**, *15*, 7522. [CrossRef] [PubMed]
94. Tambay, V.; Raymond, V.A.; Voisin, L.; Meloche, S.; Bilodeau, M. Reprogramming of Glutamine Amino Acid Transporters Expression and Prognostic Significance in Hepatocellular Carcinoma. *Int. J. Mol. Sci.* **2024**, *25*, 7558. [CrossRef] [PubMed]
95. Gessner, A.; Konig, J.; Fromm, M.F. Clinical Aspects of Transporter-Mediated Drug-Drug Interactions. *Clin. Pharmacol. Ther.* **2019**, *105*, 1386–1394. [CrossRef] [PubMed]
96. Squibb, B.-M. SmPC Eliquis 5 mg Filmtabletten. Available online: https://www.medicines.org.uk/emc/product/2878 (accessed on 28 September 2024).
97. Frost, C.E.; Byon, W.; Song, Y.; Wang, J.; Schuster, A.E.; Boyd, R.A.; Zhang, D.; Yu, Z.; Dias, C.; Shenker, A.; et al. Effect of ketoconazole and diltiazem on the pharmacokinetics of apixaban, an oral direct factor Xa inhibitor. *Br. J. Clin. Pharmacol.* **2015**, *79*, 838–846. [CrossRef] [PubMed]
98. Ingelheim, B. SmPC Pradaxa 110 mg Hartkapseln. Available online: https://www.medicines.org.uk/emc/product/6229 (accessed on 28 September 2024).
99. Sankyo, D. SmPC Lixiana 60 mg Filmtabletten. Available online: https://www.medicines.org.uk/emc/product/6905 (accessed on 28 September 2024).
100. Igel, S.; Drescher, S.; Murdter, T.; Hofmann, U.; Heinkele, G.; Tegude, H.; Glaeser, H.; Brenner, S.S.; Somogyi, A.A.; Omari, T.; et al. Increased absorption of digoxin from the human jejunum due to inhibition of intestinal transporter-mediated efflux. *Clin. Pharmacokinet.* **2007**, *46*, 777–785. [CrossRef]
101. Leahey, E.B., Jr.; Reiffel, J.A.; Drusin, R.E.; Heissenbuttel, R.H.; Lovejoy, W.P.; Bigger, J.T., Jr. Interaction between quinidine and digoxin. *JAMA* **1978**, *240*, 533–534. [CrossRef]
102. Preston, C.L. *Stockley's Drug Interactions*, 11th ed.; Pharmaceutical Press: London, UK, 2016.
103. Bayer. SmPC Xarelto 10 mg Filmtabletten. Available online: https://www.medicines.org.uk/emc/product/6402 (accessed on 28 September 2024).
104. Hartter, S.; Koenen-Bergmann, M.; Sharma, A.; Nehmiz, G.; Lemke, U.; Timmer, W.; Reilly, P.A. Decrease in the oral bioavailability of dabigatran etexilate after co-medication with rifampicin. *Br. J. Clin. Pharmacol.* **2012**, *74*, 490–500. [CrossRef]
105. Greiner, B.; Eichelbaum, M.; Fritz, P.; Kreichgauer, H.P.; von Richter, O.; Zundler, J.; Kroemer, H.K. The role of intestinal P-glycoprotein in the interaction of digoxin and rifampin. *J. Clin. Investig.* **1999**, *104*, 147–153. [CrossRef]
106. Kobayashi, K.; Abe, Y.; Kawai, A.; Furihata, T.; Endo, T.; Takeda, H. Pharmacokinetic Drug Interactions of an Orally Available TRH Analog (Rovatirelin) With a CYP3A4/5 and P-Glycoprotein Inhibitor (Itraconazole). *J. Clin. Pharmacol.* **2020**, *60*, 1314–1323. [CrossRef]
107. Schiller, H.; Huth, F.; Schuhler, C.; Drollmann, A.; Kaul, M.; Woessner, R.; Shah, B.; Weis, W.; End, P. Novel Bruton's tyrosine kinase inhibitor remibrutinib: Assessment of drug-drug interaction potential as a perpetrator of cytochrome P450 enzymes and drug transporters and the impact of covalent binding on possible drug interactions. *Eur. J. Pharm. Sci.* **2022**, *172*, 106155. [CrossRef] [PubMed]
108. Chu, J.; Panfen, E.; Wang, L.; Marino, A.; Chen, X.Q.; Fancher, R.M.; Landage, R.; Patil, O.; Desai, S.D.; Shah, D.; et al. Evaluation of Encequidar as An Intestinal P-gp and BCRP Specific Inhibitor to Assess the Role of Intestinal P-gp and BCRP in Drug-Drug Interactions. *Pharm. Res.* **2023**, *40*, 2567–2584. [CrossRef] [PubMed]
109. Desai, M.P.; Harish Patil, P.; Vullendula, S.K.A.; Birangal, S.; Shenoy, G.G.; Rao, M.; Dengale, S.J.; Bhat, K.; Channabasavaiah, J.P. Molecular Insights into the Mechanism of Modulatory Effects of Proton Pump Inhibitors on P-glycoprotein Mediated Drug Transport of Palbociclib and Ribociclib. *Curr. Drug Metab.* **2023**, *24*, 458–465. [CrossRef] [PubMed]
110. Jin, H.; Zhu, Y.; Wang, C.; Meng, Q.; Wu, J.; Sun, P.; Ma, X.; Sun, H.; Huo, X.; Liu, K.; et al. Molecular pharmacokinetic mechanism of the drug-drug interaction between genistein and repaglinide mediated by P-gp. *Biomed. Pharmacother.* **2020**, *125*, 110032. [CrossRef]
111. ML, F.M.; Loos, N.H.C.; Mucuk, S.; de Jong, D.; Lebre, M.C.; Rosing, H.; Tibben, M.; Beijnen, J.H.; Schinkel, A.H. P-Glycoprotein (ABCB1/MDR1) Controls Brain Penetration and Intestinal Disposition of the PARP1/2 Inhibitor Niraparib. *Mol. Pharm.* **2021**, *18*, 4371–4384. [CrossRef]
112. Jones, A.B.; Tuy, K.; Hawkins, C.C.; Quinn, C.H.; Saad, J.; Gary, S.E.; Beierle, E.A.; Ding, L.; Rochlin, K.M.; Lamb, L.S.; et al. Temozolomide and the PARP Inhibitor Niraparib Enhance Expression of Natural Killer Group 2D Ligand ULBP1 and Gamma-Delta T Cell Cytotoxicity in Glioblastoma. *Cancers* **2024**, *16*, 2852. [CrossRef]
113. Hanko, E.; Tommarello, S.; Watchko, J.F.; Hansen, T.W. Administration of drugs known to inhibit P-glycoprotein increases brain bilirubin and alters the regional distribution of bilirubin in rat brain. *Pediatr. Res.* **2003**, *54*, 441–445. [CrossRef]
114. Karibe, T.; Hagihara-Nakagomi, R.; Abe, K.; Imaoka, T.; Mikkaichi, T.; Yasuda, S.; Hirouchi, M.; Watanabe, N.; Okudaira, N.; Izumi, T. Evaluation of the usefulness of breast cancer resistance protein (BCRP) knockout mice and BCRP inhibitor-treated monkeys to estimate the clinical impact of BCRP modulation on the pharmacokinetics of BCRP substrates. *Pharm. Res.* **2015**, *32*, 1634–1647. [CrossRef]

115. Elsby, R.; Martin, P.; Surry, D.; Sharma, P.; Fenner, K. Solitary Inhibition of the Breast Cancer Resistance Protein Efflux Transporter Results in a Clinically Significant Drug-Drug Interaction with Rosuvastatin by Causing up to a 2-Fold Increase in Statin Exposure. *Drug Metab. Dispos.* **2016**, *44*, 398–408. [CrossRef]
116. Dong, J.; Prieto Garcia, L.; Huang, Y.; Tang, W.; Lundahl, A.; Elebring, M.; Ahlstrom, C.; Vildhede, A.; Sjogren, E.; Nagard, M. Understanding Statin-Roxadustat Drug-Drug-Disease Interaction Using Physiologically-Based Pharmacokinetic Modeling. *Clin. Pharmacol. Ther.* **2023**, *114*, 825–835. [CrossRef]
117. Heinig, R.; Fricke, R.; Wertz, S.; Nagelschmitz, J.; Loewen, S. Results from Drug-Drug Interaction Studies In Vitro and In Vivo Investigating the Inhibitory Effect of Finerenone on the Drug Transporters BCRP, OATP1B1, and OATP1B3. *Eur. J. Drug Metab. Pharmacokinet.* **2022**, *47*, 803–815. [CrossRef] [PubMed]
118. Lehtisalo, M.; Tarkiainen, E.K.; Neuvonen, M.; Holmberg, M.; Kiiski, J.I.; Lapatto-Reiniluoto, O.; Filppula, A.M.; Kurkela, M.; Backman, J.T.; Niemi, M. Ticagrelor Increases Exposure to the Breast Cancer Resistance Protein Substrate Rosuvastatin. *Clin. Pharmacol. Ther.* **2024**, *115*, 71–79. [CrossRef] [PubMed]
119. Choi, H.G.; Kwon, B.-C.; Kwon, M.J.; Kim, J.H.; Kim, J.-H.; Park, B.; Lee, J.W. Association between Gout and Dyslipidemia: A Nested Case–Control Study Using a National Health Screening Cohort. *J. Pers. Med.* **2022**, *12*, 605. [CrossRef] [PubMed]
120. Lehtisalo, M.; Keskitalo, J.E.; Tornio, A.; Lapatto-Reiniluoto, O.; Deng, F.; Jaatinen, T.; Viinamaki, J.; Neuvonen, M.; Backman, J.T.; Niemi, M. Febuxostat, But Not Allopurinol, Markedly Raises the Plasma Concentrations of the Breast Cancer Resistance Protein Substrate Rosuvastatin. *Clin. Transl. Sci.* **2020**, *13*, 1236–1243. [CrossRef] [PubMed]
121. Abed, W.; Abujbara, M.; Batieha, A.; Ajlouni, K. Statin Induced Myopathy Among Patients Attending the National Center for Diabetes, endocrinology, & genetics. *Ann. Med. Surg.* **2022**, *74*, 103304. [CrossRef]
122. Trivedi, A.; Sohn, W.; Kulkarni, P.; Jafarinasabian, P.; Zhang, H.; Spring, M.; Flach, S.; Abbasi, S.; Wahlstrom, J.; Lee, E.; et al. Evaluation of drug-drug interaction potential between omecamtiv mecarbil and rosuvastatin, a BCRP substrate, with a clinical study in healthy subjects and using a physiologically-based pharmacokinetic model. *Clin. Transl. Sci.* **2021**, *14*, 2510–2520. [CrossRef]
123. Wang, Z.; Li, Y.; He, X.; Fu, Y.; Li, Y.; Zhou, X.; Dong, Z. In vivo evaluation of the pharmacokinetic interactions between almonertinib and rivaroxaban, almonertinib and apixaban. *Front. Pharmacol.* **2023**, *14*, 1263975. [CrossRef]
124. Zhang, Y.; Shipkova, P.A.; Warrack, B.M.; Nelson, D.M.; Wang, L.; Huo, R.; Chen, J.; Panfen, E.; Chen, X.Q.; Fancher, R.M.; et al. Metabolomic Profiling and Drug Interaction Characterization Reveal Riboflavin As a Breast Cancer Resistance Protein-Specific Endogenous Biomarker That Demonstrates Prediction of Transporter Activity In Vivo. *Drug Metab. Dispos.* **2023**, *51*, 851–861. [CrossRef]
125. Xie, Y.; Shao, Y.; Gong, X.; Wang, M.; Chen, Y. Evaluation of P-glycoprotein-targeting circulating microRNAs as peripheral biomarkers for medically intractable epilepsy. *Acta Epileptol.* **2023**, *5*, 5. [CrossRef]
126. Roth, M.; Obaidat, A.; Hagenbuch, B. OATPs, OATs and OCTs: The organic anion and cation transporters of the SLCO and SLC22A gene superfamilies. *Br. J. Pharmacol.* **2012**, *165*, 1260–1287. [CrossRef]
127. Li, T.T.; An, J.X.; Xu, J.Y.; Tuo, B.G. Overview of organic anion transporters and organic anion transporter polypeptides and their roles in the liver. *World J. Clin. Cases* **2019**, *7*, 3915–3933. [CrossRef] [PubMed]
128. Bruyere, A.; Le Vee, M.; Jouan, E.; Molez, S.; Nies, A.T.; Fardel, O. Differential in vitro interactions of the Janus kinase inhibitor ruxolitinib with human SLC drug transporters. *Xenobiotica* **2021**, *51*, 467–478. [CrossRef] [PubMed]
129. Hwang, S.; Lee, Y.; Jang, Y.; Cho, J.Y.; Yoon, S.; Chung, J.Y. Comprehensive Evaluation of OATP- and BCRP-Mediated Drug-Drug Interactions of Methotrexate Using Physiologically-Based Pharmacokinetic Modeling. *Clin. Pharmacol. Ther.* **2024**, *116*, 1013–1022. [CrossRef] [PubMed]
130. Yamazaki, T.; Desai, A.; Goldwater, R.; Han, D.; Lasseter, K.C.; Howieson, C.; Akhtar, S.; Kowalski, D.; Lademacher, C.; Rammelsberg, D.; et al. Pharmacokinetic Interactions Between Isavuconazole and the Drug Transporter Substrates Atorvastatin, Digoxin, Metformin, and Methotrexate in Healthy Subjects. *Clin. Pharmacol. Drug Dev.* **2017**, *6*, 66–75. [CrossRef] [PubMed]
131. Motoki, K.; Taniguchi, T.; Ashizawa, N.; Sakai, M.; Chikamatsu, N.; Yamano, K.; Iwanaga, T. Uricosuric Agents Affect Plasma and Kidney Concentration of Adefovir via Inhibition of Oat1 and Mrp2 in Rats. *Biol. Pharm. Bull.* **2023**, *46*, 170–176. [CrossRef]
132. Wu, T.; Li, H.; Chen, J.; Cao, Y.; Fu, W.; Zhou, P.; Pang, J. Apigenin, a novel candidate involving herb-drug interaction (HDI), interacts with organic anion transporter 1 (OAT1). *Pharmacol. Rep.* **2017**, *69*, 1254–1262. [CrossRef]
133. Clarke, R.; Leonessa, F.; Trock, B. Multidrug resistance/P-glycoprotein and breast cancer: Review and meta-analysis. *Semin. Oncol.* **2005**, *32*, 9–15. [CrossRef]
134. Zhu, S.; Sun, C.; Cai, Z.; Li, Y.; Liu, W.; Luan, Y.; Wang, C. Effective therapy of advanced breast cancer through synergistic anticancer by paclitaxel and P-glycoprotein inhibitor. *Mater. Today Biol.* **2024**, *26*, 101029. [CrossRef]
135. Zhou, X.; Zhang, P.; Yang, Y.; Shi, W.; Liu, L.; Lai, Z.; Zhang, X.; Pan, P.; Li, L.; Du, J.; et al. Highly Potent and Intestine Specific P-Glycoprotein Inhibitor to Enable Oral Delivery of Taxol. *Angew. Chem. Int. Ed.* 2024; *ahead of print*. e202412649. [CrossRef]
136. Loos, N.H.C.; Martins, M.L.F.; de Jong, D.; Lebre, M.C.; Tibben, M.; Beijnen, J.H.; Schinkel, A.H. Coadministration of ABCB1/P-glycoprotein inhibitor elacridar improves tissue distribution of ritonavir-boosted oral cabazitaxel in mice. *Int. J. Pharm.* **2024**, *650*, 123708. [CrossRef]
137. Zhang, Y.; Li, C.; Xia, C.; Wah To, K.K.; Guo, Z.; Ren, C.; Wen, L.; Wang, F.; Fu, L.; Liao, N. Adagrasib, a KRAS G12C inhibitor, reverses the multidrug resistance mediated by ABCB1 in vitro and in vivo. *Cell Commun. Signal.* **2022**, *20*, 142. [CrossRef]

138. Yang, Q.; To, K.K.W.; Hu, G.; Fu, K.; Yang, C.; Zhu, S.; Pan, C.; Wang, F.; Luo, K.; Fu, L. BI-2865, a pan-KRAS inhibitor, reverses the P-glycoprotein induced multidrug resistance in vitro and in vivo. *Cell Commun. Signal.* **2024**, *22*, 325. [CrossRef] [PubMed]
139. Marcelletti, J.F.; Sikic, B.I. Continuous 72-h infusion of zosuquidar with chemotherapy in patients with newly diagnosed acute myeloid leukemia stratified for leukemic blast P-glycoprotein phenotype. *Cancer Chemother. Pharmacol.* **2024**, *93*, 595–604. [CrossRef] [PubMed]
140. Sun, D.; Liu, J.; Wang, Y.; Dong, J. Co-administration of MDR1 and BCRP or EGFR/PI3K inhibitors overcomes lenvatinib resistance in hepatocellular carcinoma. *Front. Oncol.* **2022**, *12*, 944537. [CrossRef] [PubMed]
141. Deng, F.; Sjostedt, N.; Santo, M.; Neuvonen, M.; Niemi, M.; Kidron, H. Novel inhibitors of breast cancer resistance protein (BCRP, ABCG2) among marketed drugs. *Eur. J. Pharm. Sci.* **2023**, *181*, 106362. [CrossRef]
142. Jackson, S.M.; Manolaridis, I.; Kowal, J.; Zechner, M.; Taylor, N.M.I.; Bause, M.; Bauer, S.; Bartholomaeus, R.; Bernhardt, G.; Koenig, B.; et al. Structural basis of small-molecule inhibition of human multidrug transporter ABCG2. *Nat. Struct. Mol. Biol.* **2018**, *25*, 333–340. [CrossRef]
143. Zechner, M.; Castro Jaramillo, C.A.; Zubler, N.S.; Taddio, M.F.; Mu, L.; Altmann, K.H.; Kramer, S.D. In Vitro and In Vivo Evaluation of ABCG2 (BCRP) Inhibitors Derived from Ko143. *J. Med. Chem.* **2023**, *66*, 6782–6797. [CrossRef]
144. Vesga, L.C.; Kronenberger, T.; Tonduru, A.K.; Kita, D.H.; Zattoni, I.F.; Bernal, C.C.; Bohorquez, A.R.R.; Mendez-Sanchez, S.C.; Ambudkar, S.V.; Valdameri, G.; et al. Tetrahydroquinoline/4,5-Dihydroisoxazole Molecular Hybrids as Inhibitors of Breast Cancer Resistance Protein (BCRP/ABCG2). *ChemMedChem* **2021**, *16*, 2686–2694. [CrossRef]
145. Antoni, F.; Bause, M.; Scholler, M.; Bauer, S.; Stark, S.A.; Jackson, S.M.; Manolaridis, I.; Locher, K.P.; Konig, B.; Buschauer, A.; et al. Tariquidar-related triazoles as potent, selective and stable inhibitors of ABCG2 (BCRP). *Eur. J. Med. Chem.* **2020**, *191*, 112133. [CrossRef]
146. Kanai, Y.; Clemencon, B.; Simonin, A.; Leuenberger, M.; Lochner, M.; Weisstanner, M.; Hediger, M.A. The SLC1 high-affinity glutamate and neutral amino acid transporter family. *Mol. Asp. Med.* **2013**, *34*, 108–120. [CrossRef]
147. Johnston, R.A.; Rawling, T.; Chan, T.; Zhou, F.; Murray, M. Selective inhibition of human solute carrier transporters by multikinase inhibitors. *Drug Metab. Dispos.* **2014**, *42*, 1851–1857. [CrossRef]
148. Bruyere, A.; Hubert, C.; Le Vee, M.; Chedik, L.; Sayyed, K.; Stieger, B.; Denizot, C.; Parmentier, Y.; Fardel, O. Inhibition of SLC drug transporter activities by environmental bisphenols. *Toxicol. In Vitro* **2017**, *40*, 34–44. [CrossRef] [PubMed]
149. Puris, E.; Fricker, G.; Gynther, M. The Role of Solute Carrier Transporters in Efficient Anticancer Drug Delivery and Therapy. *Pharmaceutics* **2023**, *15*, 364. [CrossRef] [PubMed]
150. Boeszoermenyi, A.; Bernaleau, L.; Chen, X.; Kartnig, F.; Xie, M.; Zhang, H.; Zhang, S.; Delacretaz, M.; Koren, A.; Hopp, A.K.; et al. A conformation-locking inhibitor of SLC15A4 with TASL proteostatic anti-inflammatory activity. *Nat. Commun.* **2023**, *14*, 6626. [CrossRef] [PubMed]
151. Chen, X.; Xie, M.; Zhang, S.; Monguio-Tortajada, M.; Yin, J.; Liu, C.; Zhang, Y.; Delacretaz, M.; Song, M.; Wang, Y.; et al. Structural basis for recruitment of TASL by SLC15A4 in human endolysosomal TLR signaling. *Nat. Commun.* **2023**, *14*, 6627. [CrossRef]
152. Rives, M.L.; Javitch, J.A.; Wickenden, A.D. Potentiating SLC transporter activity: Emerging drug discovery opportunities. *Biochem. Pharmacol.* **2017**, *135*, 1–11. [CrossRef]
153. Shchulkin, A.V.; Abalenikhina, Y.V.; Slepnev, A.A.; Rokunov, E.D.; Yakusheva, E.N. The Role of Adopted Orphan Nuclear Receptors in the Regulation of an Organic Anion Transporting Polypeptide 1B1 (OATP1B1) under the Action of Sex Hormones. *Curr. Issues Mol. Biol.* **2023**, *45*, 9593–9605. [CrossRef]
154. Buxhofer-Ausch, V.; Secky, L.; Wlcek, K.; Svoboda, M.; Kounnis, V.; Briasoulis, E.; Tzakos, A.G.; Jaeger, W.; Thalhammer, T. Tumor-specific expression of organic anion-transporting polypeptides: Transporters as novel targets for cancer therapy. *J. Drug Deliv.* **2013**, *2013*, 863539. [CrossRef]
155. Verscheijden, L.F.M.; Koenderink, J.B.; de Wildt, S.N.; Russel, F.G.M. Differences in P-glycoprotein activity in human and rodent blood-brain barrier assessed by mechanistic modelling. *Arch. Toxicol.* **2021**, *95*, 3015–3029. [CrossRef]

membranes

Review

The Influence of Cholesterol on Membrane Targeted Bioactive Peptides: Modulating Peptide Activity Through Changes in Bilayer Biophysical Properties

Juan M. Giraldo-Lorza [1], Chad Leidy [2] and Marcela Manrique-Moreno [1,*]

1 Chemistry Institute, Faculty of Exact and Natural Sciences, University of Antioquia, A.A. 1226, Medellin 050010, Colombia; juan.giraldo77@udea.edu.co
2 Biophysics Group, Physics Department, Universidad de los Andes, Bogotá 111711, Colombia; cleidy@uniandes.edu.co
* Correspondence: marcela.manrique@udea.edu.co; Tel.: +57-300-70-78-928

Abstract: Cholesterol is a biological molecule that is essential for cellular life. It has unique features in terms of molecular structure and function, and plays an important role in determining the structure and properties of cell membranes. One of the most recognized functions of cholesterol is its ability to increase the level of lipid packing and rigidity of biological membranes while maintaining high levels of lateral mobility of the bulk lipids, which is necessary to sustain biochemical signaling events. There is increased interest in designing bioactive peptides that can act as effective antimicrobial agents without causing harm to human cells. For this reason, it becomes relevant to understand how cholesterol can affect the interaction between bioactive peptides and lipid membranes, in particular by modulating the peptides' ability to penetrate and disrupt the membranes through these changes in membrane rigidity. Here we discuss cholesterol and its role in modulating lipid bilayer properties and discuss recent evidence showing how cholesterol modulates bioactive peptides to different degrees.

Keywords: cholesterol; bioactive peptides; membrane fluidity; cholesterol–peptide interactions

1. Introduction

Eukaryotic lipid membranes are structurally intricate, possessing hundreds of lipid species with a wide range of physicochemical properties [1]. They have a remarkable asymmetric lipid distribution, variations in acyl chain lengths and saturation levels, and variations in headgroup composition and geometries. The different chemical structures found in these lipid molecules influence several membrane properties, such as lateral mobility, mechanical strength, intrinsic curvature levels, and lateral organization, characterized by the presence of rafts or clusters [1]. Lipids that make up eukaryotic membranes can be divided into three categories according to their structure and function: glycerophospholipids, sphingolipids, and sterols. Glycerophospholipids are characterized by a glycerol backbone, with two acyl chains attached through ester bonds, and a headgroup characterized by the presence of a phosphate group that is negatively charged at physiological pH. Headgroup composition varies, leading to different net charges, sizes, and polarities, and phosphatidylcholine and phosphatidylethanolamine are found to be the most abundant headgroup species in eukaryotic membranes [2]. Sphingolipids share structural similarities with glycerophospholipids. However, the chemical structure does not contain a glycerol backbone. Instead, a sphingosine attaches a fatty acid through the amide group in the sphingosine, while the headgroup is attached to the hydroxyl of the sphingosine. Sphingomyelin contains a phosphate group and a choline in its headgroup, making it very similar to phosphatidylcholine. Sphingomyelin is found to be highly enriched, mainly in the extra-cellular leaflet of the plasma membrane, where it plays a role in the formation of lipid rafts [3].

Among sterols, cholesterol is the main sterol found in eukaryotic membranes, where it makes up 10–50% of total lipids in eukaryotic cells and has a range of effects on overall membrane properties [4]. It has long been noticed that the biophysical and biochemical properties of the membrane are diversely affected by the lipid composition and organization [5]. However, the unique structure of CHO with a polar head and rigid ring structure in its hydrophobic tail renders the molecule particularly important in regulating biophysical properties that affect lipid packing, fluidity, lateral organization, and the shape of phospholipid bilayers [6]. As a regulator of membrane physical properties, it is a crucial component of many cellular functions, including vesicular transport and the signal transduction pathways [7]. CHO plays important roles in the organism as the precursor of several hormones, bile acids, and vitamin D, holds the membrane proteins in a functional conformation, and is also an important mediator in cell signaling pathways [8]. Some studies also show that cholesterol takes part in the communication between intra- and extra-cellular processes as a modulator of various processes [9].

The rigid ring structure found in cholesterol induces increased lipid packing by forcing the alignment of the lipid acyl chains of neighboring phospholipids, leading to ordering of the acyl chain conformations. The hydroxyl group in the molecule positions the molecule close the hydrophilic/hydrophobic interface of the bilayer, with the ring structure extending into the acyl chain region and positioning itself parallel to the acyl chains, resulting in a reduction in rotational conformations of the chains. Acyl chain alignment induces an increase in bilayer thickness in regions enriched in cholesterol. These structural changes lead to a modulation of the mechanical properties of the membrane reflected in an increase in the bending rigidity of the membrane. However, since cholesterol acts as a spacer that separates neighboring phospholipids, it is also able to maintain lateral mobility of phospholipids and proteins in the membrane even while inducing an increase in membrane rigidity. For this reason, cholesterol has the remarkable property of increasing the mechanical strength of the lipid bilayer without strongly affecting the lateral mobility of lipid and protein components, which is crucial for maintaining essential processes, such as signaling dynamics. An additional remarkable trait of this molecule is that cholesterol can induce phase separation in the lipid membrane, which results in the formation of lipid microdomains, also known as membrane rafts, that are used by a variety of cells to segregate proteins in order to organize signaling events and endocytosis.

With regards to the modulation of mechanical properties, a prime example of the importance of cholesterol is seen in red blood cells (RBCs). RBCs contain very high cholesterol levels, close to 50 mol%, which is near the saturation concentration point of cholesterol in the membrane. This cholesterol concentration is too high to maintain phase separation, in which case the red blood-cell plasma membrane is found in a one liquid-ordered phase enriched in cholesterol. For this reason, the main purpose of these high levels appears to be an increase in the mechanical strength of the RBC membrane to prevent damage during transport in the circulatory system.

Cholesterol, the only sterol present, is found in eukaryotic membranes, where it makes up to 10–50% of total lipids and has a range of effects on overall membrane properties [4]. A molecule capable of modulating the physicochemical properties of the membrane has the potential to modulate the activity of membrane-active molecules, like bioactive peptides (BAPs). These molecules are fundamental components of the immune system of all living organisms [10]. They are small amino acid sequences with multiple effects on body function [11–16]. These physiological properties have led to BAPs being considered a promising pharmaceutical tool against several diseases [17–26]. It has been widely accepted that most BAPs exert their activity by interacting with the phospholipids of the membranes, destabilizing the membrane of the target cells. This mechanism is difficult for the target cells to counteract, and is responsible for the rapid effect of the peptides [10].

Many membranes targeted bioactive peptides are characterized by binding and inserting into biological membranes, either to traverse the membrane and enter the cytoplasm to reach intracellular targets, or to group in an inserted state to induce pore formation and leakage in the membrane. The insertion process always requires a local mechanical deformation of the membrane to take effect. Therefore, the lipid composition of target membranes, which affects the bending modulus of the lipid bilayer, is a key factor in determining the efficacy of bioactive peptides [27,28]. A regulator of membrane rigidity, cholesterol is a major structural component of the membrane, which can modulate the interactions between BAPs and lipid membranes by affecting the peptides' ability to penetrate and disrupt the membranes. In the case of antimicrobial peptides, the potential inhibition of BAPs by cholesterol will have direct relevance for how BAPs interact differentially with the host cells that do synthesize cholesterol, compared to bacterial cells that do not. This leads to intriguing questions about the potential roles of cholesterol–peptide associations as a way to protect the cells producing the peptides in order to fight pathogens. This review will focus on the general characteristics of BAPs, the cell membrane structure, and the effect of cholesterol and the consequences of its interaction with peptides. This information can be useful in the design of BAPs by helping the peptide design process to take into account information related to the role of cholesterol in the modulation of BAP activity.

2. Bioactive Peptides (BAPs)

BAPs are a heterogeneous group of molecules, expressed in organisms ranging from bacteria to humans, that provide an adaptive advantage for the innate host defense system of these organisms by exerting countless biological effects through several mechanisms [29]. BAPs are mainly known for their direct interactions with and effects on microbial membranes, but many peptides can also target an array of other key microbial components, such as lipopolysaccharides [30], teichoic acids, peptidoglycans [31], nucleic acids [32], and proteins [33], due to their exceedingly high affinity and absence of rigid binding specificity [34,35].

2.1. Physicochemical Properties of BAPs

Due to the combinatorial arrangement of amino acids within a particular primary sequence, there is an exponential number of potential primary structures. The amino acids which conform a sequence can be neutral, charged, polar, hydrophobic, and even synthetically modified. The structure–activity analysis of numerous BAPs has revealed that their biological activities are linked to a number of common features. Specifically, they are short sequences of 5–50 amino acid residues (less than 10 KDa), mainly cationic, amphipathic, and with approximately 50% of residues in the sequence being hydrophobic [36]. Despite these common characteristics, they are highly diverse with respect to their activity against bacteria [37], viruses [38], parasites [39–41], cancer [42], fungi [43,44], etc. Most BAPs possess a net positive charge at physiological pH because of their high content of basic residues, such as arginine and lysine. The charge of the peptides, located on the side chains of the residues, enables multiple interactions, such as hydrogen bonding, salt bridges and dipole interactions with the anionic charged groups of the microorganism membranes, a step that has been accepted [45,46] as necessary for them to exert their activity [47,48]. Most of the biologically active sequences reported in the databases possess a net positive charge, ranging from +3 to +9 [45,46].

The hydrophobicity and amphipathicity of bioactive peptides are characteristics of considerable interest. Extensive study has found that peptides with higher hydrophobicity penetrate the hydrophobic core of cell membranes, leading to cell disruption through the necrotic mechanism [49,50]. Several studies have attempted to replace hydrophobic and neutral amino acids with positively charged residues, such as lysine, leucine or phenylalanine, on the polar and non-polar faces of α-helical peptides [51]. Consequently, cationic peptides with moderate hydrophobicity have shown increased cytotoxicity against target cells [46,52]. Amphipathicity is defined as the distribution of polar and nonpolar residues across the backbone of the polypeptide chain. Essentially, an amphipathic peptide has hydrophilic and hydrophobic regions, resulting in water-repelling (nonpolar) and water-attracting (polar) regions. The structural arrangement of these regions allows BAPs to interact with and disrupt the target membranes, which is crucial for their biological activity. Adverse interactions with eukaryotic membrane counterparts then depend upon a delicate balance between hydrophobicity and charge. Amphipathicity, together with net charge and length, determines the disposition of the peptide towards the cell membrane and therefore the mechanism by which the peptide exerts its biological activity. Multiple BAPs, with overlapping but non-identical sequences, activities and mechanisms of action, can often act cooperatively to facilitate the killing of invading pathogens [53].

2.2. Structural Characteristics

Bioactive peptides can adopt different secondary structures (Figure 1). Some peptides are unstructured upon interaction (random/coil), while others adopt a more stable structure before or after they associate with lipids. The largest population of reported secondary structures of BAPs is α-helix, followed by random coil and β-sheet [54]. These elements often coexist in the same peptide.

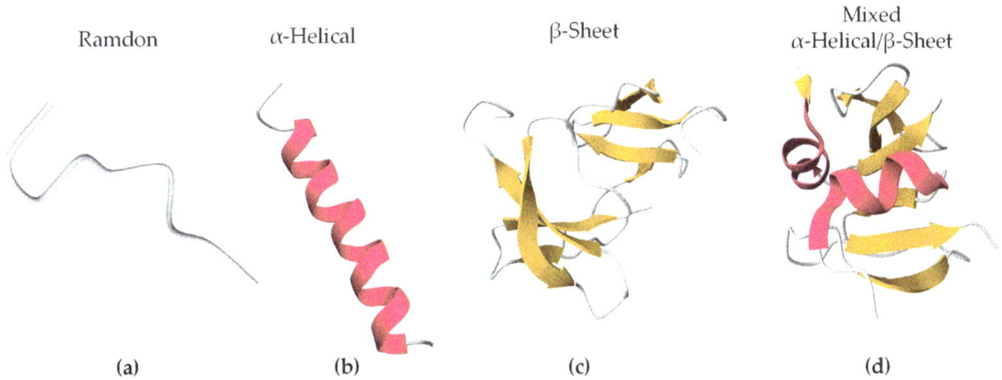

Figure 1. Secondary structures of representative BAPs using PyMOL 3.0. In brackets are the PDB ID and the origin of the peptides. The colors represent the secondary structures: (**a**) Indolicidin (PDB ID 8IS3, *Bos taurus*), (**b**) Magainin 2 (PDB ID 2MAG, *Xenopus laevis*), (**c**) Human β-Defensin-4 (PDB ID 5KI9, *Homo sapiens*), and (**d**) Human β-Defensin-2 (PDB ID 1FD4, *Homo sapiens*).

3. Eukaryotic Membranes

3.1. Structure of Membranes

In addition to populating many unicellular communities, eukaryotic cells are the building blocks of complex multicellular living systems. Eukaryotic cells require a demarcating structure that separates them from their external environment and envelopes their internal membrane-bound organelles. This structure is called plasma membrane, and is consistent in its construction across the *Eukarya* [55]. The classic model of bio-membranes was proposed by Singer and Nicolson, who named it the fluid mosaic model [56]. In this model, the membrane was described as an unstructured liquid–crystal characterized by a disorderly and chaotic arrangement of molecules in a two-dimensional liquid–crystal matrix. However, with advances in membrane biophysics, the model was complemented and improved, extending the description and the understanding of the membrane organization by including the notion of a heterogenous arrangement of lipids and proteins on the membrane surface due to the formation of lipid domains. The cell membrane is currently described as a heterogeneous two-dimensional semi-fluid barrier that acts as a matrix containing a population of lipids, proteins, and glycolipids with transitory mobility (Figure 2) [57].

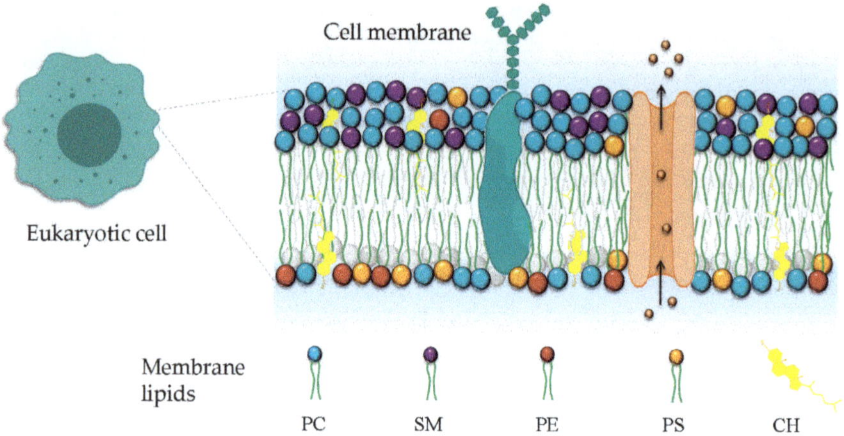

Figure 2. Representation of eukaryotic cell plasma membrane. The phospholipid bilayer contains all molecules, including phospholipids, proteins, and cholesterol. Phosphatidylcholine (PC), phosphatidylethanolamine (PE), phosphatidylserine (PS), and sphingomyelin (SM). The irregular representation of the lipid acyl chains of lipids denotes the fluid nature of the bilayer.

The main structure of the membrane consists of lipids arranged in a bilayer. These lipids are non-covalently associated in the lipid bilayer. Their interactions are weak entropic and transient dipole in nature, which leads to constant lateral motion, characteristic of a liquid state. In addition, there are other molecules, such as proteins, on the membrane (peripheral) and within it (integral), that are associated with the lipid bilayer by non-covalent interactions and therefore differing degrees of mobility. These proteins have functions as diverse as those of pore formers, signal transducers, and enzymes. Thus, the membrane is more than just two-dimensional and is a complex array of dynamic patterns involving inter- and intra-molecular interactions. The integration of proteins and glycolipids into the cell membrane reflects one of the priorities of the signal hypothesis, that the membrane permits the controlled transmission of biological information in either a secreted, (open) or transduced (closed) channel manner [58].

3.2. Lipid Composition

Cell membranes are composed mostly of proteins and lipids. There are several families of lipids present in the eukaryotic plasma membrane, showing varying compositions depending on the cell type. In human cells, the main families are phospholipids and glycolipids. Phospholipids are composed of a negatively charged phosphate group, a complex alcohol group (glycerol moiety), and two fatty acids. Phospholipids are described as amphipathic molecules because of their double-nature structure: the polar head group is hydrophilic, and the non-polar tails are hydrophobic. The properties of membrane lipids, such as their charge distribution, molecular size, and molecule-to-molecule interactions, depend on their head group and tail chemistry. The phosphate-containing portions of the molecule are widely diverse (Figure 3). This moiety can be charged or polar, and so it interacts attractively with the surrounding water through polar interactions. The long, hydrophobic fatty acid tails, on the other hand, do not present a permanent charge polarity, which impedes the fatty acids from interacting with the surrounding water through polar interactions. The acyl chains can also be diverse in terms of their length and the presence of unsaturated bonds (Figure 3), influencing the level of lipid packing. After the lipids assemble into a bilayer, it will feature a hydrophilic outer surface on both leaflets, facing the water environment, and a hydrophobic inner core characterized by a fluid hydrocarbon environment at physiological temperatures.

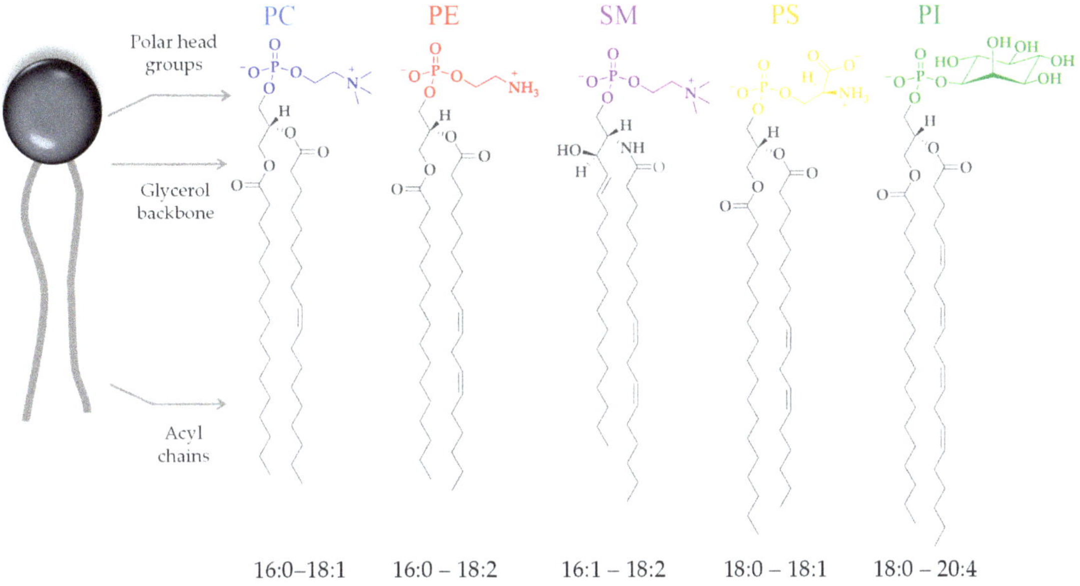

Figure 3. Chemical representation of the head groups and acyl chains of the most abundant phospholipids of eukaryotic cell membranes. Phosphatidylcholine (PC), phosphatidylethanolamine (PE), phosphatidylserine (PS), sphingomyelin (SM), and phosphatidylinositol (PI). The acyl chains range from fully saturated to multiple unsaturated.

The plasma membranes and organelle membranes of the different cell types found in the human body exhibit widely diverse lipid compositions. Lipid bilayers have many functions, and can have specific features depending on the functional characteristics that the different membranes have to fulfill [1]. Corresponding to these functions, different lipid types are present in the different cavities and leaflets of each membrane. Lipids, such as sphingomyelin and cholesterol, found in the outer leaflet of the plasma membrane of T-cells, help organize transmembrane proteins in charge of detecting and eradicating pathogens [59], while cardiolipin is found in the inner membrane of mitochondria, providing the spontaneous curvature needed to induce the characteristic folding of this membrane [60].

Lipids are important integral components of the organelles of eukaryotic cells. They are considered to determine the function of these subcellular structures [3]. The endoplasmic reticulum is equipped with a characteristic lipid metabolism with de novo synthesis of phospholipids. The metabolism of de novo phosphatidylethanolamine and phosphatidylcholine, the most synthesized phospholipids of this compartment, also involves their acylation. The secretory pathway next to the endoplasmic reticulum ends with the Golgi apparatus, which has a different lipid metabolism from the precursor [2]. In particular, the metabolism of sphingomyelin, which increases in the Golgi apparatus, and that of the phospholipid characteristic of this compartment, is remarkable. The process of intra-Golgi membrane trafficking, and other processes that Golgi is not associated with, would modify the lipid composition of this compartment, whose function would not only provide lipids and proteins for the plasma membrane but also distribute certain lipids, including sphingolipids, in a classic way toward the apical layer, coupled with the formation of a basal loop of the trans-Golgi [3]. Figure 4 summarizes the lipid composition reported for the eukaryotic cell organelles.

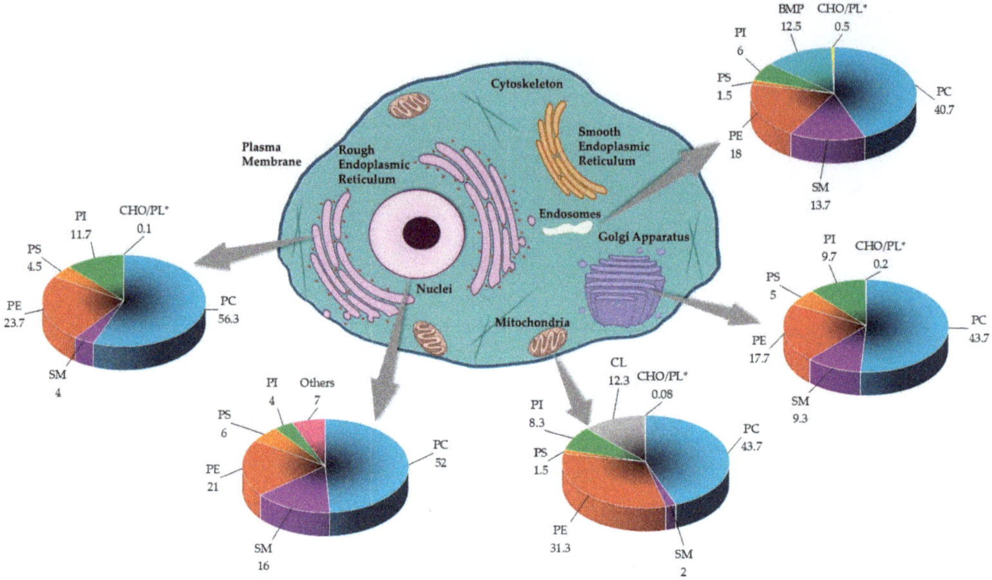

Figure 4. The lipid composition of different organelles throughout the eukaryotic cell. The lipid data in the graphs are presented as a percentage of total phospholipids (PL) in mammals. The cholesterol content is presented as the molar ratio of cholesterol (CHO) with respect to the PL [2,3,61]. * Data was reported in CHO/PL ratio.

The complexity of cell membranes suggest that the lipid models used in biophysical studies must be prepared as multi-component lipid systems based on the composition of the membrane under study. This would lead to more representative models, which can reflect more biologically relevant conditions [62–64]. Nevertheless, when deciding what proportions of lipids to use as representative of plasma membranes, the reported distribution of phospholipids is very broad. To visualize the different possibilities that are available, Figure 5 summarizes different works that reported the distributions of lipids in various cell types. Additionally, in Table 1, the original data are presented, which includes how the authors reported the using different techniques for the quantification and different methods for the extraction of the lipids of the outer layer of the membrane. Regarding the amount of cholesterol, the published data are quite imprecise. Most publications do not include the quantification of sterol leading to gaps in the information needed to determine the proportions to include in a potential artificial lipid model. As cholesterol is such an important molecule in the regulation of biophysical properties of the membrane, a clearer understanding of cholesterol content for the different biological membranes is necessary.

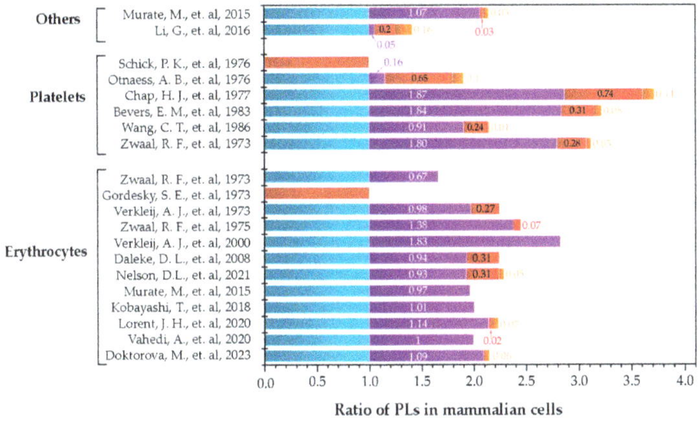

Figure 5. Summary of publications that quantified the phospholipid distribution in mammalian cells. Others are fibroblasts and cancer cells. The colors correspond to PC (■), SM (■), PE (■), and PS (■). Data are presented in ratios based on PC abundance.

Table 1. Summary of studies reporting the quantification of lipids of the outer layer (OL) of the membrane of eukaryotic cells.

Cell Type	Results OL	Unit	Technique Used for Quantification	Year	Ref.
Erythrocytes	PC: 8.5 SM: 9.3 PS: 0.48	Mol %	Enzymatic digestion/Mass spectrometry	2023	[65]
	PC: 1 SM: 1	Ratio	Cyclodextrin chemistry/Thin-layer chromatography/Mass spectrometry	2020	[66]
	PC: 42 SM: 48 PE: 1 PS: 3	Mol %	Enzymatic digestion/Mass spectrometry	2020	[67]
	PC: 98 SM:99	Total PL (%)	Sodium dodecyl sulphate-digested freeze–fracture/Specifics monoclonal antibodies	2018	[68]
	PC: 98 SM: 95	Total PL (%)	Sodium dodecyl sulphate-digested freeze-fracture/Specifics monoclonal antibodies	2015	[69]
	PC: 19.2 SM: 20 PE: 7 PS: 1.3	Total PL (%)	Not reported	2012	[70]
	PC: 22.5 SM: 21.1 PE: 7	Total PL (%)	Not reported	2008	[71]
	PC: 15.1 SM: 27.7	Total PL (%)	Not reported	2000	[72]

Table 1. *Cont.*

Cell Type	Results OL	Unit	Technique Used for Quantification	Year	Ref.
	PC: 61.5 SM: 85 PE: 4.5 PS: 0	Total PL (%)	Enzymatic degradation/Thin-layer chromatography and determined as phosphorus	1975	[73]
	PC: 21 SM: 20.5 PE: 5.7	Total PL (%)	Enzymatic degradation and freeze-etching electron microscopy/Two-dimensional thin-layer chromatography and determined as phosphorus	1973	[74]
	PE: 33 PS: 0	Total PL (%)	Titrobenzenesulfonate-diazosulfanilic acid/UV–Vis spectrometry	1973	[75]
	PC: 3 SM: 2	Ratio	Enzymatic degradation/Two-dimensional thin-layer chromatography and UV–Vis spectrometry	1973	[76]
Platelets	PC: 6.5 SM: 11.7 PE: 1.8 PS: 0.3	Total PL (%)	Enzymatic degradation	1987	[77]
	PC: 45.9 SM: 41.8 PE: 10.8 PS: 0.5	Total PL (%)	Enzymatic degradation/Thin-layer chromatography and determined as phosphorus	1986	[78]
	PC: 31 SM: 57.1 PE: 9.5 PS: 2.4	Total PL (%)	Enzymatic degradation/Thin-layer chromatography and determined as phosphorus	1983	[79]
	PC: 12.7 SM: 23.8 PE: 9.4 PS: 1.4	Total PL (%)	Enzymatic degradation/Thin-layer chromatography and determined as phosphorus	1977	[80]
	PC: 26.7 SM: 4.2 PE: 17.3 PS: 2.8	Total PL (%)	Enzymatic degradation/Thin-layer chromatography and determined as phosphorus	1976	[81]
	PE: 17.9 PS: 0	Total PL (%)	Titrobenzenesulfonate/Thin-layer chromatography and determined as phosphorus		[82]
Others	PC: 60.1 SM: 3.3 PE: 11.9 PS: 9.9	Total PL (%)	Cyclodextrin chemistry/Thin-layer chromatography/Mass spectrometry	2016	[83]
	PC: 82.3 SM: 87.8 PE: 2.5 PS: 4.2	Total PL (%)	Sodium dodecyl sulphate-digested freeze-fracture/Specific monoclonal antibodies	2015	[69]

4. Cholesterol

Cholesterol is a 27-carbon compound with a unique structure containing a central sterol core made of four hydrocarbon rings, a hydroxyl group, and a hydrocarbon tail (Figure 6a). The center sterol nucleus or ring is a feature of all steroid hormones. The cholesterol molecule is composed of three distinct regions: a small polar hydroxyl group, a rigid, plate-like steroid ring, and a flexible iso-octyl chain. The iso-octyl chain, which has a much smaller cross-sectional area than the rigid ring, is located at the lipid core of the membrane, and presents conformational freedom, similar to the lipid acyl chains (Figure 6b). When cholesterol incorporates into the membrane, it is expected that the polar hydroxyl group will be positioned between the polar headgroups of phospholipids, while the rigid ring structure extends into the hydrophobic acyl chains, aligning with the carbon atoms [84]. However, biophysical studies using polyunsaturated phosphatydilcholines (20:4–20:4) and neutron diffraction experiments showed that cholesterol can even lie in the membrane midplane. Harroun et. al. attributed the change in location to the high level of disorder of polyunsaturated fatty acids, which is incompatible with proximity to the rigid steroid moiety in its usual upright orientation [85].

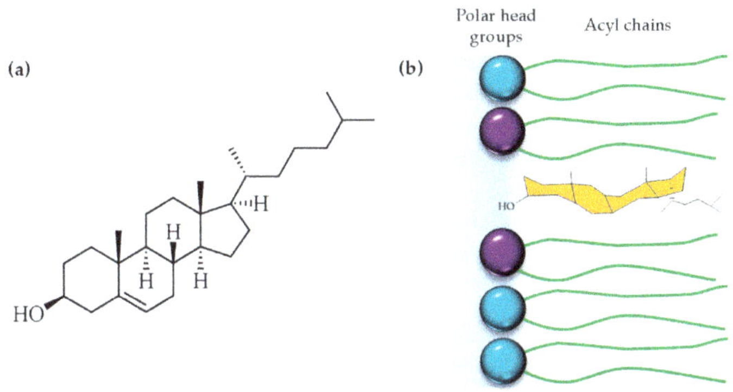

Figure 6. (**a**) Chemical structure of cholesterol and (**b**) disposition of cholesterol into the hydrophobic region of the cell membrane.

4.1. Cholesterol as a Physiological and Structural Component

Cholesterol, an essential component of eukaryotes, is a precursor for several steroid hormones, vitamin D3, and bile acids [8]. This neutral lipid is well known for being implicated in several medical issues, particularly heart diseases [86]. However, it also mediates several physiological functions, particularly in animal cells. After absorbing cholesterol from a healthy diet, the body ceases to synthesize cholesterol de novo, reducing endogenous cholesterol production and balancing the variation [87].

Organisms use cholesterol for the maintenance of cell membrane stability [88], increasing mechanical resistance [89], regulating membrane internalization [90,91], membrane trafficking following exocytosis [92], and modulating the formation of endocytic synapsis [93,94]. The cholesterol molecule plays a crucial role in the spatial arrangement of the lipid bilayer. García-Arribas reported that the addition of cholesterol to the membrane imparts stability and rigidity, preserving membrane architecture [95]. However, in addition to this mechanical function, cholesterol modulates the functional organization of the membrane by inducing the formation of mesoscopic domain regions on the plasma membrane, known as lipid rafts, that serve to segregate relevant transmembrane and peripheral proteins into functional regions [7]. These domains are formed due to thermodynamic phase separation of lipid components in the plane of the membrane.

Cholesterol, sphingomyelin, and glycosphingolipids segregate to form liquid-ordered phase regions segregated from liquid-disordered phase regions. The liquid-ordered phase receives its name from the condensed ordering of the acyl chains, which is induced by the presence of specific lipids interacting with cholesterol, which otherwise maintains a mobile liquid state of the lipids in the plane of the membrane. The liquid-disordered phase maintains a high level of conformational disorder of the acyl chains in addition to behaving as a liquid in the plane of the membrane. For this reason, this domain structure is denominated as liquid–liquid coexistence. This segregation of these phases in the plane of the membrane plays several crucial physiological roles in a diverse range of cellular processes [96]. The main structural feature that promotes this segregation is the preferential interaction of cholesterol with sphingolipids. Cholesterol occupies spaces between the acyl chains and polar headgroups in the lipid bilayer, as well as influencing lipid interdigitation between different lipid leaflets [95]. Since cholesterol is rigid and planar, cholesterol will prefer to accommodate itself next to the extended acyl chains of sphingomyelin, which lack cis-double bonds. The cis-double bonds normally present in phosphatidylcholine lead to kinks in the acyl chains that result in a steric hindrance to the cholesterol lining up with these chains [97]. This preferred interaction with sphingomyelin leads to a straightening of the sphingomyelin acyl chains, which induces a local increase in bilayer thickness. The difference in bilayer thickness between the SM-rich regions and the PC-rich regions induces line tensions that lead to domain formation. Cholesterol is also an important mediator in cell signaling pathways. Some evidence also shows that cholesterol takes part in the communication between intra- and extra-cellular processes as a modulator of various processes. The nature of cholesterol presents a contradiction, and its dual functions have been studied extensively [98].

4.2. Cholesterol as a Membrane Fluidity Regulator

Cholesterol is a unique lipid, and an essential constituent of eukaryotic membranes. Although the role of cholesterol in biological systems has been studied for more than 100 years, the complete biophysical role of this lipid is not yet fully understood. One well-established role of cholesterol in biological systems is as a regulator of membrane biophysical properties [6]. It is well known that cholesterol down-regulates overall membrane fluidity, both through the formation of tighter packing inside membranes induced by the straightening of lipid acyl chains and by directly modulating the anomalous and chain-conformational lengths. This modulation is due to its large (3-hydroxyl) group and O linkage at the C3 position, which both set constraints that restrict local chain bending and improve steric packing within the bilayer [6]. At the same time, it also softens the anomalous chain tilted permutations, due to its shorter structure. Furthermore, the large tetracyclic A-ring of cholesterol enables it to pack tightly and with a high tilt angle against the normal membrane, with its ribose ester side-chain also acting to help further pack and deform the opposing leaflet. The more local scale effect of cholesterol-induced liquid-ordered domains is its imposition of a higher local curvature that attracts charged atoms (such as the phosphate in the head-group of PC and SM, and sterols) towards the center of the interfaces. The main effect of cholesterol, therefore, is liquid phase region stiffening, increasing the measurable two-dimensional elastic properties that encapsulate both monolayer stretching and compression properties, and vesicle surface bending and glycerol angular propulsion (stretch-coupling) [99].

Cholesterol is present at different concentrations in membranes, ranging from 10 to 50 mol%. The concentration of cholesterol affects the membrane properties. Low cholesterol concentrations have been detected in the nuclear envelope, membranes of rough and smooth endoplasmic reticulum, and membranes of Golgi apparatus [4]. These membranes are highly fluid and exhibit a moderate hydrophobic barrier, which allows the intracellular transport of substances and products. At high concentrations, cholesterol causes a decrease in membrane fluidity. For typical plasma membranes containing 10 to 30% cholesterol, the presence of cholesterol induces domain formation, as discussed above [100]. The presence of cholesterol reduces the tilt mobility of the lipid acyl chains and decreases the lateral diffusion of lipids in the bilayer [4,6].

Membranes with a high cholesterol content from 30 to 50 mol%, such as erythrocytes, are characterized by the hydrophobicity of the superior membrane, which increases its rigidity by minimizing trans-gauche isomerization (promoting an extended conformation) and reducing translational and rotational motion within the membrane. These properties are consistent with those expected for erythrocytes. Red blood cells must pass through narrow capillaries without losing molecules contained in the cytosol; its membranes must be able to avoid permeation of small molecules [101]. Extensive study has found that cholesterol limits oxygen diffusion across membranes, which helps regulate its proper transport in the circulatory system [102–104]. High concentrations of cholesterol simultaneously heighten hydrophobic barriers for polar molecules and strengthen rigidity barriers for nonpolar molecules [4]. Thus, the above observations suggest that cholesterol increases the thickness of the hydrophobic region of the bilayer, with the hydroxyl group of cholesterol located on the outer edge of the bilayer [6]. On the other hand, at low concentrations, cholesterol functions to increase the fluidity of membranes by preventing hydrophobic collapse of the acyl tails [105].

4.3. Cholesterol Asymmetry

Eukaryotic membranes are known for their asymmetric distribution of lipids, with distinct compositions in their bilayer leaflets. This asymmetric distribution generates an electrostatic potential that influences protein–lipid interactions, highlighting the significance of lipid asymmetry in the plasma membranes [106]. The active maintenance of lipid asymmetry through ATP-dependent transporters underscores its importance for cellular function, as evidenced by its association with a congenital bleeding disorder and its role in cell apoptosis, where the external exposure of the cytoplasmic leaflet lipid, phosphatidylserine, is key to this pathology [107].

One of the most recent works in the field showed, using quantitative lipidomics, that the phospholipid imbalance in red blood cells is facilitated by an uneven distribution of cholesterol between the leaflets, which quickly redistributes to alleviate leaflet stress. The asymmetric abundance and composition of phospholipids work together to concentrate cholesterol in the exoplasmic leaflet of the erythrocyte plasma membrane [5]. Blumer et al. demonstrated, through atomistic and coarse-grained simulations, that the number of lipids in each leaflet of the bilayer can be adjusted to minimize leaflet stress, highlighting the coupling between the bilayer's two leaflets. Moreover, their findings suggest that cholesterol can reduce membrane asymmetry by flipping between leaflets and can influence certain properties such as the lipid packing parameter [108]. This insight underscores the intricate interplay between cholesterol and membrane asymmetry, shedding light on the complex nature of lipid distributions within the bilayer.

Additionally, different researchers have emphasized the significant influence of cholesterol on membrane proteins, particularly the functional effects of cholesterol on the superfamily of pentameric ligand-gated ion channels (pLGICs) [109,110]. Studies have demonstrated that cholesterol affects the ligand-recognition, gating, and ion permeation properties of hormone and neurotransmitter receptors, as well as ligand- and voltage-gated ion channels. Additionally, the presence of cholesterol in specific membrane domains dynamically compartmentalizes proteins, influencing cell-surface organization and protein trafficking [111]. Moreover, simulations have demonstrated the impact of bilayer asymmetry on bilayer properties, the influence of lipid components, and the adaptability of lipids to different environments [108].

The establishment of cholesterol asymmetry in eukaryotic membranes involves the action of flippases and floppases, which contribute to the uneven distribution of cholesterol across the membrane [112]. Studies have shown that the bilayer responds to asymmetry based on the composition and chemical structure of its lipids, with the number of lipids in each leaflet being adjusted to minimize leaflet surface tension. Additionally, cholesterol has been found to impact membrane asymmetry by reducing it for certain properties, such as overall membrane density and area per lipid, while increasing effects on others, such as lipid packing propensity [108,113]. Furthermore, the influence of cholesterol flip-flop on membrane properties, particularly the interleaflet coupling of cholesterol-enriched domains, has been investigated [114]. It was found that suppressing interleaflet cholesterol population significantly reduces the cholesterol density correlation between the leaflets of an average mammalian plasma membrane, suggesting an amplifying role of cholesterol in signal transduction across the leaflets [114]. These findings underscore the intricate mechanisms through which cholesterol asymmetry is established and its implications for membrane properties and functions.

5. Interaction of Cholesterol and BAPs

As explained above, cholesterol is an important modulator of the physicochemical properties of the lipid bilayer. One of the main effects that cholesterol induces in membranes is a change in their bending rigidity induced by the straightening of the lipid acyl chains, which results in an increase in membrane thickness, which is coupled to an increase in the bulk compressibility modulus of the membrane. Both of these properties, which affect the bending rigidity, are increased in the presence of cholesterol. Therefore, overall, membranes that contain cholesterol will have a higher free energy cost to bend. Since the insertion process of BAPs normally requires a high level of local deformation in the membrane, these changes in the mechanical properties of the membrane in the presence of cholesterol imply that there is a potential relationship between cholesterol concentration and the membrane activity of BAPs.

When looking at cholesterol–peptide interactions, a number of questions come to mind. Do BAPs interact specifically with cholesterol leading to inhibition of the insertion process? In this case, the inhibition would be specific to a group of peptides. Does cholesterol induce inhibition of peptide activity indirectly through shifts in the mechanical properties of the membrane? This would lead to a broad inhibition of peptide activity independent of the BAP mechanism of insertion. Table 2 summarizes some research works that studied the effect of cholesterol on the activity of membrane-active peptides. The pore-forming activity of several of the alpha-helical BAPs presented in Table 2, such as melittin and magainin, is inhibited in the presence of cholesterol. However, this inhibitory effect is also seen in Protegrin-1, which forms a beta hairpin when interacting with the bilayer. Gramicidin also shows inhibition. This peptide, in particular, requires dimerization in the membrane of two subunits to construct a membrane spanning beta helix, which acts as an ion selective channel. Dimerization normally requires local deformation of the membrane due to a hydrophobic mismatch between the thickness of the membrane and the length of the gramicidin dimer. Since cholesterol rigidifies the membrane, this would result in a greater energy cost to induce the deformation in order to account for the mismatch. The main message is that, due to the regulation of the mechanical properties of membranes by cholesterol, BAP activity, which requires membrane deformation, in general will be inhibited by the presence of this lipid molecule.

Table 2. Summary of research works that analyze the potential modulatory effect of cholesterol on the activity of bioactive peptides. Ornithine (O*), D-Phenylalanine (F*).

Peptide	Sequence	Net Charge	Technique	Result	Ref.
Melittin	GIGAVLKVLTTGLPALISWIKRKRQQ	+6	Dynamic giant uni-lamellar vesicle leakage assay and coarse-grained (CG)/Fluorescence spectroscopy/Molecular dynamics (MD)	A higher peptide concentration is necessary to induce calcein release in the DOPC/CHO system as opposed to the membrane model with DOPC alone. CHO hinders peptide-induced pore formation in the membrane model	[115]
Protegrin-1 (PG-1)	RGGRLCYCRRRFCVCVGR-NH_2	+7	Isothermal titration calorimetry (ITC) and atomic force microscopy (AFM)	Increasing CHO content decreases the favorability of the peptide–lipid interaction	[116]
Amyloid-beta (Aβ)	DAEFRHDSGYEVHHQKLVFFAED VGSNKGAIIGLMVGGVVIA	−3	MβCD-cholesterol complex/Cell viability assay (MTT)/Patch-perforated/Generalized polarization (GP)	Effect on Aβ depends on the amount of CHO in the membranes. Low CHO induced a facilitation of the membrane perforation andhigh CHO inhibited membrane disruption	[117]

Table 2. *Cont.*

Peptide	Sequence	Net Charge	Technique	Result	Ref.
Melittin	GIGAVLKVLTTGLPALISWIKRKRQQ	+6	Small-angle neutron scattering (SANS)/Circular dichroism (CD)	Increasing CHO content decreases the amount of transmembrane peptide. CHO reduces the penetration depth of melittin, which would decrease the amount by which it thins the bilayer	[118]
Melittin	GIGAVLKVLTTGLPALISWIKRKRQQ	+6	Circular dichroism (CD)/Elastic incoherent neutron scattering (EINS)/Quasi-elastic neutron scattering (QENS)	Adding CHO to DMPC vesicles mitigates peptide interaction and prevents melittin embedding in the membrane depth	[119]
MSI-78 MSI-594 MSI-367 MSI-843	GIGKFLKKAKKFGKAFVKILKK-NH_2 GIGKFLKKAKKGIGAVLKVLTTGL-NH_2 KFAKKFAKFAKKFAKFAKKFA-NH_2 Oct-OOLLOOLOOL-NH_2	+10 +7 +10 +7	Fluorescence spectroscopy	A strong reduction in membrane alteration is only observed for all peptides tested above 20% CHO content. CHO's protective, membrane stabilizing effect does not occur to an appreciable level in lipid systems containing raft domains	[120]
Cys-TP	GWTLNSAGYLLGCINLKALAALAKISIL-NH_2	+3	Spin-label electron paramagnetic resonance (EPR)	An increase in CHO in the DMPC lipid system resists the incorporation of Cys-TP into the lipid bilayer	[121]
DD K	H-GLWSKIKAAGKEAAKAAGKAAL NAVSEAV-NH_2	+5	Surface plasmon resonance (SPR)/Atomic force microscopy (AFM)/Isothermal titration microcalorimetry (ITC)/Fluorescence spectroscopy/Dynamic light scattering (DLS)	Peptide interaction with mimetic PC membranes is reduced by CHO presence in concentrations typically found in mammalian cells	[122]
Magainin 2 Indolicidin	GIGKFLHSAKKFGKAFVGEIMNS ILPWKWPWWPWRR	+4 +3	Magnetically stirred circular Teflon wells/Fluorescence spectroscopy	The CHO presence prevents the membrane-perturbing action of magainin 2 and, in the same system with CHO, indolicidin exerts a disturbing effect on the membrane model	[123]
Gramicidin S	Cyclo-(VO*LF*P)$_2$	-	Fourier transform infrared spectroscopic (FT-IR)/Circular dichroism (CD)/Nuclear magnetic resonance (^{31}P-NMR)/Fluorescence spectroscopy	The presence of CHO attenuates the interaction of GS with PC bilayers. CHO-containing POPC vesicles are more resistant to peptide-induced permeabilization than CHO-free vesicles	[124]

The activity of BAPs appears to be closely connected to the mechanical properties of the membrane, and a more resilient membrane, presenting a higher bending rigidity, will increase the cost of deformation and therefore inhibit BAP activity. Due to the connection between the energy cost associated with deformation and the chemical activity of the BAPs, a membrane with increased rigidity will require higher concentrations of BAPs in the media to achieve insertion. In addition to increasing the acyl chain order, it has been established that cholesterol can remove the tilt angle of the hydrophobic tails in the lipid bilayer, which will also induce membrane thickening, leading to inhibition of BAP activity. With a face-hinged cyclohexane ring, cholesterol can reduce the van der Waals attractive energy between adjacent lipid tails. However, cholesterol will induce its own interactions with neighboring lipids, and, since cholesterol is rigid, the lipid acyl chains will have to pay a higher energy price to bend away from cholesterol in the process of inducing membrane curvature. This would be considered an additional energy cost to bending, and would further inhibit the activity of BAPs [125].

The protective effect of cholesterol is peptide dependent. In a recent study using two synthetic peptides, it was shown that the presence of high cholesterol levels, such as those found in red blood cells, prevented peptide-induced thinning as measured by X-ray diffraction. In the same study, bacterial cell models were used and compared with minimum inhibitory concentrations in Gram-positive and Gram-negative bacteria, showing that thinning was correlated with toxicity. However, for the cholesterol containing membranes, even though thinning was prevented in the presence of cholesterol, only one of the two peptides showed strongly reduced hemolytic activity, which appeared to indicate that toxicity is not necessarily correlated to BAP-induced membrane thinning [126]. On the other hand, peptide localization within the membrane was measured through neutron diffraction, where cholesterol appeared to inhibit peptide penetration into the membrane's core region. In addition, the secondary structure of the peptides showed significant changes in the red blood cell model, presenting reduced formation of alpha helical structures. The study was complemented by NMR spectroscopy showing changes in headgroup mobility, and differential scanning calorimetry showing downward shifts in the melting temperature of the membrane in the presence of the peptides [127].

These structural studies provide very valuable information with regards to peptide positioning and changes in membrane structural parameters, comparing these with minimum inhibitory concentrations and toxicity studies on live cells, in particular when comparing bacterial and eukaryotic models. However, certain structural details cannot be teased out in full detail with these techniques. A new approach is to use scattering guided molecular dynamic simulations to be able to gain more insight into the response of the membrane in the presence of these antimicrobial peptides [127]. While molecular dynamics provides exquisite molecular detail, it falls short in some cases in accurately representing the physical parameters of the membrane. By using X-ray diffraction to stir the molecular dynamic simulations, the physical parameters can be represented more closely, which leads to a better representation of the interaction of the peptides with membrane presenting different compositions.

Charge has always been thought of as the main discriminating factor in the selectivity of antimicrobial peptides towards bacterial membranes. However, hydrophobic interactions can play an important role in peptide binding to the membrane. For this reason, it is important to consider other structural factors to induce preferential binding towards bacterial membranes versus eukaryotic membranes. The exclusive presence of cholesterol in eukaryotic membranes can provide an additional parameter to induce selectivity. However, a better understanding of how cholesterol influences BAP lytic activity is necessary to have a clear picture when designing antimicrobial peptides. This requires multiple experimental techniques to elucidate changes in the structural parameters of the membrane associated with peptide adhesion and insertion, and location of the peptide within the membrane. These experimental results should be complemented with molecular dynamics to obtain a more detailed structural picture of this interaction. These structural studies should be correlated with leakage experiments in liposomes made of model systems and cell lipid extracts and compared with minimum inhibitory concentrations in bacteria and toxicity measurements in eukaryotic cells. Cholesterol is a promising candidate to push towards a higher level of selectivity for antimicrobial peptides. The structural role that cholesterol plays in interfering with peptide activity needs to be studied in more detail to fully understand its potential in increasing peptide selectivity.

6. Conclusions

Despite the unquestionable importance of cholesterol, very few studies on the interaction between BAPs and cholesterol have been published as yet, but such interaction is important; as approximately 30% of the steroids in the mammalian plasma membrane consist of cholesterol, it is likely that a significant proportion of BAPs encounter this molecule (in addition to frequently meeting a sphingolipid component). In summary, the interaction of AMPs with previously proposed plasma membrane microdomain components (sphingolipids, glycolipids, cholesterol) is most relevant when designing experiments with a view to mimicking native membrane properties. Interactive work concerning the interaction between AMPs and cholesterol may also lead to novel insights.

Author Contributions: J.M.G.-L.: original draft preparation and formal analysis; C.L. formal analysis, writing—review and editing; and M.M.-M.: review, editing, and funding acquisition. All authors have read and agreed to the published version of the manuscript.

Funding: This research was financially supported by the University of Antioquia (CODI Grant 2024–66871) and the Faculty of Sciences of the Universidad de los Andes (INV-2023-162-2782).

Institutional Review Board Statement: Not applicable.

Informed Consent Statement: Not applicable.

Data Availability Statement: The data involved in this paper are presented in articles and supporting materials in the form of diagrams or tables.

Conflicts of Interest: The authors declare no conflicts of interest.

References

1. Harayama, T.; Riezman, H. Understanding the diversity of membrane lipid composition. *Nat. Rev. Mol. Cell Biol.* **2018**, *19*, 281–296. [CrossRef]
2. van Meer, G. Membrane lipids, where they are and how they behave: Sphingolipids on the move. *FASEB J.* **2010**, *24*, 312.1. [CrossRef]
3. Casares, D.; Escribá, P.V.; Rosselló, C.A. Membrane lipid composition: Effect on membrane and organelle structure, function and compartmentalization and therapeutic avenues. *Int. J. Mol. Sci.* **2019**, *20*, 2167. [CrossRef]
4. Subczynski, W.K.; Pasenkiewicz-Gierula, M.; Widomska, J.; Mainali, L.; Raguz, M. High cholesterol/low cholesterol: Effects in biological membranes: A review. *Cell Biochem. Biophys.* **2017**, *75*, 369–385. [CrossRef]
5. Doktorova, M.; Symons, J.L.; Levental, I. Structural and functional consequences of reversible lipid asymmetry in living membranes. *Nat. Chem. Biol.* **2020**, *16*, 1321–1330. [CrossRef]
6. Boughter, C.T.; Monje-Galvan, V.; Im, W.; Klauda, J.B. Influence of Cholesterol on Phospholipid Bilayer Structure and Dynamics. *J. Phys. Chem. B* **2016**, *120*, 11761–11772. [CrossRef]
7. Sugahara, M.; Uragami, M.; Yan, X.; Regen, S.L. The structural role of cholesterol in biological membranes. *J. Am. Chem. Soc.* **2001**, *123*, 7939–7940. [CrossRef]
8. Baron, S.F.; Hylemon, P.B. Biotransformation of bile acids, cholesterol, and steroid hormones. In *Gastrointestinal Microbiology: Volume 1 Gastrointestinal Ecosystems and Fermentations*; Springer: Heidelberg, Germany, 1997; pp. 470–510.
9. King, R.J.; Singh, P.K.; Mehla, K. The cholesterol pathway: Impact on immunity and cancer. *Trends Immunol.* **2022**, *43*, 78–92. [CrossRef]
10. Akbarian, M.; Khani, A.; Eghbalpour, S.; Uversky, V.N. Bioactive peptides: Synthesis, sources, applications, and proposed mechanisms of action. *Int. J. Mol. Sci.* **2022**, *23*, 1445. [CrossRef]
11. Nourmohammadi, E.; Mahoonak, A.S. Health implications of bioactive peptides: A review. *Int. J. Vitam. Nutr. Res.* **2019**, *88*, 223–343. [CrossRef]
12. Garcia-Mora, P.; Peñas, E.; Frías, J.; Gómez, R.; Martinez-Villaluenga, C. High-pressure improves enzymatic proteolysis and the release of peptides with angiotensin I converting enzyme inhibitory and antioxidant activities from lentil proteins. *Food Chem.* **2015**, *171*, 224–232. [CrossRef]
13. Mojica, L.; De Mejía, E.G. Optimization of enzymatic production of anti-diabetic peptides from black bean (*Phaseolus vulgaris* L.) proteins, their characterization and biological potential. *Food Funct.* **2016**, *7*, 713–727. [CrossRef]
14. Xuan, J.; Feng, W.; Wang, J.; Wang, R.; Zhang, B.; Bo, L.; Chen, Z.-S.; Yang, H.; Sun, L. Antimicrobial peptides for combating drug-resistant bacterial infections. *Drug Resist. Updates* **2023**, *68*, 100954. [CrossRef]
15. Rademacher, F.; Gläser, R.; Harder, J. Antimicrobial peptides and proteins: Interaction with the skin microbiota. *Exp. Dermatol.* **2021**, *30*, 1496–1508. [CrossRef]
16. Toosi, S.; Behravan, J. Osteogenesis and bone remodeling: A focus on growth factors and bioactive peptides. *Biofactors* **2020**, *46*, 326–340. [CrossRef]
17. Gomes, B.; Augusto, M.T.; Felício, M.R.; Hollmann, A.; Franco, O.L.; Gonçalves, S.; Santos, N.C. Designing improved active peptides for therapeutic approaches against infectious diseases. *Biotechnol. Adv.* **2018**, *36*, 415–429. [CrossRef]
18. Mammari, N.; Krier, Y.; Albert, Q.; Devocelle, M.; Varbanov, M.; OEMONOM. Plant-derived antimicrobial peptides as potential antiviral agents in systemic viral infections. *Pharmaceuticals* **2021**, *14*, 774. [CrossRef]

19. Dadar, M.; Shahali, Y.; Chakraborty, S.; Prasad, M.; Tahoori, F.; Tiwari, R.; Dhama, K. Antiinflammatory peptides: Current knowledge and promising prospects. *Inflamm. Res.* **2019**, *68*, 125–145. [CrossRef]
20. Tornesello, A.L.; Borrelli, A.; Buonaguro, L.; Buonaguro, F.M.; Tornesello, M.L. Antimicrobial peptides as anticancer agents: Functional properties and biological activities. *Molecules* **2020**, *25*, 2850. [CrossRef]
21. Shin, M.K.; Jang, B.-Y.; Bu, K.-B.; Lee, S.-H.; Han, D.-H.; Oh, J.W.; Sung, J.S. De novo design of AC-P19M, a novel anticancer peptide with apoptotic effects on lung cancer cells and anti-angiogenic activity. *Int. J. Mol. Sci.* **2022**, *23*, 15594. [CrossRef]
22. Prabha, N.; Sannasimuthu, A.; Kumaresan, V.; Elumalai, P.; Arockiaraj, J. Intensifying the Anticancer Potential of Cationic Peptide Derived from Serine Threonine Protein Kinase of Teleost by Tagging with Oligo Tryptophan. *Int. J. Pept. Res. Ther.* **2020**, *26*, 75–83. [CrossRef]
23. Gach-Janczak, K.; Biernat, M.; Kuczer, M.; Adamska-Bartłomiejczyk, A.; Kluczyk, A. Analgesic Peptides: From Natural Diversity to Rational Design. *Molecules* **2024**, *29*, 1544. [CrossRef]
24. Pavlicevic, M.; Marmiroli, N.; Maestri, E. Immunomodulatory peptides—A promising source for novel functional food production and drug discovery. *Peptides* **2022**, *148*, 170696. [CrossRef]
25. Pountos, I.; Panteli, M.; Lampropoulos, A.; Jones, E.; Calori, G.M.; Giannoudis, P.V. The role of peptides in bone healing and regeneration: A systematic review. *BMC Med.* **2016**, *14*, 103. [CrossRef]
26. Shoari, A.; Khodabakhsh, F.; Cohan, R.A.; Salimian, M.; Karami, E. Anti-angiogenic peptides application in cancer therapy; A review. *Res. Pharm. Sci.* **2021**, *16*, 559–574. [CrossRef]
27. Ciumac, D.; Gong, H.; Hu, X.; Lu, J.R. Membrane targeting cationic antimicrobial peptides. *J. Colloid Interface Sci.* **2019**, *537*, 163–185. [CrossRef]
28. Aisenbrey, C.; Marquette, A.; Bechinger, B. The mechanisms of action of cationic antimicrobial peptides refined by novel concepts from biophysical investigations. In *Antimicrobial Peptides: Basics for Clinical Application*; Springer: Berlin/Heidelberg, Germany, 2019; pp. 33–64. [CrossRef]
29. Hemshekhar, M.; Anaparti, V.; Mookherjee, N. Functions of cationic host defense peptides in immunity. *Pharmaceuticals* **2016**, *9*, 40. [CrossRef]
30. Abarca-Cabrera, L.; Fraga-García, P.; Berensmeier, S. Bio-nano interactions: Binding proteins, polysaccharides, lipids and nucleic acids onto magnetic nanoparticles. *Biomater. Res.* **2021**, *25*, 12. [CrossRef]
31. Savini, F.; Loffredo, M.R.; Troiano, C.; Bobone, S.; Malanovic, N.; Eichmann, T.O.; Caprio, L.; Canale, V.C.; Park, Y.; Mangoni, M.L.; et al. Binding of an antimicrobial peptide to bacterial cells: Interaction with different species, strains and cellular components. *Biochim. Biophys. Acta BBA Biomembr.* **2020**, *1862*, 183291. [CrossRef]
32. Benfield, A.H.; Henriques, S.T. Mode-of-action of antimicrobial peptides: Membrane disruption vs. intracellular mechanisms. *Front. Med. Technol.* **2020**, *2*, 610997. [CrossRef]
33. Wenzel, M.; Chiriac, A.I.; Otto, A.; Zweytick, D.; May, C.; Schumacher, C.; Gust, R.; Albada, H.B.; Penkova, M.; Krämer, U.; et al. Small cationic antimicrobial peptides delocalize peripheral membrane proteins. *Proc. Natl. Acad. Sci. USA* **2014**, *111*, E1409–E1418. [CrossRef]
34. Seyfi, R.; Kahaki, F.A.; Ebrahimi, T.; Montazersaheb, S.; Eyvazi, S.; Babaeipour, V.; Tarhriz, V. Antimicrobial peptides (AMPs): Roles, functions and mechanism of action. *Int. J. Pept. Res. Ther.* **2020**, *26*, 1451–1463. [CrossRef]
35. Zhang, Q.-Y.; Yan, Z.-B.; Meng, Y.-M.; Hong, X.-Y.; Shao, G.; Ma, J.-J.; Cheng, X.-R.; Liu, J.; Kang, J.; Fu, C.-Y. Antimicrobial peptides: Mechanism of action, activity and clinical potential. *Mil. Med. Res.* **2021**, *8*, 48. [CrossRef]
36. Goki, N.H.; Tehranizadeh, Z.A.; Saberi, M.R.; Khameneh, B.; Bazzaz, B.S. Structure, Function, and Physicochemical Properties of Pore-forming Antimicrobial Peptides. *Curr. Pharm. Biotechnol.* **2024**, *25*, 1041–1057. [CrossRef]
37. Erdem Büyükkiraz, M.; Kesmen, Z. Antimicrobial peptides (AMPs): A promising class of antimicrobial compounds. *J. Appl. Microbiol.* **2022**, *132*, 1573–1596. [CrossRef]
38. Liu, R.; Liu, Z.; Peng, H.; Lv, Y.; Feng, Y.; Kang, J.; Lu, N.; Ma, R.; Hou, S.; Sun, W.; et al. Bomidin: An optimized antimicrobial peptide with broad antiviral activity against enveloped viruses. *Front. Immunol.* **2022**, *13*, 851642. [CrossRef]
39. El-Dirany, R.; Shahrour, H.; Dirany, Z.; Abdel-Sater, F.; Gonzalez-Gaitano, G.; Brandenburg, K.; Martinez de Tejada, G.; Nguewa, P.A. Activity of anti-microbial peptides (AMPs) against Leishmania and other parasites: An overview. *Biomolecules* **2021**, *11*, 984. [CrossRef]
40. Patiño, A.; Hernández, L.; Patiño, E.; Ortíz, B.; Manrique-Moreno, M. Isolation and Evaluation of Galleria mellonella Peptides with Anti-leishmanial Activity. *Anal. Biochem.* **2016**, *546*, 35–42. [CrossRef]
41. Patiño-Márquez, I.A.; Manrique-Moreno, M.; Patiño González, E.; Jemioła-Rzemińska, M.; Strzałka, K. Effect of antimicrobial peptides from Galleria mellonella on molecular models of Leishmania membrane. Thermotropic and fluorescence anisotropy study. *J. Antibiot.* **2018**, *71*, 642–652. [CrossRef]
42. Manrique-Moreno, M.; Santa-González, G.A.; Gallego, V. Bioactive cationic peptides as potential agents for breast cancer treatment. *Biosci. Rep.* **2021**, *41*, BSR20211218C. [CrossRef]
43. Koehbach, J.; Craik, D.J. The vast structural diversity of antimicrobial peptides. *Trends Pharmacol. Sci.* **2019**, *40*, 517–528. [CrossRef]
44. Xu, R.; Tang, J.; Hadianamrei, R.; Liu, S.; Lv, S.; You, R.; Pan, F.; Zhang, P.; Wang, N.; Cai, Z.; et al. Antifungal activity of designed α-helical antimicrobial peptides. *Biomater. Sci.* **2023**, *11*, 2845–2859. [CrossRef]
45. Huang, Y.; Huang, J.; Chen, Y. Alpha-helical cationic antimicrobial peptides: Relationships of structure and function. *Protein Cell* **2010**, *1*, 143–152. [CrossRef]

46. Yin, L.M.; Edwards, M.A.; Li, J.; Yip, C.M.; Deber, C.M. Roles of hydrophobicity and charge distribution of cationic antimicrobial peptides in peptide-membrane interactions. *J. Biol. Chem.* **2012**, *287*, 7738–7745. [CrossRef]
47. Fillion, M.; Valois-Paillard, G.; Lorin, A.; Noël, M.; Voyer, N.; Auger, M. proteins, a. Membrane interactions of synthetic peptides with antimicrobial potential: Effect of electrostatic interactions and amphiphilicity. *Probiotics Antimicrob. Proteins* **2015**, *7*, 66–74. [CrossRef]
48. Duque, H.M.; Rodrigues, G.; Santos, L.S.; Franco, O.L. The biological role of charge distribution in linear antimicrobial peptides. *Expert Opin. Drug Discov.* **2023**, *18*, 287–302. [CrossRef]
49. Huang, Y.-B.; Wang, X.-F.; Wang, H.-Y.; Liu, Y.; Chen, Y. Studies on mechanism of action of anticancer peptides by modulation of hydrophobicity within a defined structural framework. *Mol. Cancer Ther.* **2011**, *10*, 416–426. [CrossRef]
50. Wieprecht, T.; Dathe, M.; Beyermann, M.; Krause, E.; Maloy, W.L.; MacDonald, D.L.; Bienert, M. Peptide hydrophobicity controls the activity and selectivity of magainin 2 amide in interaction with membranes. *Biochemistry* **1997**, *36*, 6124–6132. [CrossRef]
51. Datta, G.; Chaddha, M.; Hama, S.; Navab, M.; Fogelman, A.M.; Garber, D.W.; Mishra, V.K.; Epand, R.M.; Epand, R.F.; Lund-Katz, S.L.; et al. Effects of increasing hydrophobicity on the physical-chemical and biological properties of a class A amphipathic helical peptide. *J. Lipid Res.* **2001**, *42*, 1096–1104. [CrossRef]
52. Chen, Y.; Guarnieri, M.T.; Vasil, A.I.; Vasil, M.L.; Mant, C.T.; Hodges, R.S. Role of peptide hydrophobicity in the mechanism of action of α-helical antimicrobial peptides. *Antimicrob. Agents Chemother.* **2007**, *51*, 1398–1406. [CrossRef]
53. Bechinger, B.; Juhl, D.W.; Glattard, E.; Aisenbrey, C. Revealing the mechanisms of synergistic action of two magainin antimicrobial peptides. *Front. Med. Technol.* **2020**, *2*, 615494. [CrossRef]
54. Lee, H.-T.; Lee, C.-C.; Yang, J.-R.; Lai, J.Z.; Chang, K.Y. A large-scale structural classification of antimicrobial peptides. *BioMed Res. Int.* **2015**, *2015*, 475062. [CrossRef]
55. Singh, V. The origin of eukaryotic cells. *Resonance* **2021**, *26*, 479–489. [CrossRef]
56. Singer, S.J.; Nicolson, G.L. The fluid mosaic model of the structure of cell membranes. *Science* **1972**, *175*, 720–731. [CrossRef]
57. Nicolson, G.L. The Fluid—Mosaic Model of Membrane Structure: Still relevant to understanding the structure, function and dynamics of biological membranes after more than 40 years. *Biochim. Biophys. Acta BBA Biomembr.* **2014**, *1838*, 1451–1466. [CrossRef]
58. Cho, W.; Stahelin, R.V. Membrane-protein interactions in cell signaling and membrane trafficking. *Annu. Rev. Biophys. Biomol. Struct.* **2005**, *34*, 119–151. [CrossRef]
59. Jury, E.C.; Flores-Borja, F.; Kabouridis, P.S. Lipid rafts in T cell signalling and disease. *Semin. Cell Dev. Biol.* **2007**, *18*, 608–615. [CrossRef]
60. Ahmadpour, S.T.; Mahéo, K.; Servais, S.; Brisson, L.; Dumas, J.-F. Cardiolipin, the Mitochondrial Signature Lipid: Implication in Cancer. *Int. J. Mol. Sci.* **2020**, *21*, 8031. [CrossRef]
61. Vance, J.E. Phospholipid synthesis and transport in mammalian cells. *Traffic* **2015**, *16*, 1–18. [CrossRef]
62. Marquardt, D.; Geier, B.; Pabst, G. Asymmetric lipid membranes: Towards more realistic model systems. *Membranes* **2015**, *5*, 180–196. [CrossRef]
63. St. Clair, J.R.; Wang, Q.; Li, G.; London, E. Preparation and physical properties of asymmetric model membrane vesicles. In *The Biophysics of Cell Membranes: Biological Consequences*; Springer: Heidelberg, Germany, 2017; pp. 1–27.
64. Nguyen, M.H.; Rickeard, B.W.; DiPasquale, M.; Marquardt, D. Asymmetric Model Membranes: Frontiers and Challenges. In *Biomimetic Lipid Membranes: Fundamentals, Applications, and Commercialization*; Springer: Heidelberg, Germany, 2019; pp. 47–71.
65. Doktorova, M.; Symons, J.L.; Zhang, X.; Wang, H.-Y.; Schlegel, J.; Lorent, J.H.; Heberle, F.A.; Sezgin, E.; Lyman, E.; Levental, K.R.; et al. Cell Membranes Sustain Phospholipid Imbalance Via Cholesterol Asymmetry. *bioRxiv* **2023**. [CrossRef]
66. Vahedi, A.; Bigdelou, P.; Farnoud, A.M. Quantitative analysis of red blood cell membrane phospholipids and modulation of cell-macrophage interactions using cyclodextrins. *Sci. Rep.* **2020**, *10*, 15111. [CrossRef]
67. Lorent, J.H.; Levental, K.R.; Ganesan, L.; Rivera-Longsworth, G.; Sezgin, E.; Doktorova, M.; Lyman, E.; Levental, I. Plasma membranes are asymmetric in lipid unsaturation, packing and protein shape. *Nat. Chem. Biol.* **2020**, *16*, 644–652. [CrossRef]
68. Kobayashi, T.; Menon, A.K. Transbilayer lipid asymmetry. *Curr. Biol.* **2018**, *28*, R386–R391. [CrossRef]
69. Murate, M.; Abe, M.; Kasahara, K.; Iwabuchi, K.; Umeda, M.; Kobayashi, T. Transbilayer distribution of lipids at nano scale. *J. Cell Sci.* **2015**, *128*, 1627–1638. [CrossRef]
70. Nelson, D.; Cox, M. *Lehninger Principles of Biochemistry*, 8th ed.; WH Freeman: New York City, NY, USA, 2021.
71. Daleke, D.L. Regulation of phospholipid asymmetry in the erythrocyte membrane. *Curr. Opin. Hematol.* **2008**, *15*, 191–195. [CrossRef]
72. Verkleij, A.J.; Post, J.A. Membrane phospholipid asymmetry and signal transduction. *J. Membr. Biol.* **2000**, *178*, 1–10. [CrossRef]
73. Zwaal, R.F.A.; Roelofsen, B.; Comfurius, P.; Van Deenen, L.L.M. Organization of phospholipids in human red cell membranes as detected by the action of various purified phospholipases. *Biochim. Biophys. Acta BBA Biomembr.* **1975**, *406*, 83–96. [CrossRef]
74. Verkleij, A.J.; Zwaal, R.F.; Roelofsen, B.; Comfurius, P.; Kastelijn, D.; van Deenen, L.L.M. The asymmetric distribution of phospholipids in the human red cell membrane. A combined study using phospholipases and freeze-etch electron microscopy. *Biochim. Biophys. Acta BBA Biomembr.* **1973**, *323*, 178–193. [CrossRef]

75. Gordesky, S.E.; Marinetti, G.V. The asymetric arrangement of phospholipids in the human erythrocyte membrane. *Biochem. Biophys. Res. Commun.* **1973**, *50*, 1027–1031. [CrossRef]
76. Zwaal, R.F.; Roelofsen, B.; Colley, C.M. Localization of red cell membrane constituents. *Biochim. Biophys. Acta BBA Rev. Biomembr.* **1973**, *300*, 159–182. [CrossRef]
77. Zwaal, R.F.; Bevers, E.M.; Comfurius, P.C. Platelets and coagulation. In *Blood Coagulation*; Elsevier Science: Amsterdam, The Netherlands, 1986; Chapter 6; pp. 141–169.
78. Wang, C.T.; Shiao, Y.J.; Chen, J.C.; Tsai, W.J.; Yang, C.C. Estimation of the phospholipid distribution in the human platelet plasma membrane based on the effect of phospholipase A2 from Naja nigricollis. *Biochim. Biophys. Acta BBA Biomembr.* **1986**, *856*, 244–258. [CrossRef]
79. Bevers, E.M.; Comfurius, P.; Zwaal, R.F. Changes in membrane phospholipid distribution during platelet activation. *Biochim. Biophys. Acta BBA Biomembr.* **1983**, *736*, 57–66. [CrossRef]
80. Chap, H.J.; Zwaal, R.F.A.; van Deenen, L.L.M. Action of highly purified phospholipases on blood platelets. Evidence for an asymmetric distribution of phospholipids in the surface membrane. *Biochim. Biophys. Acta BBA Biomembr.* **1977**, *467*, 146–164. [CrossRef]
81. Otnaess, A.B.; Holm, T. The effect of phospholipase C on human blood platelets. *J. Clin. Investig.* **1976**, *57*, 1419–1425. [CrossRef]
82. Schick, P.K.; Kurica, K.B.; Chacko, G.K. Location of phosphatidylethanolamine and phosphatidylserine in the human platelet plasma membrane. *J. Clin. Investig.* **1976**, *57*, 1221–1226. [CrossRef]
83. Li, G.; Kim, J.; Huang, Z.; St Clair, J.R.; Brown, D.A.; London, E. Efficient replacement of plasma membrane outer leaflet phospholipids and sphingolipids in cells with exogenous lipids. *Proc. Natl. Acad. Sci. USA* **2016**, *113*, 14025–14030. [CrossRef]
84. Léonard, A.; Escrive, C.; Laguerre, M.; Pebay-Peyroula, E.; Néri, W.; Pott, T.; Katsaras, J.; Dufourc, E.J. Location of Cholesterol in DMPC Membranes. A Comparative Study by Neutron Diffraction and Molecular Mechanics Simulation. *Langmuir* **2001**, *17*, 2019–2030. [CrossRef]
85. Harroun, T.A.; Katsaras, J.; Wassall, S.R. Cholesterol Is Found To Reside in the Center of a Polyunsaturated Lipid Membrane. *Biochemistry* **2008**, *47*, 7090–7096. [CrossRef]
86. Carson, J.A.S.; Lichtenstein, A.H.; Anderson, C.A.; Appel, L.J.; Kris-Etherton, P.M.; Meyer, K.A.; Petersen, K.; Polonsky, T.; Van Horn, L. Dietary cholesterol and cardiovascular risk: A science advisory from the American Heart Association. *Circulation* **2020**, *141*, e39–e53. [CrossRef]
87. Cerqueira, N.M.; Oliveira, E.F.; Gesto, D.S.; Santos-Martins, D.; Moreira, C.; Moorthy, H.N.; Ramos, M.J.; Fernandes, P. Cholesterol biosynthesis: A mechanistic overview. *Biochemistry* **2016**, *55*, 5483–5506. [CrossRef]
88. Yamaguchi, T.; Ishimatu, T. Effects of cholesterol on membrane stability of human erythrocytes. *Biol. Pharm. Bull.* **2020**, *43*, 1604–1608. [CrossRef]
89. Chakraborty, S.; Doktorova, M.; Molugu, T.R.; Heberle, F.A.; Scott, H.L.; Dzikovski, B.; Nagao, M.; Stingaciu, L.-R.; Standaert, R.F.; Barrera, F.N.; et al. How cholesterol stiffens unsaturated lipid membranes. *Proc. Natl. Acad. Sci. USA* **2020**, *117*, 21896–21905. [CrossRef]
90. Kumar, G.A.; Chattopadhyay, A. Membrane cholesterol regulates endocytosis and trafficking of the serotonin1A receptor: Insights from acute cholesterol depletion. *Biochim. Biophys. Acta BBA Mol. Cell Biol. Lipids* **2021**, *1866*, 158882. [CrossRef]
91. Islam, M.M.; Hlushchenko, I.; Pfisterer, S.G. Low-density lipoprotein internalization, degradation and receptor recycling along membrane contact sites. *Front. Cell Dev. Biol.* **2022**, *10*, 826379. [CrossRef]
92. Rituper, B.; Guček, A.; Lisjak, M.; Gorska, U.; Šakanović, A.; Bobnar, S.T.; Lasič, E.; Božić, M.; Abbineni, P.S.; Jorgačevski, J.; et al. Vesicle cholesterol controls exocytotic fusion pore. *Cell Calcium* **2022**, *101*, 102503. [CrossRef]
93. Olkkonen, V.M.; Ikonen, E. Cholesterol transport in the late endocytic pathway: Roles of ORP family proteins. *J. Steroid Biochem. Mol. Biol.* **2022**, *216*, 106040. [CrossRef]
94. Ogunmowo, T.H.; Jing, H.; Raychaudhuri, S.; Kusick, G.F.; Imoto, Y.; Li, S.; Itoh, K.; Ma, Y.; Jafri, H.; Dalva, M.B.; et al. Membrane compression by synaptic vesicle exocytosis triggers ultrafast endocytosis. *Nat. Commun.* **2023**, *14*, 2888. [CrossRef]
95. García-Arribas, A.B.; Alonso, A.; Goñi, F.M. Cholesterol interactions with ceramide and sphingomyelin. *Chem. Phys. Lipids* **2016**, *199*, 26–34. [CrossRef]
96. Ayuyan, A.G.; Cohen, F.S. Raft Composition at Physiological Temperature and pH in the Absence of Detergents. *Biophys. J.* **2008**, *94*, 2654–2666. [CrossRef]
97. McIntosh, T.J.; Simon, S.A.; Needham, D.; Huang, C.H. Structure and cohesive properties of sphingomyelin/cholesterol bilayers. *Biochemistry* **1992**, *31*, 2012–2020. [CrossRef]
98. Olkkonen, V.M.; Ikonen, E. Inter-and intra-membrane lipid transport. In *Biochemistry of Lipids, Lipoproteins and Membranes*; Elsevier: Amsterdam, The Netherlands, 2021; pp. 457–486.
99. Doole, F.T.; Kumarage, T.; Ashkar, R.; Brown, M.F. Cholesterol stiffening of lipid membranes. *J. Membr. Biol.* **2022**, *255*, 385–405. [CrossRef]
100. McMullen, T.P.; Lewis, R.N.; McElhaney, R.N. Cholesterol–phospholipid interactions, the liquid-ordered phase and lipid rafts in model and biological membranes. *Curr. Opin. Colloid Interface Sci.* **2004**, *8*, 459–468. [CrossRef]

101. Subczynski, W.K.; Hyde, J.S.; Kusumi, A. Effect of alkyl chain unsaturation and cholesterol intercalation on oxygen transport in membranes: A pulse ESR spin labeling study. *Biochemistry* **1991**, *30*, 8578–8590. [CrossRef]
102. Khan, N.; Shen, J.; Chang, T.Y.; Chang, C.C.; Fung, P.C.; Grinberg, O.; Demidenko, E.; Swartz, H. Plasma membrane cholesterol: A possible barrier to intracellular oxygen in normal and mutant CHO cells defective in cholesterol metabolism. *Biochemistry* **2003**, *42*, 23–29. [CrossRef]
103. Menchaca, H.J.; Michalek, V.N.; Rohde, T.D.; Hirsch, A.T.; Tuna, N.; Buchwald, H. Improvement of blood oxygen diffusion capacity and anginal symptoms by cholesterol lowering with simvastatin. *J. Appl. Res. Clin. Exp. Ther.* **2004**, *4*, 410–418.
104. Widomska, J.; Raguz, M.; Subczynski, W.K. Oxygen permeability of the lipid bilayer membrane made of calf lens lipids. *Biochim. Biophys. Acta BBA Biomembr.* **2007**, *1768*, 2635–2645. [CrossRef]
105. Róg, T.; Pasenkiewicz-Gierula, M. Cholesterol effects on the phospholipid condensation and packing in the bilayer: A molecular simulation study. *FEBS Lett.* **2001**, *502*, 68–71. [CrossRef]
106. Scott, H.L.; Kennison, K.B.; Enoki, T.A.; Doktorova, M.; Kinnun, J.J.; Heberle, F.A.; Katsaras, J. Model membrane systems used to study plasma membrane lipid asymmetry. *Symmetry* **2021**, *13*, 1356. [CrossRef]
107. Clarke, R.J.; Hossain, K.R.; Cao, K. Physiological roles of transverse lipid asymmetry of animal membranes. *Biochim. Et Biochim. Biophys. Acta BBA Biomembr.* **2020**, *1862*, 183382. [CrossRef]
108. Blumer, M.; Harris, S.; Li, M.; Martinez, L.; Untereiner, M.; Saeta, P.N.; Carpenter, T.S.; Ingólfsson, H.I.; Bennett, W.D. Simulations of asymmetric membranes illustrate cooperative leaflet coupling and lipid adaptability. *Front. Cell Dev. Biol.* **2020**, *8*, 575. [CrossRef]
109. Barrantes, F.J. Cholesterol and nicotinic acetylcholine receptor: An intimate nanometer-scale spatial relationship spanning the billion year time-scale. *Biomed. Spectrosc. Imaging* **2016**, *5*, S67–S86. [CrossRef]
110. Budelier, M.M.; Cheng, W.W.; Chen, Z.-W.; Bracamontes, J.R.; Sugasawa, Y.; Krishnan, K.; Mydock-McGrane, L.; Covey, D.F.; Evers, A.S. Common binding sites for cholesterol and neurosteroids on a pentameric ligand-gated ion channel. *Biochim. Biophys. Acta BBA Mol. Cell Biol. Lipids* **2019**, *1864*, 128–136. [CrossRef]
111. Thiriet, M. Ion carriers and receptors. In *Biology Mechanics of Blood Flows: Part I: Biology*; Springer: Heidelberg, Germany, 2008; pp. 93–153.
112. Sakuragi, T.; Nagata, S. Regulation of phospholipid distribution in the lipid bilayer by flippases and scramblases. *Nat. Rev. Mol. Cell Biol.* **2023**, *24*, 576–596. [CrossRef]
113. Arashiki, N.; Saito, M.; Koshino, I.; Kamata, K.; Hale, J.; Mohandas, N.; Manno, S.; Takakuwa, Y. An unrecognized function of cholesterol: Regulating the mechanism controlling membrane phospholipid asymmetry. *Biochemistry* **2016**, *55*, 3504–3513. [CrossRef]
114. Thallmair, S.; Ingólfsson, H.I.; Marrink, S.J. Cholesterol flip-flop impacts domain registration in plasma membrane models. *J. Phys. Chem. Lett.* **2018**, *9*, 5527–5533. [CrossRef]
115. Li, J.; Lu, X.; Ma, W.; Chen, Z.; Sun, S.; Wang, Q.; Yuan, B.; Yang, K. Cholesterols Work as a Molecular Regulator of the Antimicrobial Peptide-Membrane Interactions. *Front. Mol. Biosci.* **2021**, *8*, 638988. [CrossRef]
116. Henderson, J.M.; Iyengar, N.S.; Lam, K.L.H.; Maldonado, E.; Suwatthee, T.; Roy, I.; Waring, A.J.; Lee, K.Y.C. Beyond electrostatics: Antimicrobial peptide selectivity and the influence of cholesterol-mediated fluidity and lipid chain length on protegrin-1 activity. *Biochim. Biophys. Acta Biomembr.* **2019**, *1861*, 182977. [CrossRef]
117. Fernandez-Perez, E.J.; Sepulveda, F.J.; Peters, C.; Bascunan, D.; Riffo-Lepe, N.O.; Gonzalez-Sanmiguel, J.; Sanchez, S.A.; Peoples, R.W.; Vicente, B.; Aguayo, L.G. Effect of Cholesterol on Membrane Fluidity and Association of Abeta Oligomers and Subsequent Neuronal Damage: A Double-Edged Sword. *Front. Aging Neurosci.* **2018**, *10*, 226. [CrossRef]
118. Qian, S.; Heller, W.T. Melittin-induced cholesterol reorganization in lipid bilayer membranes. *Biochim. Biophys. Acta BBA Biomembr.* **2015**, *1848*, 2253–2260. [CrossRef]
119. Sharma, V.K.; Mamontov, E.; Anunciado, D.B.; O'Neill, H.; Urban, V.S. Effect of antimicrobial peptide on the dynamics of phosphocholine membrane: Role of cholesterol and physical state of bilayer. *Soft Matter* **2015**, *11*, 6755–6767. [CrossRef]
120. McHenry, A.J.; Sciacca, M.F.; Brender, J.R.; Ramamoorthy, A. Does cholesterol suppress the antimicrobial peptide induced disruption of lipid raft containing membranes? *Biochim. Biophys. Acta BBA Biomembr.* **2012**, *1818*, 3019–3024. [CrossRef]
121. Arsov, Z.; Nemec, M.; Schara, M.; Johansson, H.; Langel, U.; Zorko, M. Cholesterol prevents interaction of the cell-penetrating peptide transportan with model lipid membranes. *J. Pept. Sci.* **2008**, *14*, 1303–1308. [CrossRef]
122. Verly, R.M.; Rodrigues, M.A.; Daghastanli, K.R.; Denadai, A.M.; Cuccovia, I.M.; Bloch, C., Jr.; Frezard, F.; Santoro, M.M.; Pilo-Veloso, D.; Bemquerer, M.P. Effect of cholesterol on the interaction of the amphibian antimicrobial peptide DD K with liposomes. *Peptides* **2008**, *29*, 15–24. [CrossRef]
123. Zhao, H.; Mattila, J.P.; Holopainen, J.M.; Kinnunen, P.K. Comparison of the membrane association of two antimicrobial peptides, magainin 2 and indolicidin. *Biophys. J.* **2001**, *81*, 2979–2991. [CrossRef]
124. Prenner, E.J.; Lewis, R.N.; Jelokhani-Niaraki, M.; Hodges, R.S.; McElhaney, R.N. Cholesterol attenuates the interaction of the antimicrobial peptide gramicidin S with phospholipid bilayer membranes. *Biochim. Biophys. Acta BBA Biomembr.* **2001**, *1510*, 83–92. [CrossRef]
125. Leeb, F.; Maibaum, L. Spatially resolving the condensing effect of cholesterol in lipid bilayers. *Biophys. J.* **2018**, *115*, 2179–2188. [CrossRef]

126. Heinrich, F.; Salyapongse, A.; Kumagai, A.; Dupuy, F.G.; Shukla, K.; Penk, A.; Huster, D.; Ernst, R.K.; Pavlova, A.; Gumbart, J.C.; et al. Synergistic biophysical techniques reveal structural mechanisms of engineered cationic antimicrobial peptides in lipid model membranes. *Chem. Eur. J.* **2020**, *26*, 6247–6256. [CrossRef]
127. Allsopp, R.; Pavlova, A.; Cline, T.; Salyapongse, A.M.; Gillilan, R.E.; Di, Y.P.; Deslouches, B.; Klauda, J.B.; Gumbart, J.C.; Tristram-Nagle, S. Antimicrobial peptide mechanism studied by scattering-guided molecular dynamics simulation. *J. Phys. Chem. B* **2022**, *126*, 6922–6935. [CrossRef]

MDPI AG
Grosspeteranlage 5
4052 Basel
Switzerland
Tel.: +41 61 683 77 34

Membranes Editorial Office
E-mail: membranes@mdpi.com
www.mdpi.com/journal/membranes

www.ingramcontent.com/pod-product-compliance
Lightning Source LLC
LaVergne TN
LVHW070448061126
851858LV00011B/187
* 9 7 8 3 7 2 5 8 8 2 8 5 4 *